218
Topics in Current Chemistry

Springer

Berlin
Heidelberg
New York
Barcelona
Hong Kong
London
Milan
Paris
Tokyo

Host-Guest Chemistry

Mimetic Approaches to Study Carbohydrate Recognition

Volume Editor: Soledad Penadés

With contributions by
J.-H. Fuhrhop, R. García, M.-F. Gouzy,
S. R. Haseley, B. T. Houseman, T. D. James,
J. P. Kamerling, G. Li, T. K. Lindhorst, J. C. Morales,
M. Mrksich, S. Penadés, J. Rojo, S. Shinkai,
C. Tromas, J. F. G. Vliegenthart

Springer

The series *Topics in Current Chemistry* presents critical reviews of the present and future trends in modern chemical research. The scope of coverage includes all areas of chemical science including the interfaces with related disciplines such as biology, medicine and materials science. The goal of each thematic volume is to give the non-specialist reader, whether at the university or in industry, a comprehensive overview of an area where new insights are emerging that are of interest to a larger scientific audience.

As a rule, contributions are specially commissioned. The editors and publishers will, however, always be pleased to receive suggestions and supplementary information. Papers are accepted for *Topics in Current Chemistry* in English.

In references *Topics in Current Chemistry* is abbreviated Top. Curr. Chem. and is cited as a journal.

Springer WWW home page: http://www.springer.de
Visit the TCC home page at http:/link.springer.de/series/tcc/
or http://link.springer-ny.com/series/tcc/

ISSN 0340-1022
ISBN 3-540-42096-7
Springer-Verlag Berlin Heidelberg New York

Library of Congress Catalog Card Number 74-644622

Springer-Verlag Berlin Heidelberg New York
a member of BertelsmannSpringer Science+Business Media GmbH

© Springer-Verlag Berlin Heidelberg 2002
Printed in Germany

The use of general descriptive names, registered names, trademarks, etc. in this publication does not imply, even in the absence of a specific statement, that such names are exempt from the relevant protective laws and regulations and therefore free for general use.

Cover design: Friedhelm Steinen-Broo, Barcelona; MEDIO, Berlin
Typesetting: Fotosatz-Service Köhler GmbH, 97084 Würzburg

SPIN: 10760678   02/3020 ra – 5 4 3 2 1 0 – Printed on acid-free paper

Topics in Current Chemistry
Now Also Available Electronically

For all customers with a standing order for Topics in Current Chemistry we offer the electronic form via LINK free of charge. Please contact your librarian who can receive a password for free access to the full articles by registration at:

http://link.springer.de/orders/index.htm

If you do not have a standing order you can nevertheless browse through the table of contents of the volumes and the abstracts of each article at:

http://link.springer.de/series/tcc
http://link.springer-ny.com/series/tcc

There you will also find information about the

– Editorial Board
– Aims and Scope
– Instructions for Authors

Preface

It is now well established that cell surface carbohydrates are involved in cell adhesion and communication. Glycoconjugates (glycoproteins and glycolipids) have been characterized as ligands for toxins, bacteria, viruses, or other cells and the role they play in morphogenesis, fertilization, antigen recognition and other processes began to be understood. All these processes imply carbohydrate protein interactions or, as it has more recently been proposed, carbohydrate-carbohydrate interactions.

The understanding of the mechanism of carbohydrate recognition at molecular level requires both the knowledge of the exact intermolecular forces at work and how these forces determine the affinity and selectivity of the recognition process.

In the early seventies the pioneer work of R. U. Lemieux and others established the bases for the understanding of carbohydrate-protein interactions. Carbohydrates are, however, very complex molecules and this complexity has represented, during the past few years, major challenges for chemists and biologists involved in carbohydrate research. These challenges have prompted the development of chemical, enzymatic and analytical tools for studying the chemistry and biology of carbohydrates.

A characteristic feature of carbohydrate binding is its low affinity. Nature overcomes this problem with a multivalent presentation of carbohydrate epitopes to achieve high binding affinities.

In the biological context carbohydrate interaction co-exist with other interactions. This complexity makes difficult to dissect and analyze the individual contribution of each interacting system. As in the study of other biological interactions, the use of model systems has being a productive approach in the study of carbohydrate recognition.

Model systems to mimic polyvalent presentation of glycoconjugates, as represented by glycopolymers, dendrimers or self-assembled monolayers (SAM), have recently been prepared for studies in glycobiology. These models with well-defined chemical composition have provided valuable information on the influence of multivalent presentation of carbohydrates in their biological activity.

The confluence of these chemical approaches with the use of classical (traditional) and new analytical techniques (biosensors, scanning force microscopy) opens fascinating ways to evaluate in more detail how carbohydrate interfere with their specific receptors.

This volume is dedicated to review some recent advances in mimetic approaches aiming to a better understanding of carbohydrate recognition. Two interesting contributions dealing with multivalent model systems open and close the volume. *B. T. Houseman* and *M. Mrksich* provide in the first contribution a mechanistic analysis of polyvalent recognition and review several multivalent model systems with special emphasis on self-assembled monolayers. The last contribution by *T. K. Lindhorst* highlights mainly the field of glycodendrimers, glycopolymers and cell surface modifications as well as the problematic of the biology of these multivalent ligands.

New evidence is confirming the importance of carbohydrate-carbohydrate interactions as mechanism for specific cell adhesion and communication. The contribution of *J. Rojo, J. C. Morales* and *S. Penadés* reviews these new emerging interactions in biological systems and presents the few examples described on quantitative studies with polyvalent model systems.

Some technical challenges in the evaluation of the energetic of carbohydrate recognition are analytical in nature. The application of new and more sensitive techniques to study weak interactions such as biosensors, scanning force microscopy, or affinity chromatography is of paramount importance to demonstrate and quantify biomolecular interactions. *S. R. Haseley, J. P. Kamerling* and *J. F. G. Vliegenthart* report on the use of Surface Plasmon Resonance (SPR) detection for investigating carbohydrate ligand interactions. *C. Tromas* and *R. García* focus on the use of Atomic Force Microscopy (AFM) for measurements of intermolecular forces and show, in the very few examples described up to date, the potential of this technique to measure very weak single molecule interactions between carbohydrate ligands. In a through contribution *T. D. James* and *S. Shinkai* highlight how simple boronic acid based artificial receptors offer the possibility of creating carbohydrate chemosensors sensitive for any chosen saccharide. Finally, *G. Li, M.-F. Gouzy* and *J.-H. Fuhrhop* describe how artificial supramolecular structures, distinct from the biological ones, can help to unravel the role of water in carbohydrate organization.

I thank all the authors for their contributions and hope that this effort will help the readers to have an insight in actual problems of carbohydrate recognition. I also would like to thank the staff of Springer Verlag for editorial assistance.

Seville, July 2001   Soledad Penadés

Contents

Contents of Volume 215

Glycoscience

Epimerisation, Isomerisation and Rearrangement Reactions of Carbohydrates

Volume Editor: Arnold E. Stütz
ISBN 3-540-41383-9

Model Systems for Studying Polyvalent Carbohydrate Binding Interactions

Benjamin T. Houseman, Milan Mrksich

Department of Chemistry, The University of Chicago, Chicago, IL 60637, USA
E-mail: mmrksich@midway.uchicago.edu

This chapter describes several classes of model systems that have been used for studies of polyvalent binding interactions between carbohydrates and proteins. This review is divided into two parts. The first section provides a mechanistic analysis of polyvalent recognition. The second section utilizes this mechanistic framework to describe and analyze several major classes of model systems. The description of each model system is accompanied by a summary of recent experimental results from the literature. The chapter concludes with a comparison of the model systems and an introduction to several novel model systems that promise to broaden the scope of future studies in polyvalent carbohydrate-protein interactions.

Keywords. Carbohydrates, Polyvalency, Model system, Monolayers

Topics in Current Chemistry, Vol. 218
© Springer-Verlag Berlin Heidelberg 2002

List of Abbreviations

Con A	concanavalin A
ELLA	enzyme-linked lectin assay
GalTase	β-(1,4)-galactosyltransferase
GlcNAc	N-acetylglucosamine
HAI	hemagglutination inhibition assay
HRP	horseradish peroxidase
ITC	isothermal titration calorimetry
LacNAc	N-acetyllactosamine
PDA	10,12-pentacosadiynoic acid
ROMP	ring-opening metathesis polymerization
SAM	self-assembled monolayer
SLe^x	sialyl lewis X
SPR	surface plasmon resonance
WGA	wheat germ agglutinin

1
Introduction

The interactions of proteins with carbohydrates define and regulate many important biological phenomena, ranging from the adhesion of leukocytes to endothelial cells to the guidance of neuronal growth cones during development [1–3]. Despite the scope and importance of carbohydrates in biology, the difficulty in studying carbohydrate-protein interactions has hindered efforts to develop a mechanistic understanding of carbohydrate structure and function. A major reason for this difficulty stems from the structural complexity of carbohydrates. While the other two classes of biopolymers – nucleic acids and proteins – have a linear arrangement of repeating units, carbohydrate building blocks have multiple points of attachment, leading to highly branched and stereochemically rich structures. This structural complexity is further increased by post-synthetic modifications, wherein unmodified hydroxyl groups can be sulfated, phosphorylated, acetylated, or oxidized to generate distinct biological activities [4]. Sulfated carbohydrates, for example, function as potent mediators of inflammation [5], while phosphosugars play important roles in cellular signal transduction and metabolism [6].

A second difficulty in studies of carbohydrate-protein interactions is that binding affinities are weak, often with dissociation constants in the millimolar range [7]. In biological contexts, this limitation is often overcome by combining multiple interactions between two or more carbohydrates and a corresponding multimeric protein. These interactions – often described as polyvalent – have several mechanistic and functional advantages over their monovalent counterparts. Among these are the ability to create conformal contact between large biological surfaces, the ability to produce graded responses with a single type of interaction, and the ability to increase the specificity of an interaction [8].

These characteristics, together with the complex nature and widespread importance of polyvalent interactions, have prompted the development of model systems to characterize carbohydrate-protein interactions. This chapter reviews several model systems that have been important in mechanistic studies of glycobiology and in applications that rely on the recognition of carbohydrates. We first describe the principles necessary for understanding polyvalent binding events. We next introduce several model systems that have been important for studies of polyvalent carbohydrate-protein interactions. A description of each model system is accompanied by a brief summary of recent experimental results from the literature. We conclude the chapter with a description of three novel model systems and a comparison of the advantages and limitations of the model systems in the review.

2
Polyvalent Recognition

Polyvalent binding is characterized by the simultaneous interaction between multiple ligands on one entity and multiple receptors on another. Scheme 1 illustrates a polyvalent interaction involving three ligands and three receptors. Note that this scheme can be generalized to N ligands and N receptors.

The following section provides a mechanistic description of polyvalent binding interactions, with an emphasis on the distinctions between these interactions and their monovalent counterparts. The remainder of the chapter utilizes these mechanistic principles to describe and analyze several classes of polyvalent model systems that present carbohydrates.

2.1
Mechanistic Features of Polyvalent Binding

The formation of an N-valent interaction is a stepwise process that involves N binding events and $(N–1)$ intermediates. This stepwise binding process has two characteristics that are absent in its monovalent counterpart: the ability to reg-

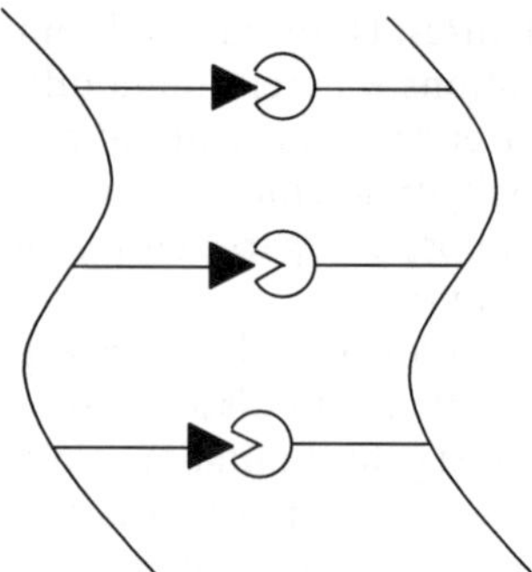

Scheme 1. Illustration of a trivalent interaction between 3 ligands and 3 receptors. Note that this scheme can be extended to an N-valent interaction between N ligands and N receptors

ulate the strength of an interaction through the number of ligand-receptor contacts and the ability to rapidly dissociate the complex with soluble ligand [9]. Whitesides and coworkers illustrated these principles with a trivalent complex between vancomycin and the peptide D-Ala-D-Ala [10, 11]. Whereas the monovalent vancomycin-peptide complex had a dissociation constant of 1×10^{-6} mol/l, the trivalent complex had a dissociation constant that was ten orders of magnitude greater ($K_d = 4 \times 10^{-17}$ mol/l). In contrast to the monovalent complex, however, the trivalent complex rapidly dissociated when soluble peptide was added to a solution containing the complex.

Although polyvalent binding is functionally attractive, it complicates the analysis of an interaction because both inter- and intramolecular binding may occur (Fig. 1). After the formation of a monovalent intermediate, intramolecular binding generates a single complex with multiple ligand-receptor interactions, while intermolecular binding affords a crosslinked network of ligands and receptors. Additional intermolecular binding events eventually lead to precipitation of the network from solution. In all cases, each binding event will alter the enthalpy, entropy, and free energy of subsequent interactions since the ligand and receptor are bound together in solution. The precipitation of ligand receptor networks from solution further increases the complexity of the system.

2.2
Factors that Influence Polyvalent Binding

The many possible ways in which N ligands and N receptors can interact makes it difficult to design systems that preferentially follow one binding pathway (i.e., intermolecular vs intramolecular binding). In order for intramolecular binding to predominate, the free energy of an N-valent interaction (ΔG^{poly}) must be more favorable than the free energy of N monovalent interactions ($N\Delta G^{mono}$). Since the free energy term includes enthalpic (ΔH^{poly}) and entropic (ΔS^{poly}) terms, it is necessary to examine the role of these components separately. In a recent review, Whitesides and coworkers presented a system of nomenclature for polyvalent

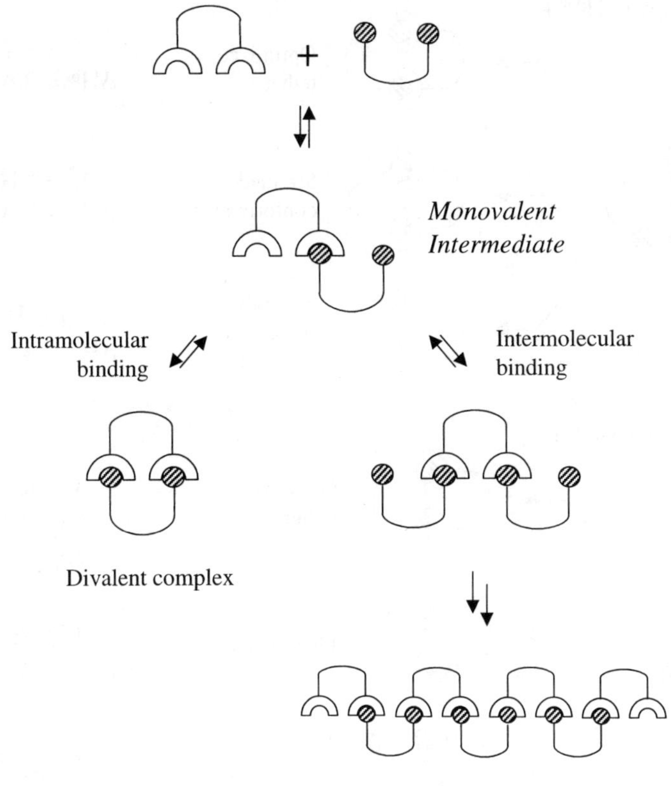

Fig. 1. Possible binding pathways in a divalent model system

interactions and described the enthalpic and entropic characteristics of a polyvalent interaction [8]. We provide a summary of this analysis for a divalent system in Fig. 2 and note that this analysis can be extended to N-valent interactions using the same principles.

The enthalpy of binding (ΔH^{poly}) reflects stabilizing interactions between ligand and receptor, less any energetic penalties for non-optimal conformations in the complex and differential solvation of bound and unbound ligand and receptor. After the formation of a monovalent complex, the enthalpy of the second binding event can be greater than, equal to, or diminished relative to the first. If the second ligand and receptor species are properly aligned, the enthalpy of the second binding event will be identical to the first and $\Delta H^{bi} = 2\Delta H^{mono}$ (Fig. 2A, Case I). This situation rarely occurs because it is difficult to design scaffolds that optimally position ligands for multipoint binding. In many cases, the second binding event introduces strain in the tether, adding an energetic penalty to the overall enthalpy of interaction (Fig. 2A, Case II). In other systems, favorable secondary interactions between the tether and the receptor during subsequent binding events stabilize the complex, enhancing the enthalpy of the second

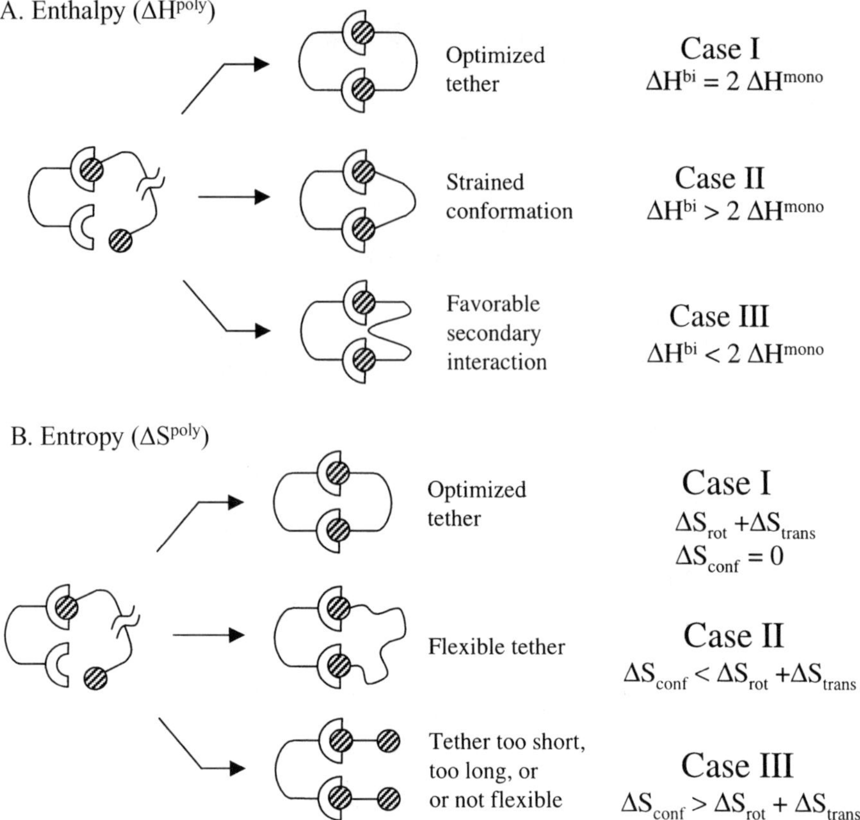

Fig. 2. Analysis of enthalpy and entropy of binding in a divalent model system

binding interaction (Fig. 2A, Case III). The binding of cholera toxin (a pentameric protein) to multiple ganglioside lipids is one possible example of this type of interaction [12]. In this system, the binding constant of the second ligand was four times more favorable than that of the first ligand. This effect was due to enthalpy alone since each ganglioside rotated and translated independently.

The entropy of binding (ΔS^{poly}) reflects changes in the molecular order that occur during a ligand-receptor interaction. Since this term is complex and often poorly understood, it is often divided into entropies of translation (ΔS_{trans}), rotation (ΔS_{rot}), conformation (ΔS_{conf}), and hydration (ΔS_{H_2O}). For a bivalent interaction, the sum of these quantities for the second binding event can again be greater than, equal to, or less than the sum observed for the first interaction. Below we consider each case in turn, assuming that the contribution from the entropy of hydration (ΔS_{H_2O}) is similar in each situation. If the two ligands and two receptors are connected by a rigid scaffold and are precisely spaced, $\Delta S_{conf} = 0$ and the interaction will occur with an entropy equivalent to a single monovalent interaction (Fig. 2B, Case I). This situation is unrealistic since most linkers do have conformational degrees of freedom and it is rare that the spacing between a ligand and receptor are matched. In general, ΔS_{conf} is unfavorable upon complexation because the binding interaction will constrain many degrees of freedom previously available to the ligand and receptor. If ΔS_{conf} is less than the sum of ΔS_{trans} and ΔS_{rot}, the binding is entropically enhanced and intramolecular interaction will be favored (Case II). If ΔS_{conf} is greater than the sum of ΔS_{trans} and ΔS_{rot}, intermolecular interactions will be favored (Case III).

This discussion illustrates the ways in which the flexibility of the spacer between ligands (and receptors) is critical to the effective design of a polyvalent model system. The optimal linker will perform two functions: increase the likelihood that interactions will occur without strain and minimize the entropic cost of multiple ligand-receptor interactions. If this balance is not achieved, the formation of cross-linked complexes – rather than a true multipoint attachment – is likely to occur. We also note that there are other strategies by which polyvalent ligands can be directed to assemble into either inter- or intramolecular complexes. For example, intermolecular complexes are favored in cases where the concentration of receptor is large, while intramolecular complexes predominate when the polyvalent ligand is immobilized on a substrate at low density (see Sect. 5.3).

2.3
Characterization of Polyvalent Binding

The models outlined above provide a molecular description of polyvalent association. In many cases, however, the mechanism by which binding occurs (see Figs. 1 and 2) is less critical than is the enhancement of binding observed in a polyvalent interaction. To aid in the description of polyvalent interactions, Whitesides and coworkers described a new parameter β [8]. This factor is defined as the ratio of the association constant for a polyvalent system (as measured by any number of methods) and the association constant for the monovalent ligand. These relationships are described quantitatively in Eqs. (1) and (2):

$$\Delta G^{poly} = \Delta G^{mono} - RT\ln\beta \tag{1}$$

$$\beta = K^{poly}/K^{mono} \tag{2}$$

Model systems that have high values of β are useful, regardless of the mechanism of action or the nature of individual interactions. In fact, many polyvalent interactions with large values of β occur with negative cooperativity – that is, the average free energy of binding for a system of N interactions is less favorable than the free energy of binding for a monovalent interaction ($\Delta G^{poly}/N < \Delta G^{mono}$). A common example in glycobiology is the binding of galactose-containing ligands to the asialoglycoprotein receptor on the surface of hepatocytes [13]. For one trivalent ligand, the value of $K^{mono} = 7 \times 10^4$ M^{-1} and $K^{tri} = 2 \times 10^8$ M^{-1}. Since $\Delta G^{tri}/3$ is less favorable than ΔG^{mono} the binding of the trivalent ligand to the receptor occurs with negative cooperativity.

In glycobiology, many model systems demonstrate values of β greater than 1. These binding enhancements are generally attributed to the "cluster glycoside effect" [7], a term that describes interactions in which the binding constant is larger than predictions based on local concentration alone. Despite a vast literature on the subject, the mechanistic basis for the cluster glycoside effect remains unclear. Isothermal titration calorimetry (ITC) has become an important tool for these studies because it measures the valency, free energy (ΔG^{poly}), and enthalpy (ΔH^{poly}) of an interaction [14]. Since at equilibrium $\Delta G^{poly} = \Delta H^{poly} - T\Delta S^{poly}$, the entropy of the interaction can be calculated from experimentally determined values. A recent study by Brewer and coworkers used ITC to investigate the thermodynamic basis for the binding enhancements that occur between concanavalin A (Con A) and a series of polyvalent mannosides [15]. They found that the valency of the interaction correlated with the number of carbohydrate ligands on a given scaffold and that the enthalpy of interaction increased linearly with the number of mannose residues on the glycoconjugate. The larger affinities (K_a) of polyvalent ligands for the lectin arose from a favorable entropy of interaction (specifically, a more positive $T\Delta S$ term). These data suggest that the polyvalent ligands bind more than one Con A molecule rather than form multiple attachments with a single lectin. These data also show that binding enhancements, while useful, may occur via multipoint binding.

3
Classes of Polyvalent Model Systems

Several classes of model systems have been developed to study polyvalent carbohydrate-protein interactions. To aid in the presentation of this body of literature, we have grouped these models into two classes based on valency (Fig. 3).

Examples in the first class, which we refer to as low valency model systems, present fewer than 20 carbohydrate ligands. Since the types of models in this class are quite diverse, we have further divided them into three groups to facilitate comparison of the different scaffold architectures. Examples in the second class, which we refer to as high valency model systems, present large numbers (often hundreds) of carbohydrate ligands. These model systems include neoglycoproteins, glycopolymers (both soluble and immobilized), hybrid bilayers, and self-assembled monolayers of alkanethiolates on gold. The last two models are functionally distinct from protein- and polymer-based systems because they present carbohydrates in a two-dimensional array.

4
Model Systems with Low Valency

Model systems that present between 2 and 20 carbohydrate ligands have been important for defining the geometric and structural requirements for tight binding to polyvalent receptors [16]. While these models are reasonably well defined and readily modified using synthetic organic chemistry, it is often necessary to prepare large numbers of compounds in order to identify polyvalent ligands that achieve significant binding enhancements. Lee and coworkers, for example, prepared over 100 compounds during the development of high affinity ligands for the hepatic asialoglycoprotein receptor [7, 13, 17]. Schnaar and

A. Systems with Low Valency

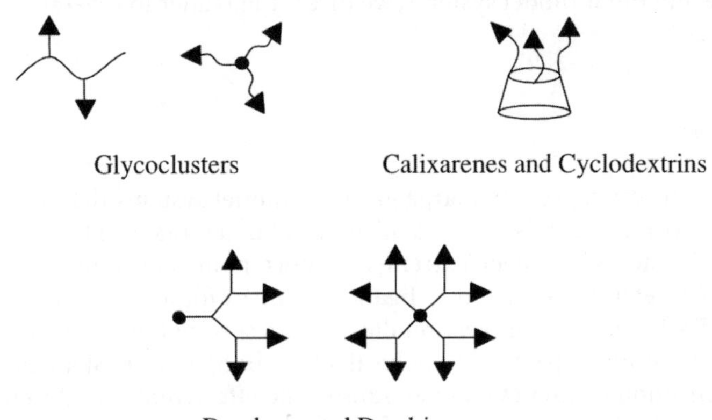

B. Systems with High Valency

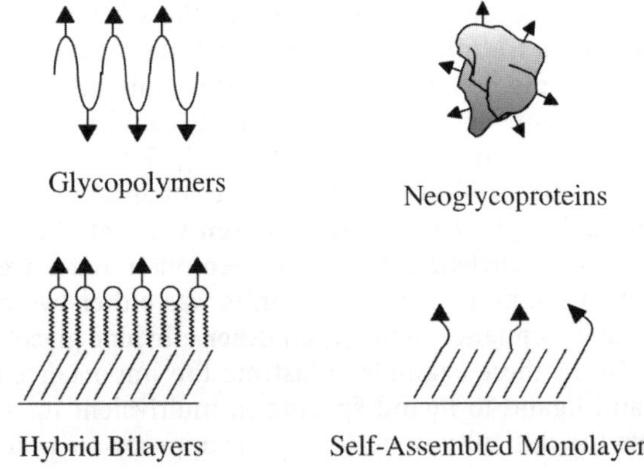

Fig. 3. Graphical summary of model systems presented in this review. Models are divided into systems with low valency and high valency

coworkers have invested similar efforts toward the development of selective ligands for a lectin from the protozoan *E. histolytica* [18, 19]. These examples reflect the difficulties in designing model systems with low valency.

The following section presents several classes of scaffolds that have been used for the presentation of carbohydrates. We illustrate the utility of these scaffolds with recent examples from the literature that highlight the functional characteristics of each model sysytem. We include two additional pieces of information with each example: an observed binding enhancement of the glycocluster (β) and a binding enhancement that is corrected for the number of carbohydrate ligands on the glycocluster (β/N). Due to space considerations, this section only presents glycoconjugates that have undergone biological evaluation and does not discuss the synthesis of these model compounds. For the preparation of these and other model systems, we direct the reader to several excellent reviews [20–24].

4.1
Clusters

Glycoclusters represent a large group of model systems that present two to five carbohydrate ligands on a small molecular template [20]. These low valency model systems have been extremely important in determining the optimal binding geometries and ligand to ligand distances for a particular multimeric protein. Each figure in this section illustrates a series of glycoclusters that has been used to study a particular polyvalent carbohydrate-protein interaction. This organization facilitates comparisons of the effectiveness of different scaffolds.

Figure 4 illustrates a group of scaffolds that present Sialyl Lewis X (SLe^x) or derivatives of SLe^x. These conjugates were designed as tight binding inhibitors for E-selectin, a carbohydrate binding protein that plays a critical role in the migration of leukocytes to sites of injury.

The examples in this figure illustrate that the success or failure of a particular model system depends on the nature of the scaffold and the length of the tether between the scaffold and the carbohydrate. Scaffolds based on butane or pentane, for example, demonstrated no binding enhancements over monovalent SLe^x [25], while more conformationally rigid scaffolds such as quaternary carbon 1 [26], carbohydrate 2 [25], peptide 3 [27], and aromatic ligand 4 [28] provided better results. In all cases, however, the enhancement of binding was quite sensitive to the length of the tether between the scaffold and SLe^x. Scaffold 5, based on a flexible ethylene glycol tether, demonstrated a fivefold enhancement over SLe^x when the number of glycol units was either two or five but demonstrated negligible enhancements when other ethylene glycol spacers were used [29]. Collectively, these examples illustrate the importance of conformational flexibility and ligand to ligand spacing in multivalent interactions. They also demonstrate the difficulty in designing the appropriate polyvalent ligand for a particular system.

It is instructive to compare these results with those in Fig. 5, which depicts a series of compounds (6a–e) that were tested for their ability to inhibit the binding of the bacterium *Staphylococcus suis* to Galα(1,4)Galβ motifs on epithelial cells of the urinary tract [30]. These data represent the minimum concentration of multivalent compound required to inhibit the crosslinking of red blood cells by the bacterium.

In this series, the nature and valency of the scaffold is less important than the length of the spacer between the scaffold and the carbohydrate ligands. Compound 6b, for example, was 20-fold more potent than monovalent carbohydrate 6a in inhibiting binding. The introduction of a three carbon segment (compound 6c) resulted in an additional 50-fold enhancement. Trivalent system 6d, based on a quaternary carbon scaffold, was a less potent inhibitor than 6c, while tetravalent ligand 6e inhibited bacterial adhesion at a twofold lower concentration than 6c.

1

R	IC_{50}	β
SLe^x-$O(CH_2)_6NH_2$	1000 μM	1
SLe^x	380 μM	7.5 (2.5)
$(CH_2)_6OSLe^x$	130 μM	50 (16.3)

2

R	IC_{50}	β
$R_1 = SLe^x$, $R_2 = H$	600 μM	1
$R_1 = R_2 = SLe^x$	200 μM	3 (1.5)

3

R	IC_{50}	β
SLe^x-OH	600 μM	1
SLe^x, n = 2	140 μM	4.3 (2.15)
SLe^x, n = 3	370 μM	1.2 (0.6)

4

R	IC_{50}	β
$R_1 = H$ $R_2 = SLe^x$ (Naphth)	200 μM	1
$R_1 = R_2 = SLe^x$ (Naphth), n=1	30 μM	6.7 (3.35)
$R_1 = R_2 = SLe^x$ (Naphth), n=2	60 μM	3.3 (1.65)

5

R	IC_{50}	β
SLe^x-OH	393 μM	1
SLe^x, n=1	228 μM	1.7 (0.85)
SLe^x, n=2	70 μM	5.6 (2.8)
SLe^x, n=3	179 μM	2.2 (1.1)
SLe^x, n=4	176 μM	2.2 (1.1)
SLe^x, n=5	81 μM	4.9 (2.45)

Fig. 4. Glycoclusters for studies of the interaction between Sialyl Lewis X and E-selectin. Absolute and valency-corrected binding enhancements over monovalent analogs are provided

An important, and often overlooked, factor in the evaluation of multivalent models is the assay used to determine binding enhancements. A number of assays are available for these measurements in glycobiology [8], and the choice of a particular assay depends on the interaction being studied, the quantity of available material, and access to specialized equipment. To illustrate the importance of the assay used to measure a polyvalent interaction, Toone and coworkers evaluated a series of carbohydrates with a hemagglutination inhibition assay

R = Galα(1,4)Galβ	Inhibitory Conc. (nM)	β
6a RO-TMS	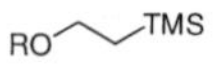 1800	1
6b	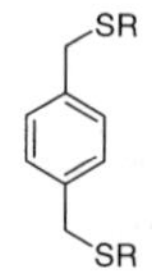 90	20 (10)
6c	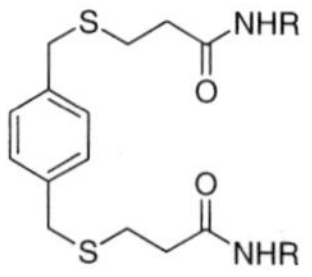 6	300 (150)
6d	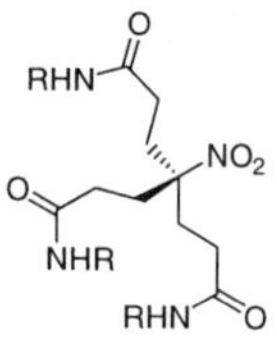 25	72 (24)
6e	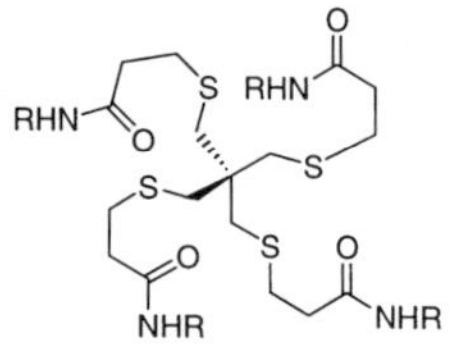 3	600 (150)

Fig. 5. Glycoclusters for studies of the interaction between galabiose and *S. suis*. Absolute and valency-corrected enhancements of binding over monovalent analogs are provided

(HAI) and enzyme-linked lectin assay (ELLA) [31]. The former assay measures the ability of a multivalent inhibitor to prevent the cross-linking of erythrocytes (red blood cells) by a lectin (in this case Con A). The latter method measures the ability of a multivalent inhibitor to prevent binding of a lectin to an immobilized multivalent ligand.

The results in Fig. 6 show that **7d** and **7e** are potent inhibitors of hemagglutination but only modest inhibitors of lectin binding by ELLA. The authors attribute this dramatic difference to the lectin used in the assays. In the ELLA, the Con A lectin was conjugated to horseradish peroxidase (HRP; a 40-kD protein) while in the hemagglutination assay, an unconjugated lectin was used. The large HRP moiety is believed to prevent effective crosslinking of the protein by the

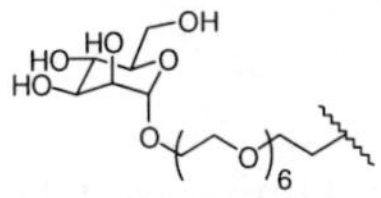	Potency relative to α-Me Mannose	
R =	HAI	ELLA
7a	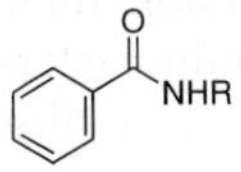 0.81	3.4
7b	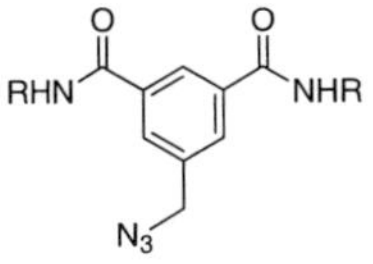 3.6 (1.8)	4.7 (2.3)
7c	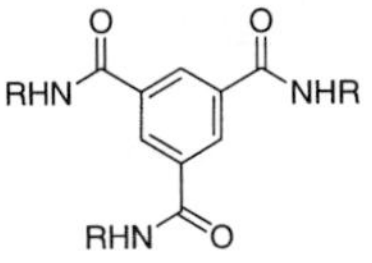 8.6 (2.9)	5.4 (1.8)
7d	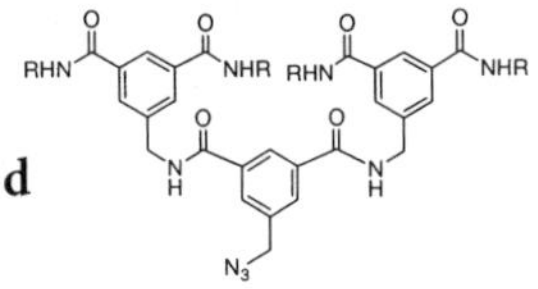 186 (47)	12 (3.0)
7e	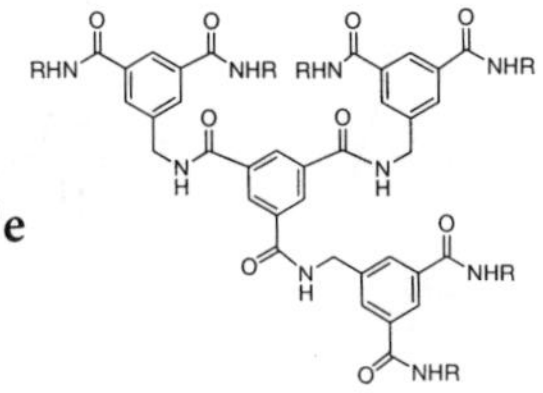 672 (112)	15 (2.5)

Fig. 6. Comparison of binding assays for polyvalent glycoclusters that present mannosides. Absolute and valency-corrected enhancements of binding over monovalent analogs are provided

multivalent ligand, reducing the potency of the glycoconjugate. This hypothesis was confirmed with ITC, which showed that the free energy of association between unconjugated Con A and **7d** and **7e** did increase (relative to α-methyl mannose) and that this increase arose primarily from a more positive entropy of interaction (see Sect. 2.3). This example illustrates the importance of the assay in the evaluation and interpretation of binding enhancements in polyvalent glycoconjugates. For the remainder of the chapter we will include the assay used for binding in our description of each model system.

4.2
Cyclodextrins and Calixarenes

Cyclodextrins are cyclic oligosaccharides containing α-(1,4)-linked glucopyranosyl units. These structures present a hydrophobic interior that is capable of binding small organic molecules and a hydrophilic exterior [32, 33]. Hydroxyl groups on the exterior of the cyclodextrin can be modified so that ligands are presented on a single face of the structure with control over ligand-to-ligand spacing (Fig. 7 A).

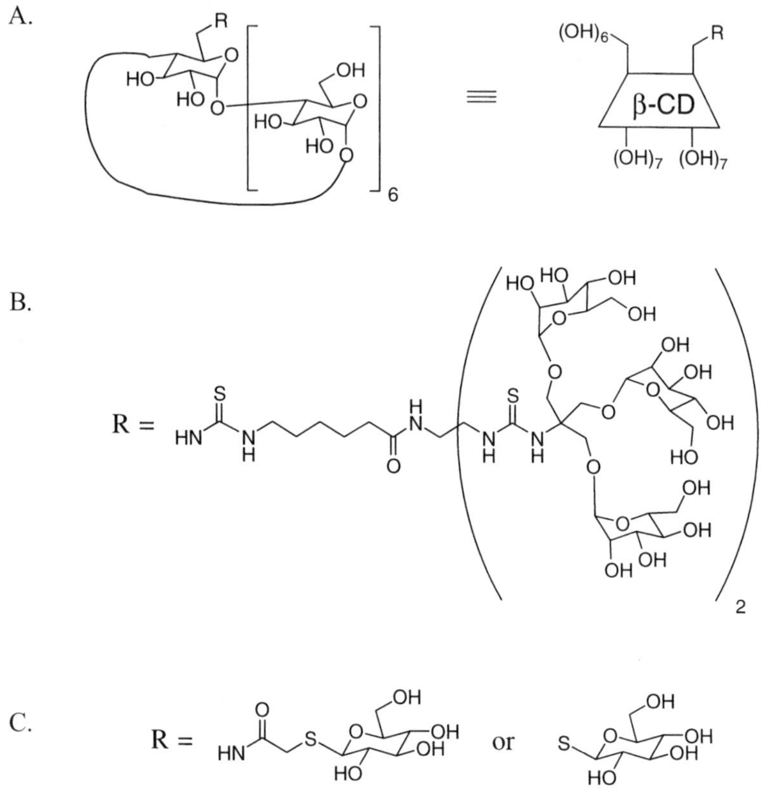

	IC$_{50}$	Enhancement
methyl-β-D-Glc	9470 μM	1
direct attachment	870 μM	10.9 (1.6)
acetamido spacer	90 μM	105.2 (15.0)

Fig. 7 A–C. Cyclodextrins that present carbohydrates. (A) general structure of β-cyclodextrin; (B) structure of the monosubstituted hexavalent structure prepared by Defaye and coworkers [37]; (C) structure of persubstituted hexavalent cyclodextrin presenting β-D-glucosides and data for the inhibition of binding of pea lectin to poly(acrylamide-*co*-allyl-α-D-mannoside) [38]

Although several synthetic methods for the preparation of glycosylated cyclodextrins have been reported [34, 35], few have evaluated the activity of these conjugates. In a recent report, Nishimura and coworkers prepared a series of perglycosylated cyclodextrins and showed that a hexavalent *N*-acetylglucosamine conjugate inhibited the agglutination of human erythrocytes at a 240-fold lower concentration than its monomeric counterpart [36]. Defaye and coworkers prepared a cyclodextrin derivatized with dendritic mannosides and showed that the hexavalent conjugate inhibited the binding of Con A to yeast mannan with an IC_{50} of 8 µmol/l, a 100-fold increase in potency over α-methyl mannoside [37] (Fig. 7B). In another study, Vargas-Berenguel and coworkers synthesized a series of β-cyclodextrins modified with either β-D-glucopyranose, β-D-galactopyranose, β-D-*N*-acetylglucopyranosylamine, or α-D-1-deoxy-1-thiomannopyranoside [38]. The inhibitory potency of these compounds was evaluated using an ELLA with the appropriate plant lectin and immobilized polymers presenting multiple carbohydrate ligands. Figure 7 C illustrates two glucosylated cyclodextrins that inhibited the binding of pea lectin to poly(acrylamide-*co*-allyl-α-D-mannoside). When the carbohydrate was attached directly to the cyclodextrin scaffold, a modest binding enhancement over β-methyl glucose was observed. The introduction of an acetamido spacer group between the carbohydrate and the cyclodextrin significantly enhanced the potency of the glycoconjugate.

Calixarenes are cyclic oligomers of substituted aromatic rings. Like cyclodextrins, these scaffolds are able to bind hydrophobic organic molecules and can be modified with carbohydrate ligands on one face of the molecule with control over ligand to ligand spacing (Fig. 8 A). These conjugates have been useful for both site-directed drug delivery [39] and for studies of water-monolayer surface interactions [40]. Roy and Kim, for example, prepared a series of dendritic *p-tert*-butylcalix[4]arenes presenting from 4 to 16 *N*-acetylgalactosamine ligands [41] (Fig. 8B). In an ELLA, the hexadecamer was 12 times more potent than allyl-α-GalNAc in inhibiting the binding of the agglutinin from *Vicia villosa* to asialoglycophorin. More recently, Aoyama and coworkers prepared a calix[4]arene presenting eight galactosides and showed that this glycoconjugate could deliver a fluorescent dye to the surface of rat hepatoma cells (Fig. 8 C) [42].

4.3
Dendrimers and Dendrons

Dendrimers represent a class of polymer composed of a core structure that is modified with several regularly hyperbranched units (Fig. 3 A). Dendrons are a subclass of dendrimers in which the core structure is modified with only one hyperbranched unit [43]. These structures are useful templates for the design of polyvalent ligands for two reasons. First, dendrimers and dendrons permit access to valencies between glycoclusters and model systems that present very large numbers of carbohydrates. Second, synthetic chemistry permits wide flexibility in adjusting the valency, size, and even shape of a dendron [20–24].

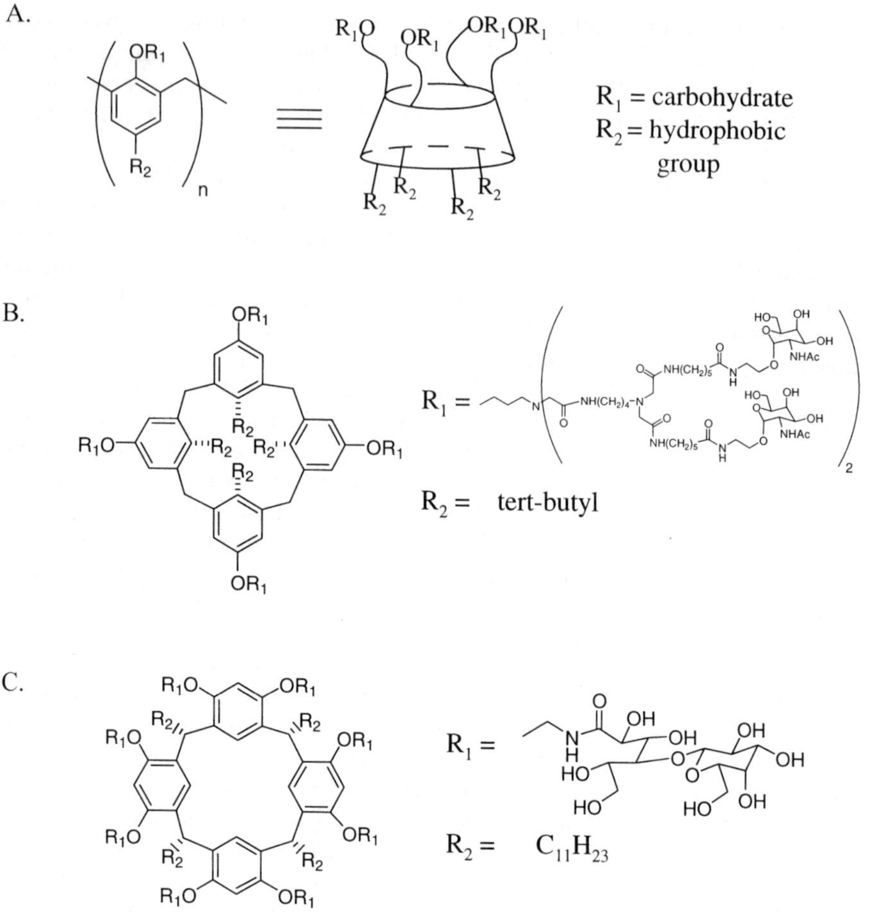

A.

R_1 = carbohydrate
R_2 = hydrophobic
group

B.

$R_1 =$

$R_2 =$ tert-butyl

C.

$R_1 =$

$R_2 =$ $C_{11}H_{23}$

Fig. 8 A–C. Calix[n]arenes that present carbohydrates. (**A**) General structure of calixarenes and (**B**) structure of calix[4]arenes that present GalNAc-containing dendrons [41] and (**C**) structure of calix[4]arenes for the delivery of fluorescent molecules to hepatoma cells [42]

Despite this flexibility in design, dendrimers and dendrons sometimes fail to display large enhancements in binding affinity toward carbohydrate-binding proteins. Roy and coworkers, for example, found that dendrons presenting *N*-acetylglucosamine displayed a valence-dependent enhancement of binding to the agglutinin from wheat germ (WGA) (Fig. 9). The same scaffold presenting *N*-acetyllactosamine (LacNAc), by contrast, showed little binding enhancement toward the lectin from *E. cristagalli*.

Many authors reason that the poor binding enhancements reflect an improper spacing or geometry of the carbohydrates presented on these scaffolds [13, 45]. It is possible, however, that larger dendrons and dendrimers (with valency greater than four) sterically prevent the cross-linking of carbohydrate-binding proteins. Further mechanistic studies using ITC will provide answers to this and

A.

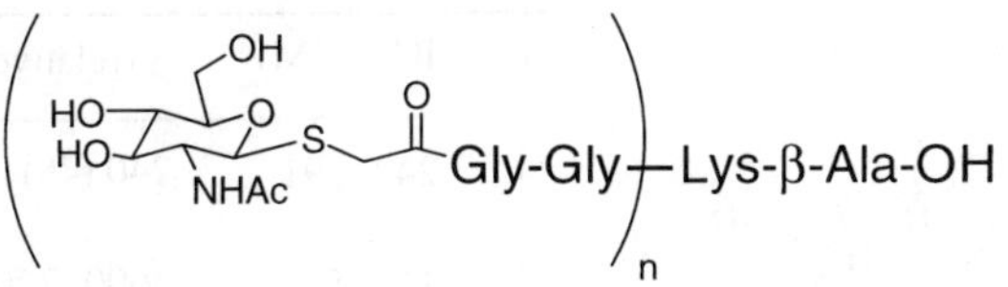

	IC_{50}	Enhancement
allyl-α-D-GlcNAc	15000 μM	1
$n = 2$	3100 μM	4.8 (2.4)
$n = 4$	510 μM	25.4 (6.4)
$n = 8$	88 μM	170 (21.2)

B.

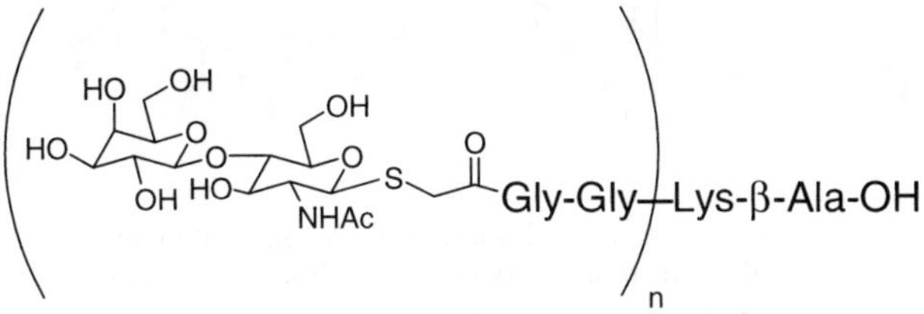

	IC_{50}	β
azido-β-D-LacNAc	430 μM	1
$n = 2$	340 μM	1.3 (0.65)
$n = 4$	140 μM	3 (0.75)
$n = 8$	86 μM	5 (0.63)

Fig. 9A, B. Comparison of dendrons presenting. (**A**) *N*-acetylglucosamine and (**B**) *N*-acetyllactosamine. Each series of compounds was evaluated for its ability to inhibit binding of wheat germ agglutinin (A) or *Erythrina cristagalli* (B) to microtiter plates coated with porcine stomach mucin. Values in parentheses represent enhancements that have been corrected for the valency of the dendron. While the compounds in (A) shows binding enhancements (per ligand) over the corresponding monomeric carbohydrate, the compounds in (B) do not [44]

related mechanistic questions. For more discussion of these model systems, we direct the reader to other chapters in this volume and to several recent reviews [24, 46, 47].

4.4
Designs Based on Structural Information

Some of the most potent polyvalent systems with low valency have been designed using detailed structural information of the carbohydrate binding protein. While this approach is limited to systems for which crystallographic or NMR structural data are available, it significantly reduces the time required to develop a polyvalent system that presents ligands with the appropriate geome-

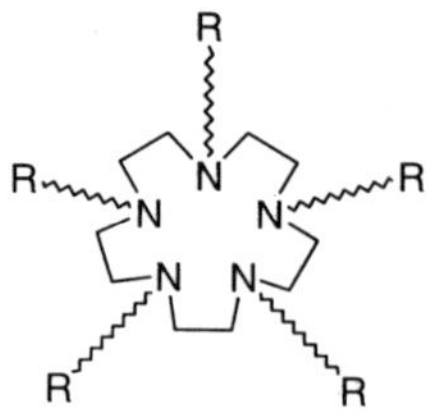

n	IC$_{50}$ (μM)	β (relative to galactose)
1	242 ± 91	240 (48)
2	16 ± 8	3600 (720)
3	6 ± 4	10 000 (2000)
4	0.56 ± 0.06	104 000 (20 800)

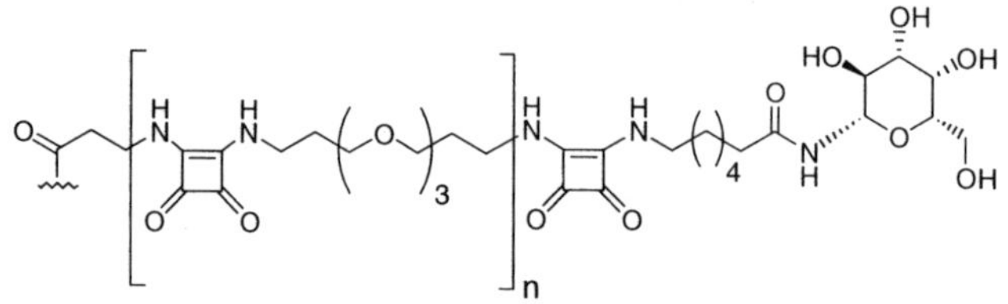

Fig. 10. Structure of a pentameric inhibitor of the heat-labile enterotoxin from *E. coli* [49]. Increasing the length of the tether between the carbohydrate and the macrocycle increases the inhibitory potency of the compound. Values in parentheses represent enhancements that have been corrected for the valency of the dendron

try and spacing. Two recent reports have utilized this approach in the design of synthetic inhibitors for members of the AB$_5$ family of bacterial toxins. These toxins, which include the heat-labile enterotoxin from *E. coli*, shiga toxin, shiga-like toxin, and pertussis toxin, consist of five identical carbohydrate-binding subunits arranged around a core subunit in a star-like fashion [48]. To mimic the structure of heat-labile enterotoxin, Hol and coworkers prepared an azamacrocycle presenting *β*-galactose and varied the length of the spacer between the macrocycle and carbohydrate groups (Fig. 10) [49]. The most potent analog demonstrated an IC$_{50}$ of 560 nmol/l, which represents an enhancement of 10^5 over monomeric galactose. This degree of enhancement is remarkable since the linker between the carbohydrate and the scaffold is quite flexible.

Bundle and coworkers recently prepared a decavalent sugar cluster for inhibition of the binding of Shiga toxin to gangliosides on the cell surface (Fig. 11) [50]. In this work, the authors used D-glucose as a scaffold for the attachment of ligands and varied the number of diethyl squarate moieties in the spacer arm to achieve optimal binding. The most successful inhibitor, named STARFISH, had an IC$_{50}$ of 4×10^{-10} mol/l. This value is comparable to the estimated affinity of the native ganglioside-pentamer interaction ($K_d = 10^{-9}$ mol/l) and is 10^6 more potent than the monomeric carbohydrate ($\beta = 10^6$). Interestingly, the authors obtained a crystal structure of the toxin-inhibitor complex and found that each STARFISH ligand bound two toxin pentamers. They attributed this finding to the lack of flexibility in the STARFISH oligomer. This example illustrates that even the most carefully designed interaction may behave unpre-

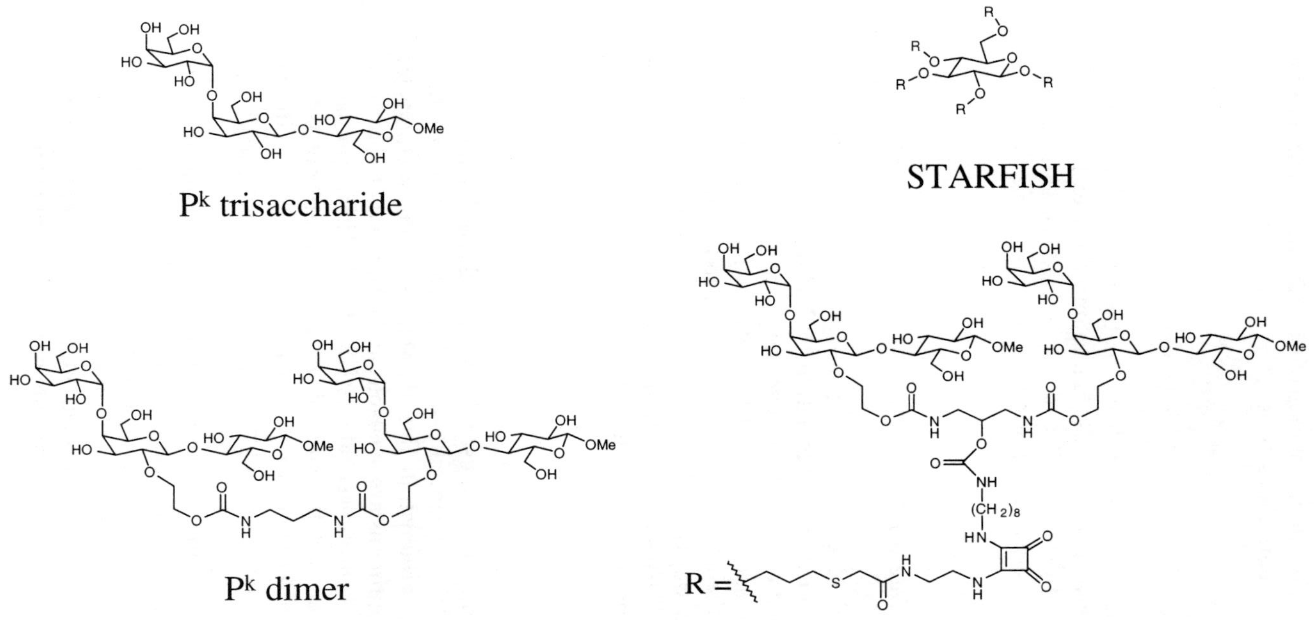

Compound	IC$_{50}$ (µM)	β
P^k trisaccharide	2100	1
P^k dimer	55	38 (19)
STARFISH	0.00024	8.75×10^6 (875000)

Fig. 11. Structure of a potent decameric inhibitor of shiga-like toxin [50]. Values in parentheses represent enhancements that have been corrected for the valency of the dendron

dictably. It also illustrates that the interpretation of experimental results is often ambiguous, even in the most controlled setting.

4.5
Other Models

A variety of other scaffolds have been utilized for the preparation of glyco-conjugates. These include glycopeptoids [51, 52], azamacrocycles [53], linear peptides [54, 55], cyclic peptides [56, 57] and others [58, 59]. Due to space considerations, we refer the reader to the original literature for further information.

5
Model Systems with High Valency

Many applications in glycobiology require model systems that present large numbers of carbohydrates, including vaccination [60], chromatography [61], and the inhibition of interactions that occur over large areas (such as cell-cell and cell-pathogen interactions). Examples of the latter include the binding of leukocyte selectins to endothelial cells (cell-cell) [5], the adhesion of influenza virus to cells of the upper airway (cell-virus) [62], and the adhesion of *E. coli* to cells of the urinary tract (cell-bacteria) [63]. In the following section we outline several model systems that present large numbers of carbohydrates.

5.1
Neoglycoproteins

Neoglycoproteins remain one of the most widely used classes of polyvalent glycoconjugates. These models are naturally occurring proteins that have been synthetically modified with carbohydrate ligands. Many synthetic methods for the preparation of neoglycoproteins are available, and a large number of these methods provide reasonable control over valency [18, 21 – 23]. These approaches have the limitation, however, that the sites of glycosylation are not strictly defined and therefore yield heterogeneous glycoforms. To overcome this limitation, several methods have been developed to produce defined glycoforms (Scheme 2).

Davis and coworkers, for example, have used site-directed mutagenesis to introduce cysteine residues in defined locations, which are then selectively coupled with methanethiosulfonate-derivatized carbohydrates [64, 65]. While useful for determining the role of glycosylation in modifying the activity of a protein, this and other chemoselective ligation methods [66 – 69] are not yet routine and have not been used to generate neoglycoproteins with large numbers of carbohydrates.

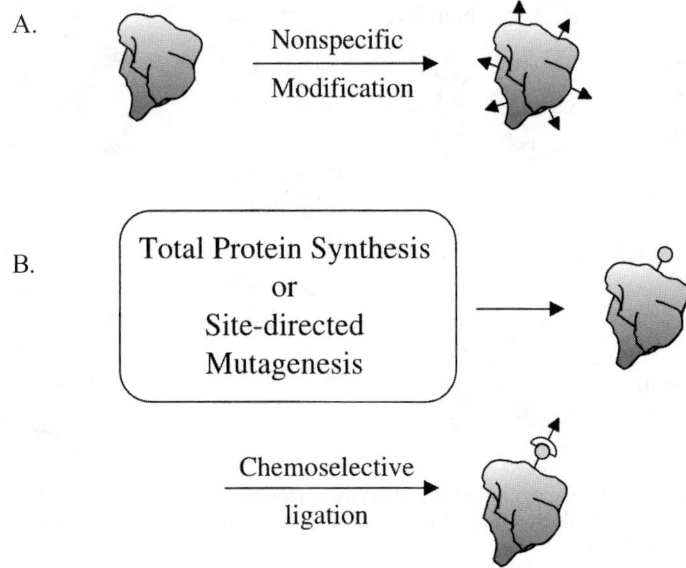

Scheme 2A, B. Preparation of neoglycoproteins by (**A**) nonspecific modification and (**B**) chemoselective ligation. The conjugates in B are prepared by introducing a chemical handle through total protein synthesis or site-directed mutagenesis

5.2
Neoglycopolymers

Polymers derivatized with carbohydrates continue to be important tools in mechanistic studies of polyvalent carbohydrate protein interactions. Polymers have been attractive scaffolds because synthetic approaches for their preparation permit control over the molecular weight of the polymer and the density of one or more ligands on the polymer backbone [22, 23, 70, 71]. This model system has the limitation that ligands are not presented in defined environments and that the number of interactions is often unknown. For studies that do not require strict mechanistic interpretations, these considerations are less important. We have separated the discussion of these glycoconjugates into two classes of model systems: soluble and immobilized polymers. The former have emerged as potent inhibitors of polyvalent carbohydrate-protein interactions, while the latter have found extensive analytical applications in glycobiology.

5.2.1
Soluble Polymers

The properties of a soluble polymer depend on a number of factors, including the structure of the backbone and the molecular weight. Several backbones have been used in the literature, including polyacrylamide, polylysine [72], polystyrene [73, 74], dextran [75], and several others [76–79]. Of these, poly-

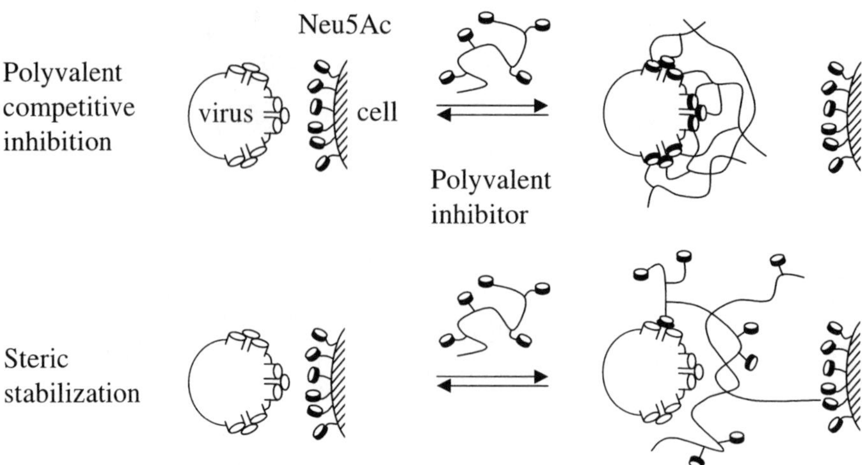

Fig. 12. Inhibition of hemagglutination of erythrocytes by polyacrylamide gels presenting sialic acid (Neu5Ac). Inhibition is attributed to both polyvalent competitive inhibition and steric stabilization [81–84]

acrylamide has been the most widely used because of its biocompatibility and ease of preparation [80]. Whitesides and coworkers, for example, have prepared polyacrylamides presenting a C-glycoside of sialic acid and shown that these conjugates strongly inhibited hemagglutination of erythrocytes by the influenza virus ($\beta = 10^{10}$ M^{-1}) [81–83]. The potency of these polymeric inhibitors was attributed to two mechanisms of action: polyvalent competitive inhibition and steric stabilization (Fig. 12). The latter effect is due to the ability of the water-swollen polymer to interfere sterically with the erythrocyte-virus interaction.

In subsequent work, these polymers were found to be more potent inhibitors of hemagglutination in the presence of a monomeric inhibitor of the enzyme neuraminidase [84]. This enzyme is also present on the surface of the virus particle and competes with hemagglutinin for sialic acid groups on the polymer (Fig. 13). Interestingly, the enhancement was proportional to the inhibition constant of the monomeric inhibitor (Fig. 14A). The authors suggested that the monomeric inhibitor displaces the polymer from the erythrocyte, enhancing its ability to inhibit sterically the virus-erythrocyte interaction (Fig. 14B). More potent monomeric inhibitors were more effective at displacing the polymer and increasing the observed value of K^{HAI} for the two-component system.

Wang and coworkers utilized polyacrylamides derivatized with Galα(1,3) Galβ(1,4)Glcβ epitopes to inhibit the immune recognition of porcine xenografts [85]. Using an inhibition ELISA with purified human anti-Gal antibody, they found that the polymers showed greater enhancements for classes of antibody that display greater numbers of binding sites (IgM > IgA > IgG). In addition, the polymers were shown to inhibit binding of anti-Gal IgM to pig kidney cells in a

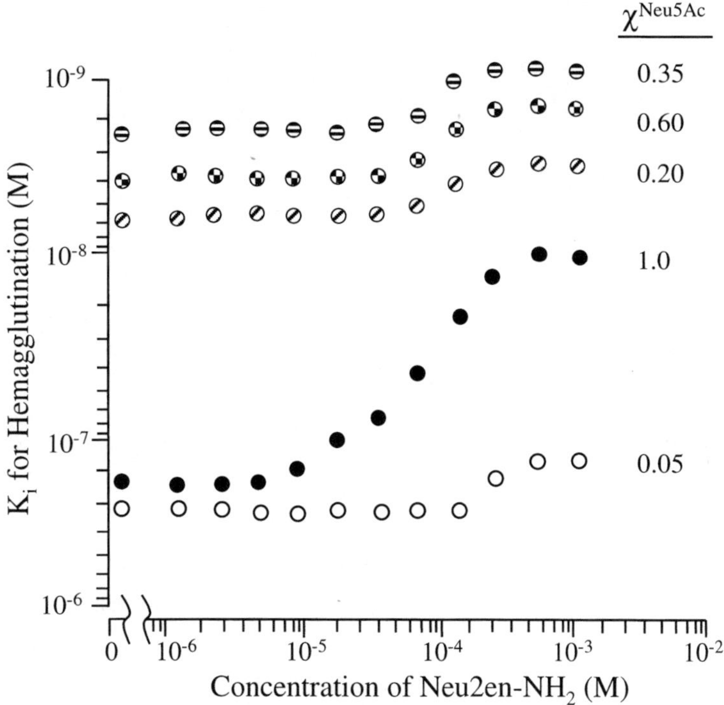

Fig. 13. The inhibitory potency of polyacrylamides presenting sialic acid is enhanced by Neu2en-NH$_2$, a monomeric inhibitor of neuraminidase [84]. Polymers presenting different sialic acid contents (χ_{NeuAc}) are shown in the figure. The monomeric inhibitor enhances inhibition most effectively with polymers that present only sialic acid ($\chi_{\text{NeuAc}} = 1$)

concentration-dependent manner whereas the trisaccharide alone had no effect, even at a concentration of 1 mmol/l.

Polyacrylamides decorated with sialic acid and sialic acid derivatives have also been used for studies of selectin-mediated interactions [86–88]. Hayashi and coworkers, for example, prepared a homopolymer from a modified SLe^x acrylamide conjugate and examined its activity against HL-60 cells in vitro and in a murine model in vivo [88]. The polymer inhibited adhesion more readily than the monomeric sugar (IC$_{50}$ = 1.5 mg/ml for Sialyl Lewis X; IC$_{50}$ = 0.01 mg/ml for the polymer), but the ability of the polymer to inhibit inflammation in vivo was only slightly greater than the monomer.

The use of ring-opening metathesis polymerization (ROMP) has also found application in the construction of glycopolymers. ROMP is a method for generating "living" polymers and permits the preparation of block copolymers with excellent control over the composition and length of each component. The living nature of the polymerization also permits the preparation of polymer chains that are end-labeled with different functional groups (Scheme 3) [89, 90].

Kiessling and coworkers have used ROMP to prepare several classes of polyvalent glycoconjugates. Initial work developed short polymers presenting sul-

A.

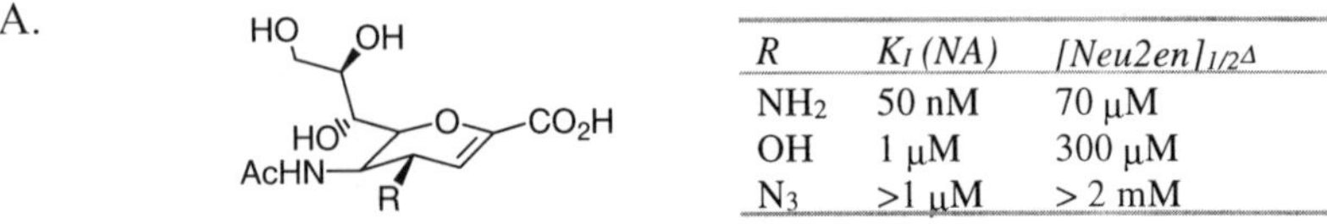

R	K_I (NA)	$[Neu2en]_{1/2\Delta}$
NH_2	50 nM	70 µM
OH	1 µM	300 µM
N_3	>1 µM	> 2 mM

B.

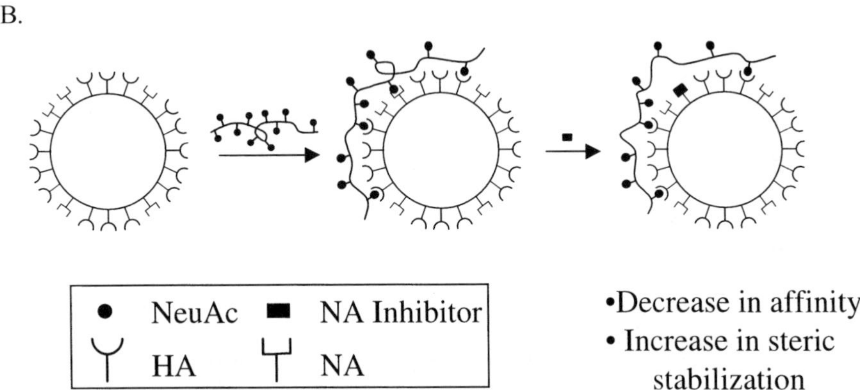

Fig. 14. A More potent monomeric inhibitors enhance inhibition of hemagglutination more readily than weaker inhibitors [84]. $[Neu2en]_{1/2\Delta}$ represents the concentration of monomeric inhibitor required to enhance the inhibition of hemagglutination by 50%. **B** Proposed mechanism for enhancement of inhibition of hemagglutination by monomeric inhibitors

Scheme 3. Preparation of glycopolymers using ROMP

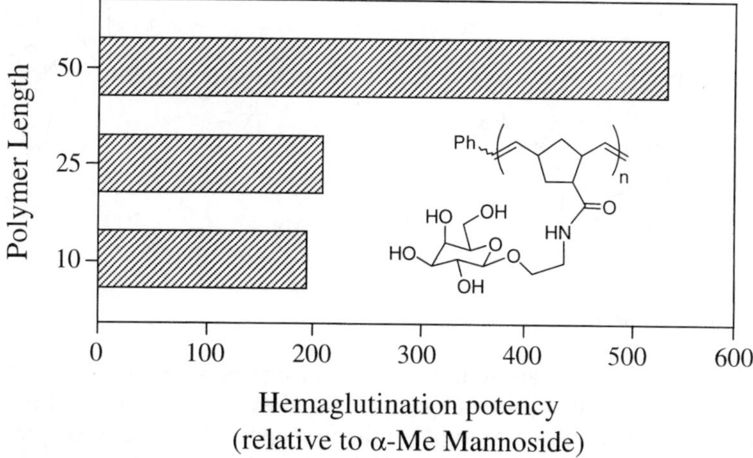

Hemaglutination potency
(relative to α-Me Mannoside)

Fig. 15. Relationship between the length of a mannoside polymer and its ability to inhibit hemagglutination by Con A [95, 96]. The 50-mer exhibits a significant enhancement of inhibition relative to the 25-mer and the 10-mer, perhaps due to its ability to span two binding sites of the lectin

fated galactose moieties as inhibitors of the binding of leukocyte selectins to glycoconjugates on the surface of high endothelial venules [91–93]. They later found that these polyvalent oligomers, but not their monovalent counterparts, were able to induce the release of L-selectin from the surface of lymphocytes [94]. Subsequent work expanded this approach to the study of binding to concanavalin A [95, 96]. In these studies, the authors found that longer polymers were more effective than shorter polymers at inhibiting hemagglutination of erythrocytes by Con A (Fig. 15).

Recent work has taken advantage of the living nature of the polymerization to generate polymers presenting carbohydrate ligands and fluorescent reporter groups. Polymer **9** has been used to visualize cell-surface selectins [97], while polymer **10** has been used to examine the chemotactic response of *E. coli* [98].

9

10

5.2.2
Immobilized Polymers

Immobilized polymers have found use as solid supports for studies in cell biology and as scaffolds that present ligands for studies of polyvalent binding interactions. The most common example of the first application uses glycopolymers for the culture of hepatocytes [99–102]. Weisz and Schnaar, for example, used polyacrylamide derivatized with galactose to study the adhesion of hepatocytes and the localization of the asialoglycoprotein receptor during adhesion [100]. Schnaar and coworkers also used gels presenting gradients of the carbohydrate N-acetylglucosamine to study the migration of melanoma cells [101]. More recently, Griffith and Lopina prepared a class of "star" poly(ethylene oxide) gels presenting galactose [102]. These gels presented the sugar in clusters and enabled the investigators to estimate the minimum ligand to ligand distance required for efficient function of the hepatocytes.

Immobilized polymers have been particularly important model systems for studying the kinetics of carbohydrate-protein interactions. Surface plasmon resonance (SPR) is an analytical technique that is well suited for these studies because it measures binding interactions in situ and without the need for molecular or enzymatic labels. SPR has been reviewed elsewhere [103, 104], and its use for studies of carbohydrate-protein interactions is the subject of another chapter in this volume. It is, however, important to describe this technique in brief here. SPR measures changes in the refractive index of the medium near a thin film of gold or silver. Since the association of proteins or other small molecular weight ligands with groups presented near the interface influences the local refractive index of the medium, this technique can monitor binding interactions in real time (Fig. 16).

The majority of studies using SPR are conducted by coupling protein or small molecules to an immobilized carboxymethylated dextran polymer. In the case of carbohydrate studies, carbohydrate-specific antibodies, lectins, or neoglycoconjugates have been coupled to the polymer. An appropriate binding partner is flowed over the polymer and binding interactions are measured. Vliegenthart and coworkers, for example, immobilized sialic acid binding lectins from *S. nigra* and *M. amunrensis* to the dextran matrix and examined the interaction of these proteins with a series of oligosaccharides [105]. Other studies have grafted preformed glycopolymers onto self-assembled monolayers on gold or silver [106, 107]. We will discuss the use of SPR to study interactions between lectins and well-defined organic surfaces later in the review.

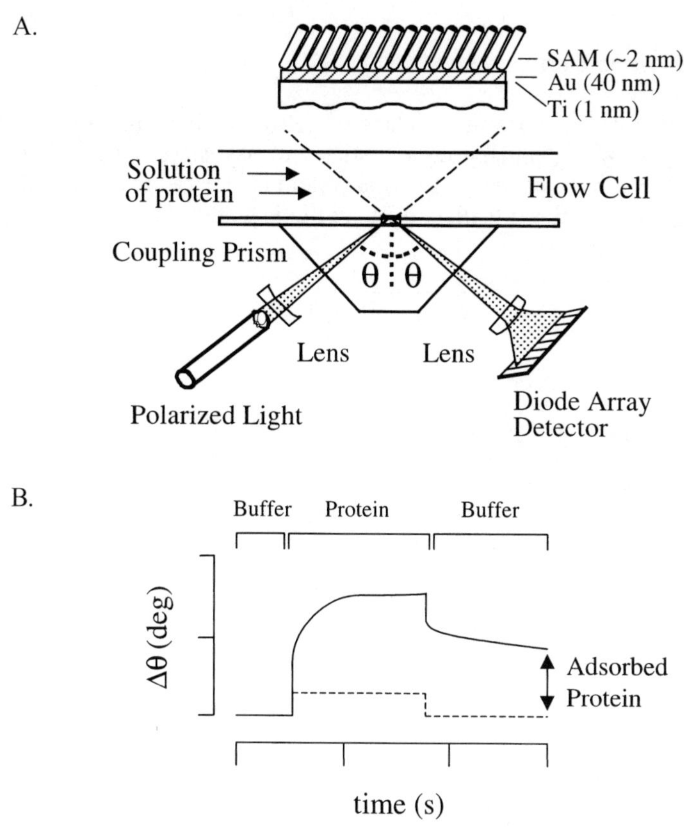

Fig. 16A, B. Surface plasmon resonance. (A) schematic of the optical biosensor and (B) example of a sensorgram that shows the association of a protein with an immobilized ligand

5.3
Self-Assembled Monolayers

Self-assembled monolayers are model systems that present one or two layers of amphiphilic molecules in a two-dimensional array [108]. Several classes of monolayers have found application in the study of biological recognition, including hybrid bilayers on glass, hybrid bilayers on gold and self-assembled monolayers of alkanethiolates on gold. Supported lipid bilayers are also an important model system in this class, but this model has not been used extensively for studies of multivalent carbohydrate-protein recognition. Each of these model systems presents carbohydrate ligands with control over the density of the ligand (but not necessarily over ligand-ligand spacing). An important feature of these models is that the immobilization of the ligand at a surface prevents intermolecular associations of multiple polyvalent ligands and receptors (as shown in Fig. 1). This characteristic simplifies the mechanistic analysis of polyvalent interactions. The following section presents recent developments in the use of these monolayers for studies of carbohydrate-protein interactions.

5.3.1
Hybrid Bilayers on Glass

Hybrid bilayers on glass are a class of self-assembled monolayers that are formed by allowing phospholipid vesicles to associate with a monolayer of hydrophobic alkylsiloxanes [109]. This system is referred to as a hybrid because it contains both a bilayer leaflet and a synthetic leaflet (Scheme 4).

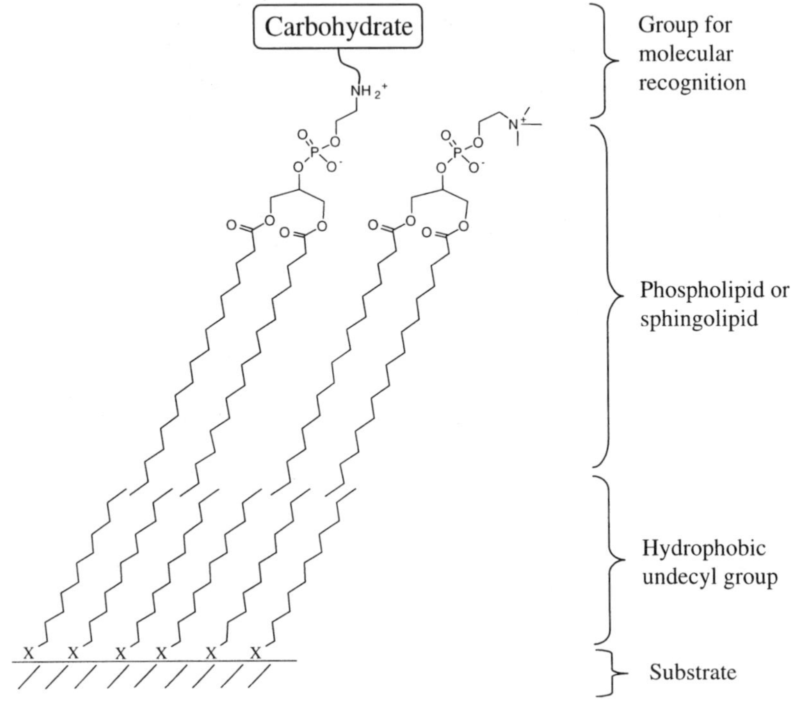

Scheme 4. Architecture of a hybrid bilayer. For a bilayer on gold, X = sulfur. For a bilayer on glass, X = Si(O)

Hybrid bilayers are good model systems for studying cell-cell and cell-pathogen interactions because ligands presented in the bilayer are free to diffuse laterally, much like ligands found in cellular membranes [110]. While this property allows the formation of ligand clusters, it may also facilitate the exchange and nonspecific adsorption of cell surface proteins. The extent to which changes in the composition of these substrates changes over time has not been determined. The utility of supported bilayers in mechanistic studies of carbohydrate-protein interactions was demonstrated by Stevens and coworkers in the development of a sensor for bacterial toxins [111]. This work utilized a hybrid bilayer presenting 10,12-pentacosadiynoic acid (PDA), a sialic acid-PDA conjugate, and a ganglioside lipid (Fig. 17).

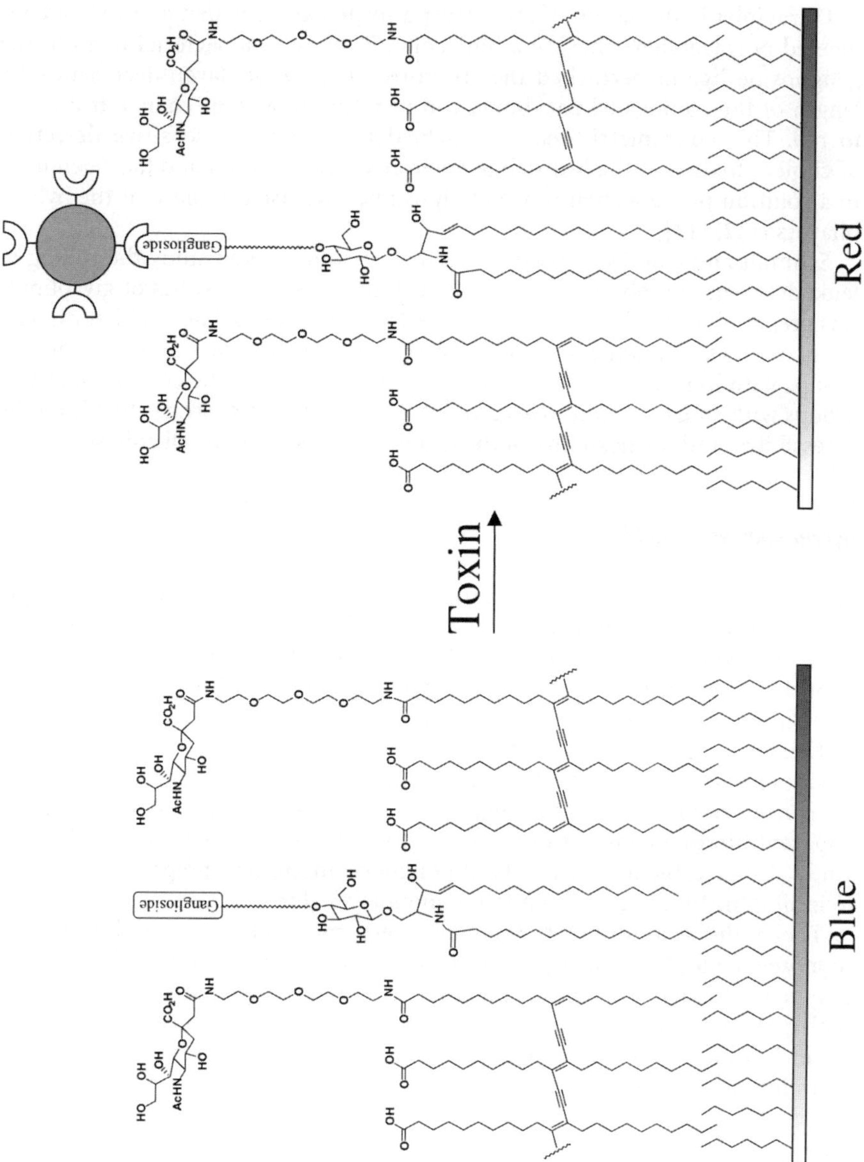

Fig. 17. Colorimetric detection of toxins using a supported bilayer [111]. The extensively conjugated poly(acetylene) network on the left exhibits a blue color. Binding of a toxin to the ganglioside in the bilayer decreases the length of the conjugated network and changes the color of the film from blue to red. This figure shows binding of the toxin but not changes in the structure of the bilayer

Ultraviolet illumination of the bilayer afforded an extended network of conjugated polyacetylenes having a blue color. Binding of a bacterial toxin to the ganglioside ligand perturbed the structure of the monolayer, decreasing the length of the conjugated backbone, and resulting in a color change from blue to red. This colorimetric readout enabled the rapid and sensitive detection of cholera toxin and the heat-labile enterotoxin from *E. coli* and has been used in a solution phase format in which liposomes are used in place of the hybrid bilayers [112, 113].

Schmidt and coworkers used hybrid bilayers on glass to study the rolling of leukocytes on immobilized SLe^x [114, 115]. In one study, a series of glycolipids was prepared in which SLe^x was connected to a glycerol backbone either directly or via tri-, hexa-, or nona(ethylene glycol) groups [114]. At low densities of ligand, groups with longer spacer arms were still able to support the cell rolling, while groups with shorter spacer arms did not. This result demonstrates that both the accessibility and the flexibility of the ligand are important in cell adhesion.

5.3.2
Hybrid Bilayers on Gold

Hybrid bilayers on gold are prepared by allowing a phospho- or sphingolipid vesicle to associate onto a hydrophobic self-assembled monolayer on gold (Scheme 4 where X = sulfur) [116]. These models are often preferred over the hybrid bilayers on glass because the underlying gold film permits the use of surface plasmon resonance for studies of real time ligand-receptor interactions.

Kiessling and coworkers used SPR to investigate the binding of Con A to mannosides presented in a hybrid bilayer [117]. By incubating the lectin with different concentrations of α-methyl mannose, the authors were able to determine the apparent binding constant of the lectin for the immobilized mannoside. This study also investigated the ability of polymeric inhibitors (prepared by ROMP) to inhibit the binding of Con A to the monolayer (Fig. 18).

The authors found that polymers were more effective inhibitors than monomers and that longer polymers were more potent inhibitors than shorter polymers. Preincubation of the polymers with the lectin significantly improved their potency. It is still unclear, however, whether the enhanced inhibition of lectin binding by longer polymers was due to polyvalent attachment or to crosslinking of the protein in solution.

5.3.3
Self-Assembled Monolayers of Alkanethiolates on Gold

Self-assembled monolayers of alkanethiolates on gold (SAMs) are a class of model substrates that are especially well suited for studies of polyvalent binding at interfaces [118, 119]. These monolayers form spontaneously by the adsorption of alkanethiols from solution onto a clean surface of gold. Since the properties of the monolayer depend upon the terminal functional group of the precursor alkanethiol, virtually any surface can be prepared using organic synthesis (Scheme 5). SAMs that present oligo(ethylene glycol) groups have the addition-

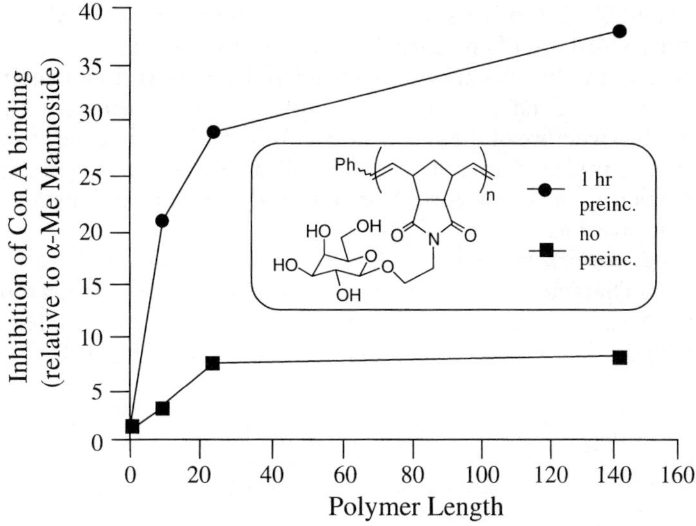

Fig. 18. Inhibition of binding of Con A to a hybrid bilayer presenting mannose groups [117]. Longer polymers were more effective than shorter polymers at inhibiting binding to the bilayer. All polymers were more potent inhibitors when they were preincubated with the lectin

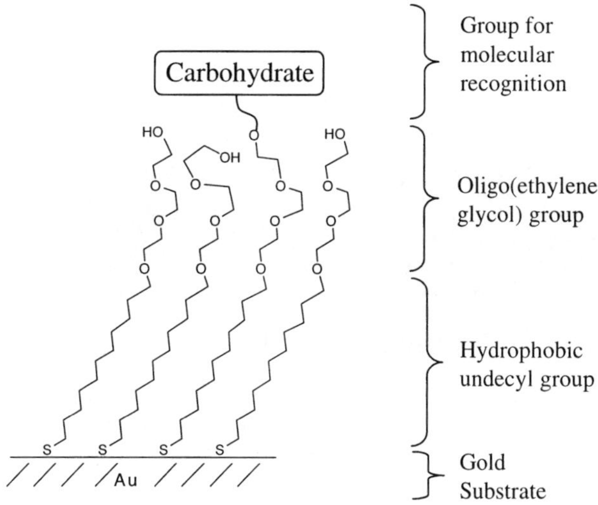

Scheme 5. Architecture of a self-assembled monolayer of alkanethiolates on gold

al advantage that they are effective at resisting the nonspecific adsorption of protein, which allows an unambiguous characterization of specific carbohydrate-protein interactions [120].

For mechanistic studies of polyvalent carbohydrate interactions, monolayers offer three advantages over other model systems. First, the properties of monolayers enable molecular level control over the structure, density, and environment of ligands. This feature is particularly important for carbohydrate-protein interactions, which are often sensitive to subtle changes in the presentation of ligand. Second, the ligands are immobilized and presented in an environment that is inert to nonspecific adsorption of protein. This characteristic eliminates changes in the structure of the substrate over time and is particularly important for studies that utilize mammalian cells. Finally, these substrates are compatible with SPR, making the characterization of ligand structure and ligand-receptor interactions possible.

Kahne and coworkers used monolayers presenting different densities of carbohydrate to characterize a lectin "specificity switch" [121]. Two carbohydrate ligands were tested in this study: a native ligand for the *B. purpura* lectin and a lead compound that bound this ligand in a library screen (Fig. 19).

A.

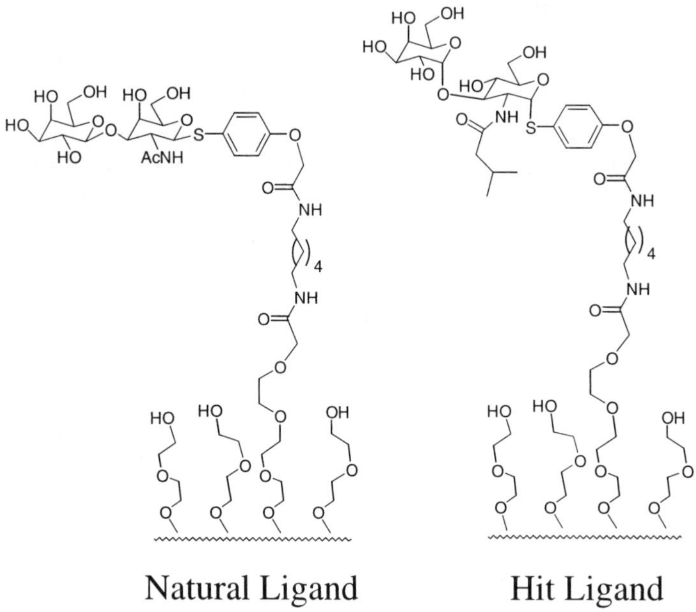

Natural Ligand   Hit Ligand

B.

IC_{25} with

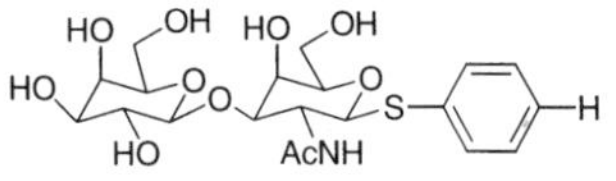

	Natural Ligand	Hit Ligand
$\chi_{sugar} = 0.1$	150 μM	130 μM
$\chi_{sugar} = 0.6$	80 μM	190 μM

Fig. 19A, B. Density-dependent binding of the lectin from *B. purpura* to two carbohydrate ligands [121]. (**A**) the structure of the two monolayers presenting the two ligands; (**B**) at low densities of sugar ($\chi_{sugar} = 0.1$), the immobilized native ligand interacts more strongly with the lectin, while at higher densities ($\chi_{sugar} = 0.6$), the hit ligand from a library screen binds more tightly

Using SPR, they found that the lectin preferred the native ligand at low densities but bound the lead compound more tightly at higher density. This finding suggests that cells may regulate their interaction with extracellular proteins through changes in the density of a particular carbohydrate in the glycocalyx.

We have used monolayers to study the interaction between cellular glycosyltransferases and glycosylated proteins of the extracellular matrix. One example of this interaction occurs during the migration of malignant cell lines that express β-(1,4)-galactosyltransferase (GalTase) in their plasma membrane [122, 123]. This enzyme, traditionally found in the Golgi apparatus, plays a crucial role in the post-translational modification of proteins. When expressed at the cell surface, GalTase transfers galactose to N-acetylglucosamine (GlcNAc) residues on the extracellular matrix protein laminin. The enzyme modifies the matrix only during migration, suggesting that it may play a mechanistic role in the migration process.

We utilized monolayers that present GlcNAc as a model system to investigate the enzymatic modification of surfaces by GalTase [124]. In initial work, we prepared monolayers presenting different mole fractions of GlcNAc and investigated the enzymatic incorporation of ^{14}C-labeled galactose onto the monolayer (Fig. 20). The modification of the substrate by GalTase showed a striking dependence on the density of ligand on the monolayer (Fig. 20B). At mole fractions less than 0.7 ($\chi_{GlcNAc} < 0.7$), galactose incorporation increased linearly with the density of carbohydrate. At higher densities, however, incorporation of ^{14}C galactose decreased rapidly. Monolayers presenting carbohydrate alone, for example, incorporated the same density of ^{14}C as monolayers with $\chi_{GlcNAc} = 0.2$. This result suggests that ligand crowding interferes with further enzymatic modification at high values of χ_{GlcNAc}.

In order to characterize the enzymatic modification, we investigated the binding of two lectins to monolayers presenting either GlcNAc or the product of the enzymatic reaction N-acetyllactosamine (LacNAc) (Fig. 21). Before the modification only the lectin from *B. simplicifolia II*, which recognizes GlcNAc, associated with the monolayer. After the modification, however, only the lectin from *E. cristagalli*, which recognizes LacNAc, bound to the surface. To determine if the lectins could be used to monitor the reaction, we examined the binding of *B. simplicifolia* over time and found that there was indeed a time-dependent decrease in binding as the enzymatic modification proceeded (Fig. 22).

These results show that the combination of monolayers and SPR is a powerful technique for mechanistic studies of carbohydrate-protein interactions at interfaces. These methods can also be extended to the study of other glycosyltransferases and glycosidases. We are currently studying these processes as well as the modification of the carbohydrate by migrating cells.

A.

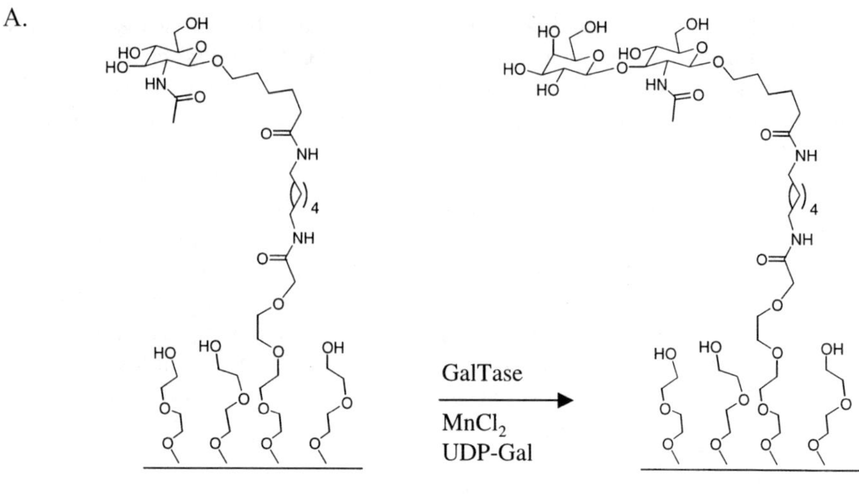

B.

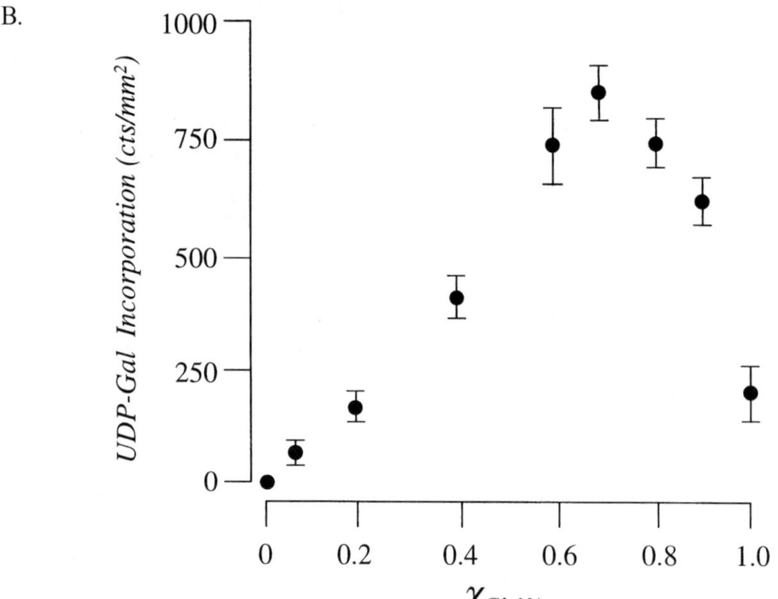

Fig. 20. (**A**) Enzymatic modification of monolayers that present GlcNAc [124]. (**B**) Relationship between the density of carbohydrate on the monolayer (χ_{GlcNAc}) and the transfer of ^{3}H galactose onto the monolayer

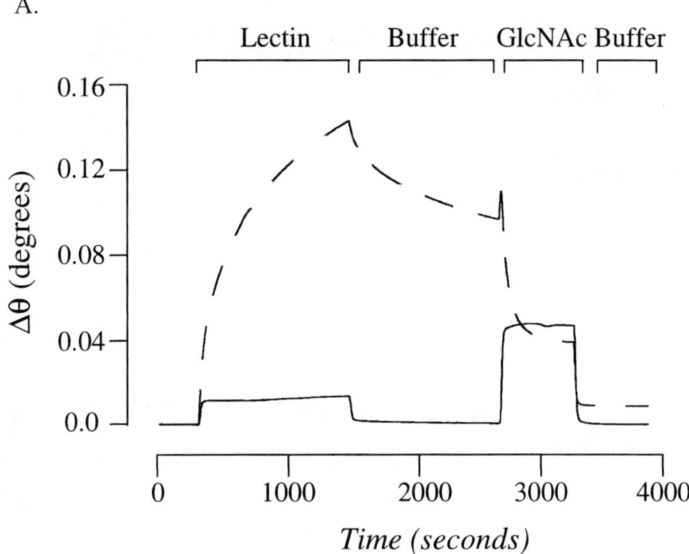

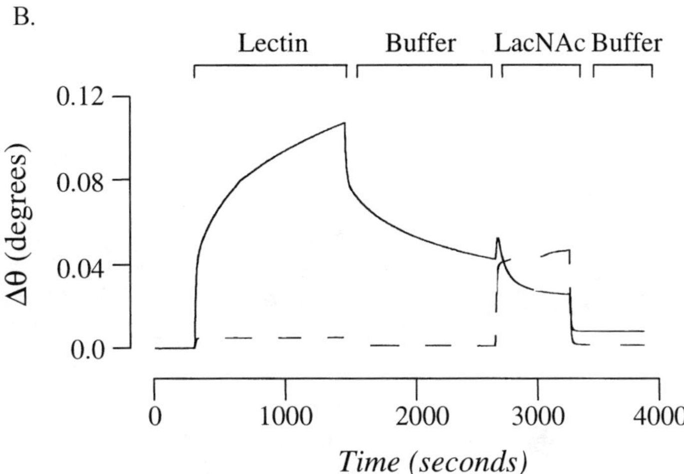

Fig. 21. Data from SPR for the binding of *Bandeiraea simplicifolia BS-II* lectin (*dashed curve*) and *Erythrina cristagalli* lectin (*solid curve*) to monolayers presenting (**A**) GlcNAc and (**B**) LacNAc among tri(ethylene glycol) groups. The *BS II* lectin bound monolayers presenting GlcNAc groups, while the *E. cristagalli* lectin bound only monolayers presenting LacNAc moieties. Binding of both lectins could be competitively inhibited by soluble carbohydrate ligand [124]

A.

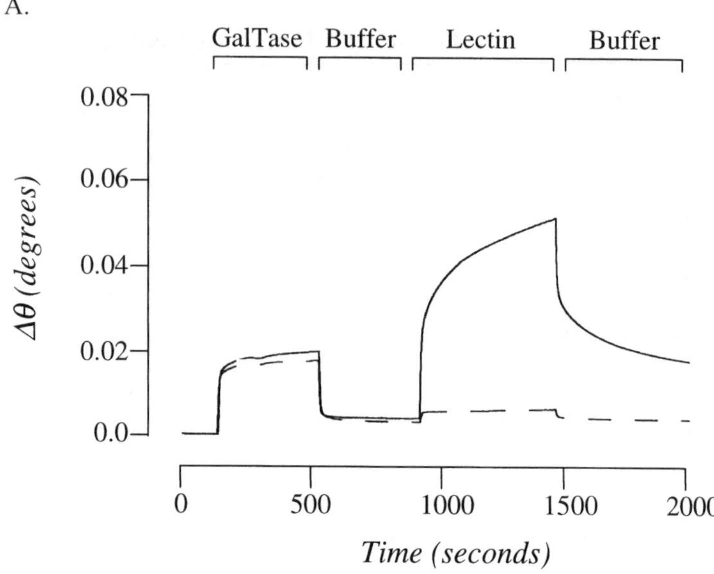

B.

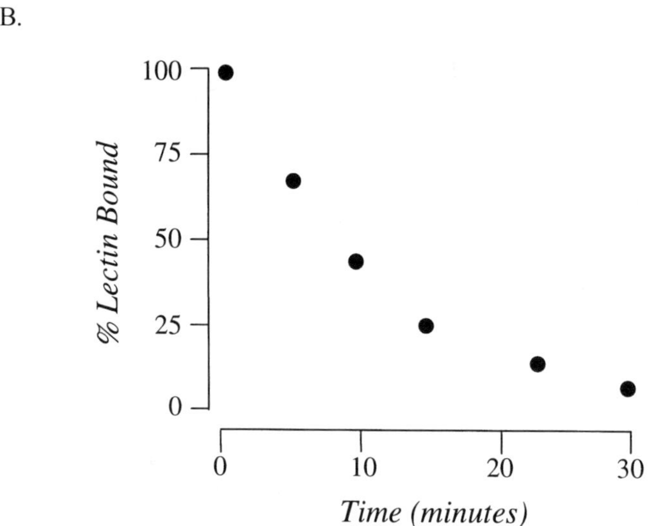

Fig. 22. (A) The *E. cristagalli* lectin (*solid curve*), but not the *BS II* lectin (*dashed curve*), binds LacNAc groups resulting from treatment of GlcNAc groups with GalTase. (B) Plot of *BS II* lectin binding vs time. All data are reported as a percentage of the binding response observed between the lectin and an untreated monolayer presenting GlcNAc [124]

6
Novel Model Systems

Several groups have begun to develop new model systems that extend the scope of current research on polyvalent carbohydrate-protein interactions. These include dynamic substrates, combinatorial methods, and the use of mammalian cells that present defined glycoforms on their surface. Each has unique capabilities and limitations that are described below.

6.1
Dynamic Self-Assembled Monolayers

Dynamic substrates are a class of model substrates that can change, in real time, the identity or density of ligands that participate in a polyvalent association. These substrates are important for modeling interactions that are altered by an external stimulus, such as the initiation of cellular or bacterial migration in response to a change in the density or structure of an immobilized carbohydrate [101]. Self-assembled monolayers on gold are ideal platforms for the development of dynamic substrates because the underlying metal film serves as an electrode that can affect oxidation-reduction chemistry at the interface [125]. Our group has employed the hydroquinone-quinone redox couple because the quinone group reacts efficiently and selectively as a dienophile in the Diels-Alder reaction, while the hydroquinone group is unreactive under the same reaction conditions [126, 127] (Fig. 23 A).

By tethering a carbohydrate to cyclopentadiene, it is possible to turn on the immobilization of the sugar (Fig. 23 B). We have used this methodology to immobilize GlcNAc to monolayers and have shown that the lectin from *B. simplicifolia* interacts with the monolayer only after the immobilization. This methodology makes possible the development of substrates for investigating a wide range of carbohydrate-mediated phenomena, including the up- or down-regulation of carbohydrate binding proteins on cell surfaces or in solution.

6.2
Combinatorial Approaches

Advances in combinatorial chemistry have permitted the rapid discovery and optimization of ligands for carbohydrate-binding proteins [128–130], minimizing the significant time investments often demanded by traditional synthetic approaches (Sect. 4). Kahne and coworkers, for example, prepared a library of disaccharides on solid support and screened this library for binding to the lectin from *Bauhinia purpura* [131, 132]. The screen identified a novel ligand that bound the lectin more tightly than the native ligand when presented on beads.

Whitesides and coworkers have used a combinatorial approach to libraries of polyacrylamides presenting derivatives of sialic acid [133]. Evaluation of these compounds identified a ter-polymer that inhibited the hemagglutination of chicken erythrocytes at subnanomolar concentration. In another study, Roy and coworkers generated a series of polyacrylamide gels presenting mono- and di-

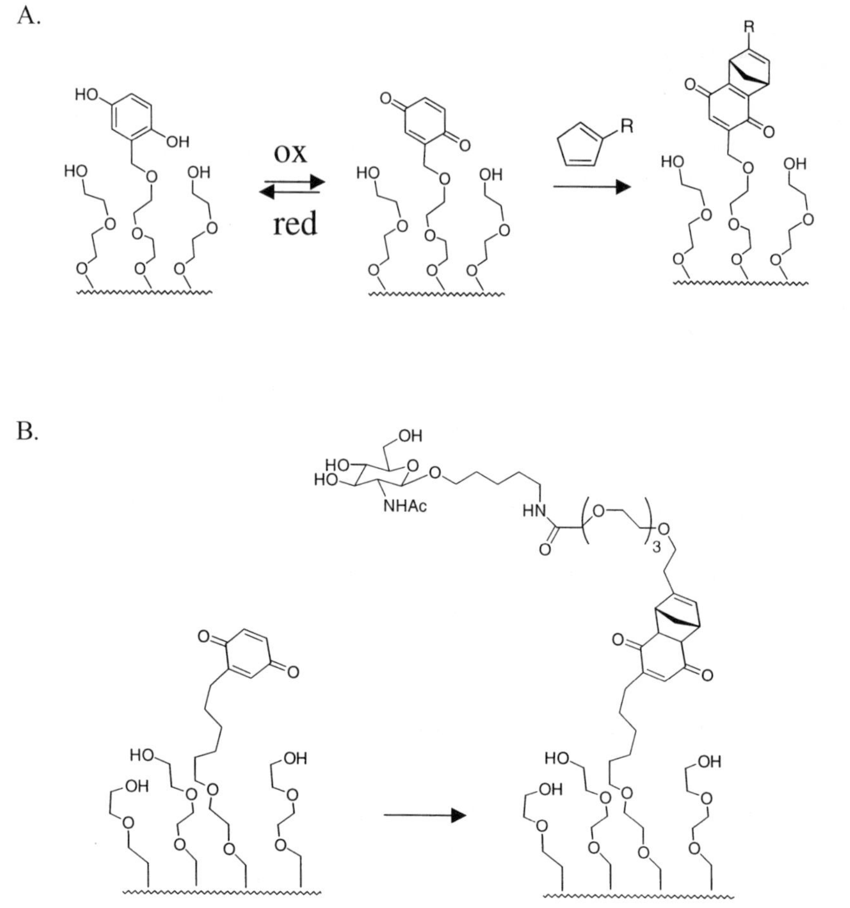

A.

B.

Fig. 23 A, B. Diels-Alder reaction for the immobilization of carbohydrates to monolayers. (A) Hydroquinone groups can be reversibly oxidized to the quinone group, which undergoes a Diels-Alder reaction with a substituted cyclopentadiene. (B) Immobilization of GlcNAc by the Diels-Alder reaction of quinone and a conjugate of the carbohydrate and cyclopentadiene

saccharides and evaluated the modification of these polymers with several gly-cosyltransferases in a 96-well format [134]. They demonstrated that this assay was both accurate and more efficient than its solution-phase counterpart. Recently, Ramstrom and Lehn utilized disulfide exchange to generate a small library of divalent carbohydrates and screened this library for its ability to bind Con A [135]. A bis-mannoside was selected from the mixture as the most potent ligand for the lectin.

The preparation of carbohydrate libraries on two-dimensional solid support also holds great promise for studies of polyvalent carbohydrate-protein interactions [128–130, 136, 137]. Jobron and Hummel, for example, generated a series of glycopeptides on cellulose using traditional methods in peptide synthesis [138]. This study did not, however, utilize the array to investigate multivalent recognition at the solid support. Our group is currently using self-assembled monolayers on gold for the preparation of carbohydrate arrays. By immobilizing a series of carbohydrates onto distinct regions of the monolayer, we are generating a "sugar chip" that can be modified either chemically or enzymatically. Target structures on the chip can be evaluated by immunofluorescence microscopy. Since monolayers permit control over the presentation and density of ligands, this approach makes it possible to define unambiguously ligands for bacterial and cellular adhesion.

6.3
Cells, Bacteria, and Viruses

An intrinsic feature of the model systems described above is that they do not present carbohydrates in their native cellular environment. Several approaches have been used to overcome this limitation. First, mammalian cells have been enzymatically tailored to present non-native carbohydrate structures [139–141]. Palcic and coworkers, for example, used Lewis $\alpha(1 \rightarrow 3,4)$-fucosyltransferase to transfer a preassembled trisaccharide onto the surface of erythrocytes [141]. This approach can, in principle, be extended to both bacterial and viral systems. A more recent methodology developed by Bertozzi and coworkers uses the cell's own metabolic machinery to present modified carbohydrates at the cell surface [142–144]. In this approach, the enzymes responsible for the synthesis of sialic acid process a derivative of N-acetylmannosamine substituted with a ketone group. The modified sialic acid is transported to the cell surface where it can be selectively modified using nucleophilic hydrazide, aminooxy, or thiosemicarbazide groups. While this method is selective and biologically relevant, it is limited to a small number of cellular pathways that tolerate non-native functional groups. We believe, however, that these cellular models will continue to become more useful and powerful in the near future.

7
Comparison of Model Systems

An overview and comparison of model systems is presented in Fig. 24.

8
Future Directions and Conclusions

This chapter establishes a mechanistic framework for understanding polyvalent carbohydrate-protein interactions and describes several classes of model systems for the evaluation of these interactions. While the examples in this review demonstrate the diversity and complex nature of carbohydrate-protein interactions, they also emphasize the need for the development of new analytical tools, novel synthetic methods, and more sophisticated model systems. We believe that the approaches described in this review, together with combinatorial methods and advances in materials science, promise to expand further the rate and scope of new discoveries in glycobiology.

A.   Overview of Model Systems

Low Valency Model Systems
Principle Use: Determination of structural and geometric requirements for tight
                          binding to a multivalent protein
Examples: Glycoclusters, cyclodextrins, calixarenes, dendrons and dendrimers

High Valency Model Systems
Principal Use: Chromatography, vaccination, inhibition of interactions that occur
                          over large surface areas
Examples: Neoglycoproteins, soluble polymers

Principal Use: Mechanistic studies of interactions that occur at interfaces
Examples: Immobilized polymers, hybrid bilayers on glass or gold, self-
                          assembled monolayers

B.   Comparison of Model Systems[a]

Model System	*Valency*	*ELLA*	*HAI/ ITC/ ACE*	*SPR*	*Stability[e,f]*	*Homogeneity[f,g]*	*Synthetic Effort*
glycoclusters	<7	Yes	Yes	No[c]	++++	++++	Significant
cyclodextrins	2-20	Yes	Yes	No[c]	++++	++++	Significant
calixarenes	2-20	Yes	Yes	No[c]	++++	++++	Significant
dendrimers	2-20	Yes	Yes	No[c]	++++	+++	Significant
neoglycoproteins	2-40	Yes	Yes	No[c]	++++	+[h]	Minimal[h]
soluble polymers	>10	Yes	Yes	No	++++	++	Minimal
immobilized polymers	>100	Yes[b]	No	Yes[d]	+++	++	Minimal
hybrid bilayers on glass	>100	Yes[b]	No	No	++	+++	Moderate
hybrid bilayers on gold	>100	Yes[b]	No	Yes	++	+++	Moderate
self-assembled monolayers	>100	Yes[b]	No	Yes	+++	++++	Moderate

[a]HAI = hemagglutination inhibition assay; ELLA = enzyme-linked lectin assay; SPR = surface
plasmon resonance; ITC = isothermal titration calorimetry; ACE = affinity capillary electrophoresis.
[b]An ELLA can be used if the model surface, rather than a ligand in solution, is evaluated.
[c]SPR can be used if the conjugate is immobilized directly or indirectly on a thin metal film.
[d]If the polymer is immobilized directly or indirectly on a thin meta film
[e]Stability depends on the nature of chemical bonds in the structure. For example, esters and
thioureas are less stable than amides.
[f]++++ = excellent; +++ = good; ++ = fair; + = poor
[g]A homogeneous system is one in which all ligands are equally accessible to a multimeric receptor
[h]These structures are more homogeneous, but less easily prepared, if chemoselective ligation
methods are used.

Fig. 24. **A** Summary of high valency and low valency model systems. **B** Comparison of model
systems presented in the review

9
References

1. Dwek RA (1996) Chem Rev 96:683
2. Koeller KM, Wong CH (1999) Nat Biotech 18:835
3. Varki A (1993) Glycobiology 3:97
4. Varki A, Cummings R, Esko J, Freeze H, Hart G, Marth J (eds) (1999) Essentials of glycobiology. Cold Spring Harbor Laboratory Press, New York
5. Rosen SD, Bertozzi CR (1996) Curr Biol 6:261
6. Vaulont S, Kahn A (1994) FASEB J 8:28
7. Lee YC, Lee RT (1995) Acc Chem Res 28:321
8. Mammen M, Choi S-K, Whitesides GM (1998) Angew Chem Int Ed Engl 37:2754
9. Williams DH, Maguire AJ, Tsuzuki W, Westwell MS (1998) Science 280:711
10. Rao JH, Lahiri J, Weis RM, Whitesides GM (2000) J Am Chem Soc 122:2698
11. Rao JH, Lahiri J, Isaacs L, Weis RM, Whitesides GM (1998) Science 280:708
12. Hoffman RC, Horvath SJ, Klevit RE (1993) Protein Sci 2:951
13. Lee YC, Lee RT (1997) Bioconj Chem 8:762
14. Jelesarov I, Bosshard HR (1999) J Mol Recognit 12:3
15. Dam TK, Roy R, Sanjoy KD, Oscarson S, Brewer CF (2000) J Biol Chem 275:14,223
16. Lee YC, Lee RT (1994) Neoglycoconjugates: preparation and applications. Academic Press, San Diego, p 53
17. Lee YC, Lee RT, Rice K, Ichikawa Y, Wong TC (1991) Pure Appl Chem 63:499
18. Yi D, Lee RT, Longo P, Boger ET, Lee YC, Petri WA, Schnaar RL (1998) Glycobiology 8:1037
19. Adler P, Wood SJ, Lee YC, Lee RT, Petri WA, Schnaar RL (1995) J Biol Chem 270:5164
20. Zanini D, Roy R (1998) Architectonic neoglycoconjugates: effects of shapes and valencies in multiple carbohydrate-protein interactions. In: Chapleur Y (ed) Carbohydrate mimics: concepts and methods. VCH, Germany, p 385
21. Davis BG (1999) J Chem Perk Trans I 3215
22. Roy R (1994) The chemistry of neoglycoconjugates. In:Boons GJ (ed) Carbohydrate chemistry. Chapman and Hall, Glasgow, p 243
23. Roy R (1997) Top Curr Chem 187:242
24. Roy R (1996) Curr Opin Struct Biol 6:692
25. Uchiyama T, Vassilev VP, Kajimoto T, Wong WC, Huang HM, Lin CC, Wong CH (1995) J Am Chem Soc 117:5395
26. Kretaschmar G, Sprengard U, Junz H, Bartnik E, Schmidt W, Toepfer A, Horsch B, Krause M, Seiffge D (1995) Tetrahedron 51:13,015
27. Baisch G, Ohrlein R (1996) Angew Chem Int Ed Engl 35:1812
28. Miyauchi H, Yuri M, Tanaka M, Kawamura N, Hayashi M (1997) Bioorg Med Chem Lett 7:989
29. Wittmann V, Takayama S, Gong SW, Wetiz-Schmidt G, Wong C-H (1998) J Org Chem 63:5137
30. Hansen HC, Haataja S, Finne J, Magnusson G (1997) J Am Chem Soc 119:6974
31. Corbell JB, Lundquist JJ, Toone EJ (2000) Tetrahedron Asymmetry 11:95
32. Duchene D, Ponchel G, Wouessidjewe D. (1999) Adv Drug Deliv Rev 36:29
33. Gattuso G, Nepogodiev SA, Stoddart JF (1998) Chem Rev 98:1919
34. Fulton DA, Stoddart JF (2000) Org Lett 2:1113
35. Garcia-Fernandez JM, Ortiz-Mellet C (2000) Adv Carbo Chem Biochem 55:35
36. Furuike T, Aiba S, Nishimura S-I (2000) Tetrahedron 56:9909
37. Baussanne I, Benito JM, Ortiz-Mettel C, Garcia-Fernandez JM, Law H, Defaye J (2000) Chem Commun 1489
38. Garcia-Lopez JJ, Hernandez-Matero F, Isac-Garcia J, Kim JM, Roy R, Santoyo-Gonzalez F, Vargas-Berenguel A (1999) J Org Chem 64:522
39. Aoyama Y, Matsuda Y, Chuleeraruk J, Mishiyama K, Fujimoto K, Fujimoto T, Shimizu T, Hayashida O (1998) Pure Appl Chem 70:3279

40. Hayashida O, Shimizu C, Fujimoto T, Aoyama Y (1998) Chem Lett 13
41. Roy R, Kim JM (1999) Angew Chemie Int Ed Engl 38:369
42. Fujimoto T, Miyata T, Aoyama Y (2000) J Am Chem Soc 122:3558
43. Kim Y, Zimmerman SC (1998) Curr Opin Chem Biol 2:733
44. Zanini D, Roy R (1997) Bioconj Chem 8:187
45. Andre S, Ortega PJC, Perez MA, Roy R, Gabius HJ (1999) Glycobiology 9:1253
46. Colonna B, Harding VD, Nepogodiev SA, Raymo FM, Spencer N, Stoddart JF (1998) Chem
    Eur J 4:1244
47. Turnbull WB, Pease AR, Stoddart JF (2000) ChemBiochem 1:70
48. Lord JM, Smith DC, Roberts LM (1999) Cell Microbiol 1:85
49. Fan E, Zhang Z, Minke WE, Hou Z, Verlinde CLMJ, Hol WGJ (2000) J Am Chem Soc
    122:2663
50. Kitov PI, Sadowska JM, Mulvey G, Armstrong GD, Ling H, Pannu NS, Read RJ, Bundle DR
    (2000) Nature 403:669
51. Saha UK, Roy R (1997) Tetrahedron Lett 39:253
52. Kim JM, Roy R (1997) Tetrahedron Lett 38:3487
53. Konig B, Fricke T, Wabmann, Krallmann-Wenzel U, Lindhorst TK (1998) Tetrahedron
    Lett 39:2307
54. Arya P, Kutterer KMK, Qin H, Roby J, Barnes ML, Lin S, Lingwood CA, Peter MG (1999)
    Bioorg Med Chem 7:2823
55. Arya P, Kutterer KMK, Qin H, Roby J, Barnes ML (1998) Bioorg Med Chem Lett 8:1127
56. Sprengard U, Schudok M, Schmidt W, Kretzschmar G, Kunz H (1996) Angew Chemie Int
    Ed Engl 35:321
57. Sprengard U, Dretzschmar G, Bartnik E, Huls C, Kunz H (1995) Angew Chemie Int Ed
    Engl 34:990
58. Burke SD, Shao Q, Schuster MC, Kiessling LL (2000) J Am Chem Soc 122:4518
59. Reuter JD, Myc A, Hayes MM, Gan ZH, Roy R, Qin DJ, Yin R, Piehler LT, Esfand R, Toma-
    lia DA, Baker JR (1999) Bioconj Chem 10:271
60. Kuberan B, Lindhardt RJ (2000) Curr Org Chem 4:653
61. Caron M, Seve AP, Bladier D, Joubert-Caron R (1998) J Chromatogr B 715:153
62. Varki A (1997) FASEB J 11:248
63. An YH, Friedman RJ (1998) J Biomed Mater Res 43:338
64. Davis BG, Lloyd RC, Jones JB (2000) Bioorg Med Chem 8:1527
65. Davis BG, Jones JB (1999) Synlett 1495
66. Macmillan D, Bertozzi CR (2000) Tetrahedron 9515
67. Borgia JA, Fields GB (2000) Trends Biotech 18:243
68. Yarema K, Bertozzi CR (1998) Curr Opin Chem Biol 2:49
69. Kochendoerfer GG, Kent SBH (1999) Curr Opin Chem Biol 3:665
70. Roy R (1996) Trends Glycosci Glyc 8:79
71. Kadokawa J, Tagaya H, Chiba K (1999) Synlett 1845
72. Thoma G, Patton JT, Magnani JL, Ernst B, Ohrlein R, Duthaler RO (1999) J Am Chem Soc
    121:5919
73. Matsuura K, Ketakouji H, Sawada N, Ishida H, Kiso M, Kitajima K, Kobayashi K (2000)
    J Am Chem Soc 122:7406
74. Tsuchida A, Kobayashi K, Matsubara N, Muramatsu T, Suzuki T, Suzuki Y (1998) Glyco-
    conj J 15:1047
75. Yoshitani N, Takasaki S (2000) Anal Biochem 277:127
76. Kobayashi K, Tsuchida A, Usui T, Akaike T (1997) Macromolecules 30:2016
77. Dohi H, Nishida Y, Mizuno M, Shinkai M, Kobayashi T, Takeda T, Uzawa H, Kobayashi K
    (1999) Bioorg Med Chem 7:2053
78. Yamada K, Minoda M, Miyamoto T (1997) J Polym Sci Pol Chem 35:751
79. Sallas F, Nishimura SI (2000) J Chem Soc Perkin Trans 1 2091
80. Bovin NV (1998) Glycoconj J 15:431
81. Mammen M, Helmerson R, Kishore R, Choi S-K, Phillips WD, Whitesides GM (1996)
    Chem Biol 3:757

82. Mammen M, Dahmann G, Whitesides GM (1995) J Med Chem 38:4179
83. Sigal GB, Mammen M, Dahmann G, Whitesides GM (1996) J Am Chem Soc 118:3789
84. Choi S-K, Mammen M, Whitesides GM (1996) Chem Biol 3:97
85. Wang J-Q, Chen X, Zhang W, Zacharek S, Chen Y, Wang PG (1999) J Am Chem Soc 121:8174
86. Wu WY, Jin B, Krippner GY, Watson KG (2000) Bioorg Med Chem Lett 10:341
87. Roy R, Park WKC, Srivastava OP, Foxall C (1996) Bioorg Med Chem Lett 6:1399
88. Miyauchi H, Tanaka M, Koike H, Kawamura N, Hayashi M (1997) Bioorg Med Chem Lett 7:985
89. Grubbs RH, Chang S (1998) Tetrahedron 54:4413
90. Grubbs RH, Miller SJ, Fu GC (1995) Acc Chem Res 28:446
91. Manning DD, Hu X, Beck P, Kiessling LL (1997) J Am Chem Soc 119:3161
92. Manning DD, Strong LE, Hu X, Beck PJ, Kiessling LL (1997) Tetrahedron 53:11,937
93. Sanders WJ, Gordon EJ, Dwir O, Beck PJ, Alon R, Kiessling LL (1999) J Biol Chem 274:5271
94. Gordon EJ, Strong LE, Kiessling LL (1998) Bioorg Med Chem 6:1293
95. Strong LE, Kiessling LL (1999) J Am Chem Soc 121:6193
96. Kanai M, Mortell KH, Kiessling LL (1997) J Am Chem Soc 199:9931
97. Gordon EJ, Gestwicki JE, Strong LE, Kiessling LL (2000) Chem Biol 7:9
98. Gestwicki JE, Strong LE, Kiessling LL (2000) Chem Biol 7:583
99. Kim SH, Goto M, Cho CS, Akaike T (2000) Biotechnol Lett 22:1049
100. Weisz OA, Schnaar RL (1991) J Cell Biol 115:495
101. Brandley BK, Shaper JH, Schnaar RL (1990) Dev Biol 140:161
102. Griffith LG, Lopina S (1998) Biomaterials 19:979
103. Hasegawa Y, Shinohara Y, Hiroyuki S (1997) Trends Glycosci Glycotechnol 9:S15
104. Green RJ, Frazier RA, Shakesheff KM, Davies MC, Roberts CJ, Tendler SJB (2000) Biomaterials 21:1823
105. Haseley SR, Talaga P, Kamerling JP, Vliegenthart JFG (1999) Anal Biochem 274:203
106. Yoshizumi A, Kanayama N, Maehara Y, Ide M, Kitano H (1999) Langmuir 15:482
107. Zeng X, Murata T, Kawagishi H, Usui T, Kobayashi K (1998) Carbo Res 312:209
108. Ullman A (1991) An introduction to ultrathin organic films: from Langmuir-Blodgett to self-assembly. Academic Press, San Diego
109. Kunitake T (1992) Angew Chem Int Ed Engl 31:709
110. Boullanger P (1997) Topics Curr Chem 187:276
111. Charych D, Cheng Q, Reichert A, Kuziemko G, Stroh M, Nagy JO, Spevak W, Stevens RC (1996) Chem Biol 3:113
112. Pan JJ, Charych D (1997) Langmuir 13:1365
113. Spevak W, Foxall C, Charych DH, Dasgupta F, Nagy JO (1996) J Med Chem 39:1018
114. Vogel J, Bendas G, Bakowsky U, Hummel G, Schmidt RR, Kettmann U, Rothe U (1998) Biochem Biophys Acta 1372:205
115. Gege C, Vogel J, Bendas G, Rothe U, Schmidt RR (2000) Chem Eur J 6:111
116. Plant AL (1999) Langmuir 15:5128
117. Mann DA, Kanai M, Maly DJ, Kiessling LL (1998) J Am Chem Soc 120:10,575
118. Mrksich M (1998) Cell Mol Life Sci 54:653
119. Mrksich M (2000) Chem Soc Rev 29:267
120. Mrksich M, Whitesides GM (1997) ACS Sym Ser 680:361
121. Horan N, Yan L, Isobe H, Whitesides GM, Kahne D (1999) Proc Natl Acad Sci USA 96:11,782
122. Wassler MJ, Shur BD (2000) J Cell Sci 113:237
123. Shur BD, Evans SC, Lu Q (1998) Glycoconj J 15:537
124. Houseman BT, Mrksich M (1999) Angew Chemie Int Ed Engl 38:782
125. Hickman JJ, Ofer D, Laibinis PE, Whitesides GM, Wrighton MS (1991) Science 252:688
126. Yousaf MN, Mrksich M (1999) J Am Chem Soc 121:4286
127. Yousaf MN, Houseman BT, Mrksich M (2000) Angew Chemie Int Ed Engl (in press)
128. Kanemitsu T, Kanie O (1999) Trends Glycosci Glyc 11:267

129. St Hilaire PM, Meldal M (2000) Angew Chemie Int Ed 39:1162
130. Haase WC, Seeberger PH (2000) Curr Org Chem 4:481
131. Liang R, Loebach J, Horan N, Ge M, Thompson C, Yan L, Kahne D (1997) Proc Natl Acad Sci USA 95:10,554
132. Liang R, Yan L, Loebach J, Ge M, Uozumi Y, Sekanina K, Horan N, Gildersleeve J, Thompson C, Smith A, Biswas K, Still WC, Kahne D (1996) Science 274:1520
133. Choi SK, Mammen M, Whitesides GM (1997) J Am Chem Soc 119:4103
134. Donovan RS, Datti A, Baek MG, Wu QQ, Sas IJ, Korczak B, Berger EG, Roy R, Dennis JW (1999) Glycoconj J 16:607
135. Ramstrom O, Lehn J-M (2000) ChemBiochem 1:41
136. Frank R (1992) Tetrahedron 48:9217
137. Lebl M (1998) Biopolymers 47:397
138. Jobron R, Hummel G (2000) Angew Chemie Int Ed Engl 39:1621
139. Ohrlein R (2000) Top Curr Chem 200:227
140. Takayama S, Wong CH (1997) Curr Org Chem 1:109
141. Srivastava G, Kaur KJ, Hindsgaul O, Palcic M (1992) J Biol Chem 267:22,356
142. Jacobs CL, Yarema KJ, Mahal LK, Nauman DA, Charters NW, Bertozzi CR (2000) Methods Enzym 327:260
143. Mahal LK, Yarema KJ, Bertozzi CR (1997) Science 276:1125
144. Saxon E, Bertozzi CR (2000) Science 287:2007

Carbohydrate-Carbohydrate Interactions
in Biological and Model Systems

Javier Rojo, Juan Carlos Morales, Soledad Penadés

Grupo de Carbohidratos, Centro de Investigaciones Científicas, CSIC, Isla de la Cartuja,
41092 Sevilla, Spain
E-mail: penades @cica.es

Carbohydrate-carbohydrate interactions are emerging as a novel and versatile mechanism for
cell adhesion and recognition. Although this interaction has not deserved much attention of
cell biologists, biochemists, or carbohydrate chemists, new evidence and studies are confirming the importance of this mechanism for specific cell adhesion and communication. The
study and evaluation of carbohydrate-carbohydrate interactions is still in its infancy. Their
development will go hand in hand with the development of new and more sensitive techniques
to study weak interactions such as biosensors, atomic force microscopy, or weak affinity chromatography. In this contribution we will review these new emerging carbohydrate-carbohydrate related interactions including those already established and those where the carbohydrate involvement, at this moment, may be considered possible as in the case of DNA-carbohydrate interaction. The two main research lines on carbohydrate-carbohydrate interaction in
biological systems and the efforts, using model systems, to demonstrate and evaluate these
interactions are reviewed here in some detail. Considerations about the intermolecular forces
and the mechanism that may be involved in carbohydrate-carbohydrate interactions are also
presented. The chapter ends with the review of the few examples existing in the literature on
quantitative studies of carbohydrate-carbohydrate interaction with model systems.

Keywords. Carbohydrate-carbohydrate interactions, DNA-carbohydrate interactions, Polyvalence, Model systems

Topics in Current Chemistry, Vol. 218
© Springer-Verlag Berlin Heidelberg 2002

List of Abbreviations

A	adenosine
AF	adhesion factor
AFM	atomic force microscopy
AMP	adenosin monophosphate
C	cytidine
CBS	cerebroside sulfate
CD	circular dichroism
Cer	ceramide
CID	collision induce decomposition
CREB	cyclic AMP response element binding protein
DNA	deoxyribonucleic acid
ECM	extra cellular matrix
EDTA	ethylenediaminetetraacetic acid
EM	electron microscopy
ESI-MS	electrospray ionization mass spectrometry
FAB-MS	fast atom bombardment mass spectrometry
FAK	focal adhesion kinase
Fuc	fucose
G	guanosine
GAG	glycosamineglycan
Gal	galactose

Glc   glucose
GlcNAc   *N*-acetyl glucosamine
GSD   glycosignaling domain
GSL   glycosphingolipid
HSEA   hard sphere exo-anomeric
IS-MS   ion spray mass spectrometry
Lac   lactose
LeX   Lewis X
LNFP   lacto-*N*-fucopentaose
Lys   lysine
NMR   nuclear magnetic resonance
NOE   nuclear Overhauser effect
NOESY   nuclear Overhauser effect spectrometry
PG   paragloboside
PNP   *p*-nitrophenol
PNPG   1-(*p*-nitrophenyl) glycerol
PNPGly   *p*-nitrophenyl glycoside
RBL   rat basophilic leukemia
RNA   ribonucleic acid
SAM   self-assembled monolayer
SPG   sialyl paragloboside
SPR   surface plasmon resonance
T   thymidine

1
Introduction

The surface of cell membranes presents a carbohydrate covering named glyco-calix. For a long time the role of this glycocalix was adscribed to the protection of the cell against the outside environment by repulsive interactions. However, just this role cannot explain the complexity of the glycoconjugates. Their location at the outer leaflet indicates that carbohydrates have to be involved in some way in cell adhesion and recognition processes based on both attractive and repulsive interactions [1]. Cells, toxins, bacteria, and other organisms interact with their receptors, which in most cases are glycoconjugates (glycolipids and/or glycoproteins), at the cell surface. The complexity of these conjugates has made their study very hard and for many years they have been kept aside in comparison with other biomolecules as nucleic acids or proteins. This complexity lays fundamentally on their diversity. Over the last years, an impressive number of glycoproteins and glycolipids have been characterized and the role they play in several processes as embryonic tissue formation, morphogenesis, fertilization, antigen recognition, etc., began to be understood. It was observed that, in addition to the well-established protein-protein interaction in cell adhesion, a weaker protein-carbohydrate interaction was involved in all these biological processes [2]. This interaction, that most of the times is multivalent and divalent cation dependent, is more specific than protein-protein interaction. After the isolation and characterization of some membrane receptor proteins (i.e., lectins) and their corresponding carbohydrate ligands, the protein-carbohydrate interaction has been well documented and several reviews can be found in the literature [3–5] (a review on the study of carbohydrate-protein interaction with model systems is found in Chaps. 1 and 7 of this volume).

Recently, new evidence has emerged in favor of a new type of interaction implicated in cell adhesion processes, the carbohydrate-carbohydrate interaction. This interaction is even weaker than the protein-carbohydrate interaction presented above and it is also divalent cation dependent. The interaction between carbohydrates is so weak compared with others that it has to be multivalent to be detected and observed. These features make it very difficult to show and evaluate this interaction and only a few cases have been described in certain detail dealing with this interaction. In spite of this, a few excellent reviews on this topic have been already published [6–8].

Interactions between carbohydrates were well known and accepted in the context of structural network formation of the extracellular matrix (ECM) and in cell walls of bacteria and plants. We will not refer in this review to this type of interaction, but we will focus on those carbohydrate interactions involved in molecular recognition processes where specificity and flexibility are essential to the healthy development of biological systems. The involvement of specific interactions between carbohydrates in recognition processes has been demonstrated by Misevic and Burger [9] in the species-specific cell aggregation of marine sponges, and by the group of Hakomori in morula compaction and metastasis of lymphoma and melanoma cells in mice [10].

In addition to the cell adhesion carbohydrate-carbohydrate interaction, we would also like to focus attention on other simpler systems where carbohydrate-carbohydrate interaction may play an important role. We refer to the interaction between nucleic acids and drugs based on carbohydrates [11–13]. A well-documented system is the interaction between aminoglycoside antibiotics and RNA. In this case, carbohydrates are aminosugars that interact with the RNA backbone mainly through electrostatic interactions between the positive charged amino groups and the negative charged phosphate of the furanose moiety. In addition to electrostatic interactions, other kinds of forces involving non-polar interactions between carbohydrates cannot be excluded. There is now a considerable amount of literature about this interaction and it will not be reviewed here. Interaction between antitumoral antibiotics and DNA, a system where the likelihood of carbohydrate-carbohydrate interactions is rather high, will be presented in Sect. 4.

Most of the antitumoral and antibiotic agents that interact with DNA are glycosidated molecules. Several examples have been described in the literature; however, not many details have been analyzed about the contribution of the carbohydrate moiety involved in these processes. The mechanism of this interaction is not well understood.

This chapter begins with the review of several systems where carbohydrate-carbohydrate interaction may be involved. Following this, the efforts carried out in biological and artificial systems to identify and characterize these interactions are discussed. The chapter concludes with the description of the few model systems that have been used to determine the energetics of carbohydrate-carbohydrate interactions. Because of their importance in determining affinity and selectivity in all interactions, we have considered it convenient to analyze the intermolecular forces involved in the carbohydrate interactions.

2
Intermolecular Forces in Carbohydrate-Carbohydrate Interactions. The Role of Water

Oligosaccharides are complex biomolecules constituted by monosaccharides as monomers. Due to the presence of different monosaccharides, and especially to the different position and stereochemistry of the possible linkages between the monomers, the number of combinations compared with other biomolecules is much higher. As an example, one nucleotide or one amino acid can form one dinucleotide and one dipeptide respectively, while one monosaccharide is capable of forming 11 different disaccharides. When the number of monomers is higher, the differences are more impressive. Four nucleotides or 4 amino acids can give 24 different tetranucleotides and 24 different tetrapeptides respectively, while 4 different monosaccharides can form up to 33,560 different tetrasaccharides. These figures illustrate the potential complexity of these oligomers. The specific sequence of monosaccharides that constitutes an oligosaccharide contains a huge amount of information to be decoded through the interaction and recognition with their receptors. This interaction is based on non-covalent bonds: van der Waals contacts, hydrogen bonds, electrostatic interactions, interactions with cations, etc.

Structural studies of carbohydrates have shown important features that are responsible for their conformational behavior in solution and therefore for their interaction processes. Restrictions around the glycosidic linkages strongly determine the conformation of carbohydrates in solution. These features control the sugar surface exposed to the environment. In other words, the sequence and their preferred conformations set a well-defined surface implicated in the interaction with other molecules. This surface is rich in hydroxyl groups, which can operate as H-bond donors and acceptor centers. It is also endowed with non-polar groups so that this surface may be considered polyamphiphilic when interacting via non-covalent forces with its specific receptor [14, 15].

How these surfaces are presented for interaction is of major importance. Carbohydrates show weak interaction forces with their receptors. These interactions are considered as one of the weakest among biological processes. To overcome this problem, glycoconjugates display a polyvalent configuration at the cell surface. Glycoproteins present repetitive epitopes on their structures while glycosphingolipids (GSL) are associated in clusters or patches as was demonstrated using ultrafast freezing and freeze-etch electron microscopy techniques [16, 17]. The organization of glycosphingolipids into domains creating a polyvalent display (similar to that represented by the repetition of epitopes in glycoproteins) enhances the adhesion forces.

This polyvalent presentation and their epitope density define clearly the carbohydrate surface involved in recognition processes and hence the affinity and the selectivity of biological processes where glycoconjugates are involved [18–20].

In these biological processes, carbohydrate-protein interactions are the better studied and well-established [21–23]. This chapter does not review this well documented interaction. Nevertheless, the forces which have been described to participate in it will be analyzed in certain detail because they summarize the forces that determine affinity and selectivity in different carbohydrate interaction processes. How these forces determine this affinity and selectivity of the process has been the subject of a long debate.

First, carbohydrates were considered as highly polar molecules where hydrogen bonds, formed through the hydroxyl groups, were the dominant interaction in carbohydrate-protein complexes [24, 25]. All hydroxyl groups of the carbohydrate act as hydrogen bond donors and/or acceptors establishing a well organized network where cooperative effect reinforce the strength of these bonds. Also, van der Waals contacts can be formed through interaction of aromatic groups of amino acids of the protein and the carbon framework of the carbohydrate and can participate in the stabilization of the final complex. The highly directional features of hydrogen bonds explained the selectivity found in recognition processes of carbohydrates by proteins.

In 1978 Lemieux et al. proposed that the main driving force for carbohydrate interactions arises from the interaction between hydrophobic patches of the sugar surface and the complementary hydrophobic cavity of the protein after desolvation [26]. Polar interactions were not expected to contribute significantly to the stability of the complex. The directional character of hydrogen bonds, however, could confer the specificity to the interactions. The hypothesis that hydrophobic effect dominates the stability of the complex evolved into the concept of carbohydrates as molecules that present a surface with a polyamphiphilic character [15]. This character plays an important role in the organization of the hydration layer on the surface. The assumption that the intermolecular attraction forces between complementary surfaces are the only driving force for recognition is part of the total picture. Affinity and selectivity between complementary molecular surfaces is achieved through a combination of non-covalent interactions in the presence of many co-solutes and in a very large molar excess of water. Because of that, water has to play an important role in this picture. In fact, the water molecules that are organized around polyamphiphilic surfaces define a perturbed situation well explained by Lemieux and other authors [15, 27–31]. The presence of hydrophobic and hydrophilic areas nearby on the same surface explains why the hydration layer contains high-energy water molecules. When these water molecules return to the bulk (desolvation or reorganization) during recognition processes, a large amount of energy is liberated. Depending on the surface, the desolvation process contributes to the free energy changes in different ways. Lemieux explains these differences considering by two different effects on desolvation: the *hydrophobic effect* and what he has named the *hydraphobic effect*. At the non-polar surfaces, when the organized layer of water molecules is released to bulk, an increase in entropy is observed (*hydrophobic effect*). In contrast, the liberation of water molecules from polyamphiphilic surfaces to the bulk causes an important decrease in both enthalpy and entropy terms (*hydraphobic effect*) [15].

From the thermodynamic parameters obtained for carbohydrate-protein interaction it has been established that, in most cases, a favorable enthalpy and a decrease in entropy are involved in the association of carbohydrates to proteins [32]. Part of the decrease in enthalpy due to the desolvation process is compensated by a decrease in the entropy of the system. Water molecules organized on polyamphiphilic surfaces go to the bulk forming stronger and better organized hydrogen bonds with the subsequent loss in entropy. It seems that the interaction of water with either a non-polar or a polyamphiphilic surface creates a perturbed water layer, the main source of energy for complexation. Dispersion forces between the complementary surfaces are an additional source of energy. Summarizing, a complex combination of these factors to the free energy of binding determines the energetics of carbohydrate binding in water. The confluence of the interaction between complementary surfaces with the release of perturbed water molecules from these surfaces is Nature's way of providing the energy required to complexation. This assumption is valid independently of the process where the carbohydrate may be involved and can be extended to all biological processes.

More than two decades ago, part of the carbohydrate community focused their interest on a new class of interaction where only carbohydrates were involved, what we refer to as carbohydrate-carbohydrate interaction [33]. The forces described above for protein-carbohydrate can be applied to this new interaction but some details are a little different.

Carbohydrate-carbohydrate recognition processes do not implicate the interaction through cavities as happens in the case of proteins. The interaction of carbohydrates takes place through surfaces determined by the carbohydrate epitopes. On these surfaces, hydrophilic and hydrophobic patches are well defined by the position of functional groups, which determine the tri-dimensional structure adopted. Van der Waals contacts can take place through non-polar patches of carbohydrates. The surface complementarity should be the most important contribution to the selective formation of carbohydrate complexes. After the approach of these surfaces, water molecules which form the solvated layer have to go to the bulk providing as described above what Lemieux has defined as "*the impetus for molecular recognition in aqueous solution*" [15]. The hydroxy groups that are close tend to form hydrogen bonds to increase the stability of the complex. Divalent ions such as Mg^{2+} and especially Ca^{2+} may act as a glue, coordinating these hydroxy groups and participating in the stability of the complex. Examples of biological and model systems where this carbohydrate-carbohydrate interaction can take place are described in certain detail in the next sections.

3
Carbohydrate-Carbohydrate Interactions in Cell Adhesion

3.1
Species-Specific Sponges Cell Adhesion

Cell adhesion of dissociated marine sponge cells was studied as early as 1907 [34] and was shown to be species-specific. Marine sponge cells have been chosen as a model system to study the mechanism involved in cellular recognition and adhesion processes.

In 1963, a macromolecule responsible for aggregation of cell sponges, named aggregation factor (AF), was isolated for the first time from *Microciona Parthena* and other sponges [35]. This AF was purified and chemically characterized as a proteoglycan containing protein and polysaccharide in approximately equal parts. The polysaccharide analysis showed 60% neutral sugars, 20% amino sugars, and 20% uronic acids. The electron microscopy examination of this extracellular proteoglycan presented a "sunburst" structure with an 800 Å diameter circle containing around 15 arms 1100 Å in length radiating from the central circle. The molecular weight was calculated to be 2.2×10^7 Daltons [36, 37].

The cell adhesion process through the aggregation factor (AF) of *Microciona* sponge involves two separated mechanisms: (1) Ca^{2+}-independent species-specific binding AF to cell surface receptor and (2) Ca^{2+}-dependent AF-AF interaction between adjacent cells; Ca^{2+} concentration should be 10 mmol/l to observe cell adhesion. Both mechanisms are necessary for the species-specific aggregation of sponge cells. AF-AF interaction was found to be inhibited by ethylenediaminetetraacetic acid (EDTA), heat, or treatment with periodate whereas the AF-cell receptor interaction remained intact under these conditions, suggesting two different binding sites or functional domains in the proteoglycan [38] (Fig. 1).

In order to deduce which were the functional domains and their chemical nature, different chemical and enzymatic degradation methods were per-

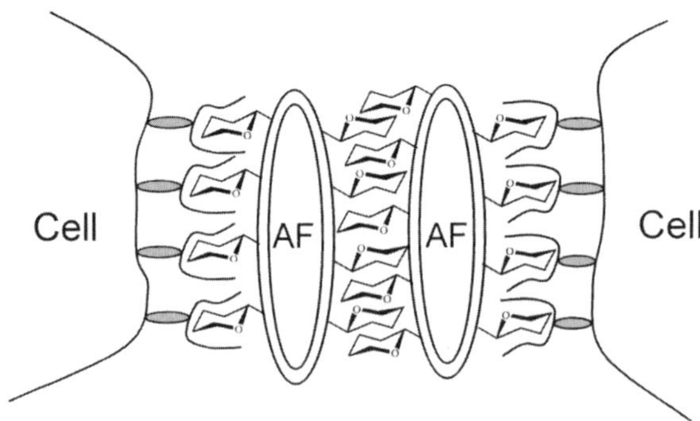

Fig. 1. Cartoon of the adhesion process of two sponges cells based on the aggregation factor (AF) [8]

formed. AF proteoglycan was digested with trypsin and the glycopeptide fragments were isolated and purified [39]. The binding affinity of these small fragments was found to be 13,000 times lower than that of intact AF. In an attempt to reconstitute the activity of AF, these glycopeptide fragments were cross-linked polymerized with diepoxybutane and glutaraldehyde. The binding constants of these polymers increased with their molecular weight although, in the best case, values 4–6 times lower than the binding constant of intact AF were obtained. Two important conclusions were deduced from these experiments: (a) the spatial arrangement of the fragments could influence in the association constants and (b) highly polyvalent low affinity sites could be the basis of strong and specific cellular interactions.

As a first step to test where the association activity lies, monoclonal antibodies were created against pure isolated AF of *Microciona prolifera* sponge cells. One of these antibodies, Block 1 or its Fab fragment was able to inhibit the AF-AF association activity. The number of antigenic sites per one AF molecule was found to be around 1100, demonstrating that the epitope, the nature of which was at that moment unknown, was highly repetitive. The possibility that this antigenic determinant presented a carbohydrate nature was analyzed.

Protein-free glycans were obtained by *Pronase* degradation of AF molecule and the role of the carbohydrate moiety of AF was analyzed. An immunobinding assay showed that Block 1 monoclonal antibody bound 19% of AF glycans. Competition experiments were performed with different monosaccharides and glycosaminoglycans (GAG). Only in the case of AF glycans and intact AF was inhibition of Block 1 antibody achieved, confirming that the epitope was localized on the carbohydrate portion, but the antigen was not one of the former carbohydrates tested [40]. For the first time, multiple low affinity carbohydrate-carbohydrate interaction was postulated to be the basis of cell aggregation in sponges more than one decade after the isolation of the proteoglycan [41].

One of the fragments isolated after the treatment with *Pronase* was purified and characterized. This larger glycan, $M_r = 200 \times 10^3$ (g-200) contained fucose, glucuronic acid, galactose, mannose, and *N*-acetyl glucosamine and its structure was different from other known glycosaminoglycans. Another monoclonal antibody, Block 2 (obtained in the same way that Block 1), also inhibited AF self-interaction and therefore cell adhesion. Block 2 antibody bound to g-200, as was shown by immunobinding gel electrophoresis retardation assay. Competition experiments with several sugars was performed with Block 2, and only AF, protein-free glycans, and g-200 inhibited its activity [40]. Binding assays of radiolabel Block 2 Fab fragments with AF determined approximately 2500 binding sites of the g-200 glycan per AF molecule. The polyvalency of the interaction was demonstrated using cross-linked g-200 polymers in the presence of 10 mmol/l Ca^{2+} solution.

In summary, self-association activity of the AF in the presence of a 10 mmol/l Ca^{2+} solution using the glycan fragments was reconstituted only when a cross-linked polymeric form was prepared with these glycans and glutaraldehyde. As was expected, glycans themselves did not undergo self-association in those conditions. This negative result could be explained based on the low affinity of single interaction sites. Since protein-free glycans were enough to self-associate and to bind intact AF, a multiple low affinity carbohydrate-carbohydrate interaction was suggested as the base of the AF-AF association.

In an attempt to characterize which were the constituent carbohydrate units, glycans isolated after *Pronase* digestion were submitted to partial acid hydrolysis and the resulting oligosaccharides were conjugated by reductive amination to lipids for immunochemical analysis. One of these oligosaccharide conjugates was identified as one of the strongest antigens of Block 1 antibody and its structure was unmistakably characterized by ^{1}H-NMR, FAB-MS, enzymatic, and chemical degradation as Pyr-4,6-Galβ1–4GlcNAcβ1–3Fuc (Fig. 2) [42]. Following the same protocol as that for Block 1, an oligosaccharide inhibitor of Block 2 was isolated and characterized by ^{1}H-NMR, FAB-MS as a disaccharide of formula 3-SO$_3$GlcNAcβ1–3Fuc (Fig. 2) [43].

It is important to notice here that the fucose ring was opened when the lipid glycoconjugate was formed for the immunobinding assays, so the structures recognized by the monoclonal antibodies Block 1 and Block 2 were in fact a disaccharide and a monosaccharide respectively. Also, other structures which include different monosaccharides in the non-reductive terminus could be envisaged as epitopes with or without the fucose ring in the reductive terminus. A better knowledge of the glycan composition using different techniques could give more details about the structure of the real epitopes. Preliminary ^{1}H-NMR studies of the AF showed the structural diversity of the backbone and made the analysis of this glycan very complicated.

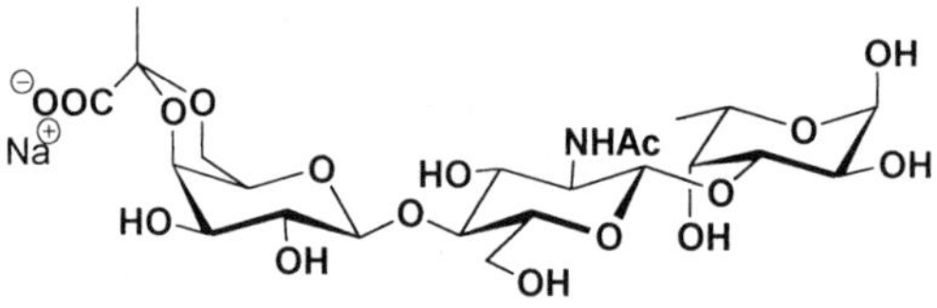

PyrGalβ(1→4)GlcNAcβ(1→3)Fuc

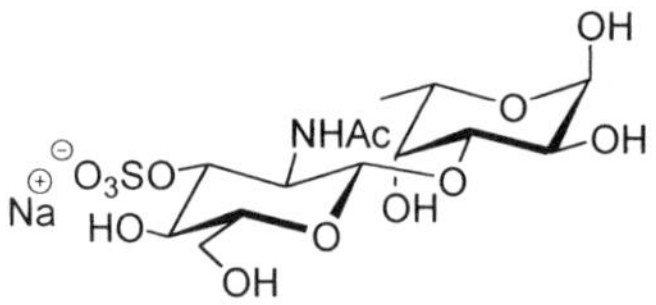

3-SO$_3$GlcNAcβ(1→3)Fuc

Fig. 2. Structure of the tri- and disaccharide epitopes of the adhesion proteoglycan responsible for the cell adhesion in *Microciona prolifera* sponge

3.2
Carbohydrate-Carbohydrate Interactions in Eukaryotic Cells

Hakomori has studied the expression of different type of carbohydrate antigens (normally glycosphingolipids) at the surface of eukaryotic cells and their implication in cell adhesion processes through carbohydrate-carbohydrate interactions. Due to the importance of biological processes in which it is involved, Le^x has been one of the antigens studied in more detail. This and other glycolipids supposedly implicated in carbohydrate-carbohydrate interactions are analyzed in this section.

3.2.1
Le^x-Le^x Interaction

Le^x antigen (Galβ1–4[Fucα1–3]GlcNAcβ1) has been identified in a large number of different eukaryotic cells [44]. This epitope is highly expressed during early stages of mice embryo development after the third cleavage division at the 8–16 *morula* cell stage. The appearance of Le^x at this stage correlates approximately in time with the onset of compaction with its expression decline rapidly after that, indicating a role of Le^x during compaction, when the intercellular interactions are maximized (Fig. 3) [45].

To demonstrate that Le^x determinant participates in this process, decompaction analysis was performed. Decompaction of early embryo mouse cells by Le^x was not observed even at concentrations as high as 5 mmol/l. However, multivalent Le^x obtained by coupling Le^x with LysLys [46] allowed decompaction at concentrations between 0.125 and 1 mmol/l. Other oligosaccharides such as LNFP II (Le^a), *N*-acetyl chitotriose LNFP I (H hapten) conjugated with LysLys (Fig. 4), did not present the same activity. These results suggested that a multivalent Le^x system might participate in the compaction process and the receptor involved in recognizing it could be a lectine, a glycosyltranferase, or even a carbohydrate [47].

An analogous system studied was the F9 teratocarcinoma cells. These cells resemble the early mouse embryo cells and express Le^x at the cell surface showing autoaggregation in the presence of bivalent cations. This self-aggregation was inhibited by EDTA, by the presence of Le^x pentasaccharide, and especially by Le^x LysLys conjugate. Several experiments were designed in an attempt to find which were the molecules involved in this interaction process. In this sense, F9 cells were fixed on plastic surface and their interaction with liposomes presenting different glycolipids was studied. Only liposomes which contained Le^x epitopes showed a preferential adsorption on these surfaces, indicating that this epitope was involved in F9 cells aggregation. Liposomes were incubated with plastic surface coated with different glycolipids (Le^x, paragloboside (PG), or sialylparagloboside (SPG)). Only liposomes containing Le^x interacted with plastic surface coated with Le^x. Moreover, liposomes with other oligosaccharides as PG or SPG interacted with this Le^x surface more weakly than liposomes without glycolipids showing for the first time a repulsive activity. The interaction of liposomes containing Le^x with Le^x coated plastic surface was inhibited by Le^x itself.

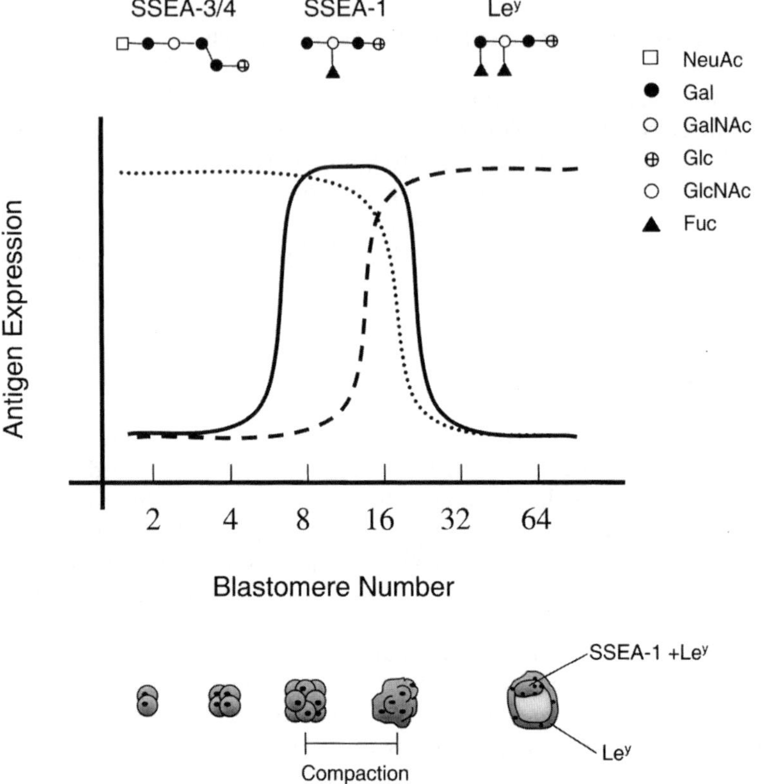

Fig. 3. Changes in cell surface expression of carbohydrate antigens during pre-implantation embryo development. Modified from [10]

Also, Le^x liposomes self-aggregate when they were incubated in the presence of bivalent cations and inhibited by EDTA, as was shown by measurements using spectrophotometer and aggregometer [48]. Autoaggregation of Le^x-coated beads was observed in the presence of Ca^{2+} but was not seen when beads were coated with other glycolipids as PG. Again, the aggregation was inhibited by EDTA [49]. This aggregation was analyzed by turbidometry. Similar results were obtained with HRT-18 cells, a human colorectal carcinoma cell line expressing high concentration of Le^x at the cell surface [10].

Aggregation of cells mediated by Le^x-Le^x interaction was confirmed using Le^x modified rat basophilic leukemia cells (RBL). These cells do not express Le^x and also do not exhibit autoaggregation. However, Le^x-glycolipids were easily incorporated into the membrane of these cells and were labeled with Hoechst 33258 dye. It was demonstrated that Le^x modified cells autoaggregate, but they were not able to interact with cells which had not been modified, indicating that the interaction was only between Le^x-Le^x and not Le^x with other structures on the plasma membrane such as Le^x-lectine like receptors, other oligosaccharides, etc. [50].

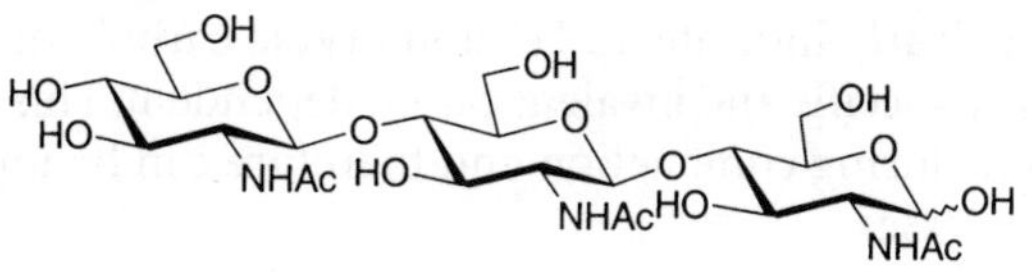

N,N',N''-tri-O-acetyl chitotrioside

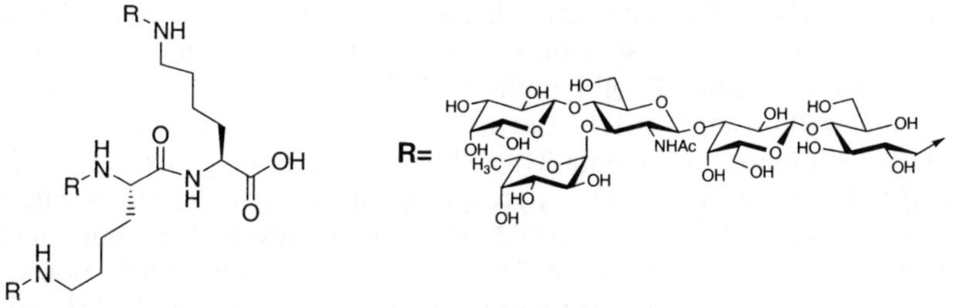

$LysLysLe^x_3$

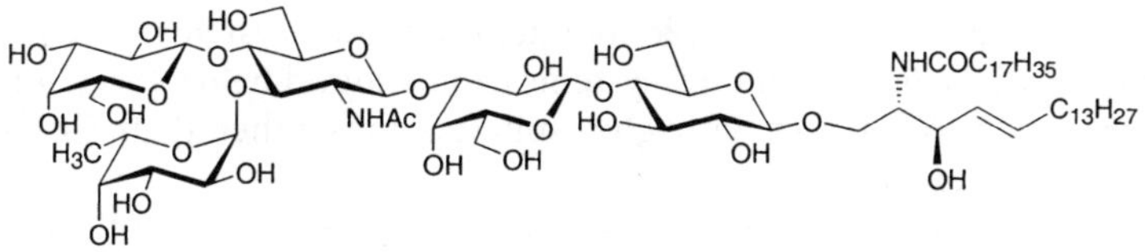

Le^xCer

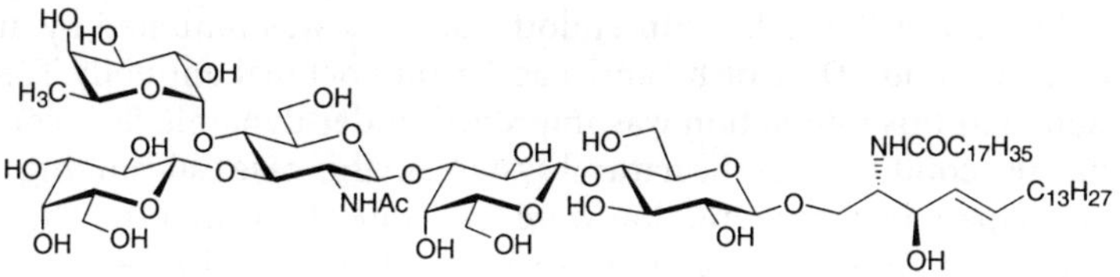

Le^aCer

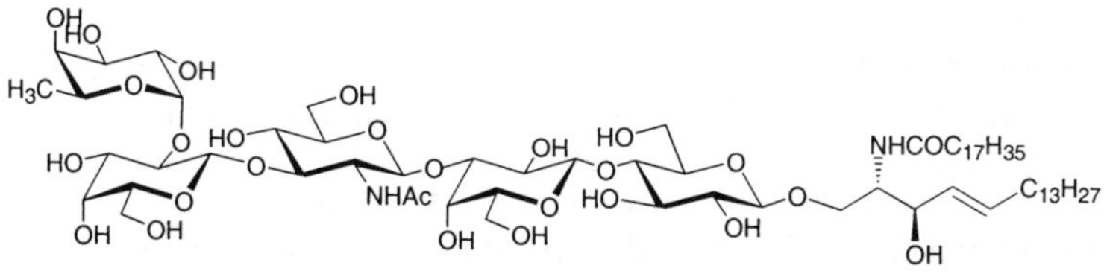

HCer

Fig. 4. Structure of Le^x, Le^a, H-hapten, *N*-acetyl chitotriose, and trivalent Le^x LysLys glycoconjugate

All these data clearly indicate an Le^x homotypic carbohydrate-carbohydrate interaction, highly specific and bivalent cation dependent. This interaction plays an important role during compaction and therefore can be implicated in other cell adhesion processes.

3.2.2
G_{M3}-Gg3 and LacCer Interaction

Highly specific GSL-GSL interaction between G_{M3} and Gg3 or lactose has been proposed as a basis for specific cell recognition between lymphoma and melanoma cells. Mouse T cell lymphoma L5178AA12 expresses high levels of Gg3 (GalNAcβ1–4Galβ1–4Glcβ1-Cer), a glycosphingolipid of globo series, and interacts with mouse melanoma B16 cell line which is a high expressor of G_{M3} (sialylβ2–3Galβ1–4Glcβ1-Cer), a glycosphingolipid of ganglio series (Fig. 5). A variant clone of the former, L5178AV27, does not express Gg3 and does not present any sort of interaction with melanoma B16. This specific interaction seems to be based on molecular interactions between glycosphingolipids Gg3 and G_{M3} [51]. The use of monoclonal antibodies highly specific for the corresponding antigens of Gg3 and G_{M3} inhibits the interaction completely. Also the absence of bivalent cations or the presence of EDTA inhibit this association.

To demonstrate this heterotypic interaction, model systems have also been used. More details on these model systems can be found below (Sect. 5.1).

G_{M3} was able to interact with glycosphingolipids other than Gg3 such as LacCer or Gb$_4$ (Fig. 5) [52].

B16 melanoma cells, which express high levels of G_{M3} at the outlet cell membrane, interact with non-activated mouse and human endothelial cells that contain LacCer and Gg3 as the major glycosphingolipid exposed at the cell surface. This cell adhesion could therefore be mediated by G_{M3}-LacCer or G_{M3}-Gg3 interaction. Adhesion of B16 cells with endothelial cells was inhibited by anti-G_{M3} monoclonal antibody DH2 or by anti-LacCer monoclonal antibody T5A7. It is remarkable that this interaction was improved under dynamic flow conditions that mimic the conditions of the natural system. In this case Gal-binding lectines which participate in the interaction were less important than the G_{M3}-LacCer (Gg3) association. In fact, the authors suggested that the initial event in tumor cell metastasis under physiological dynamic flow conditions could be this carbohydrate-carbohydrate interaction. After this interaction, endothelial cells would be activated and selectin expression could be induced as a second step in cell adhesion.

The role of carbohydrate-carbohydrate interaction in comparison with adhesion proteins (fibronectin, laminin, etc.) and their transmembrane receptors (integrin family) was analyzed in detail [53]. Plastic surface was coated with adhesion proteins in different concentrations. When this concentration was high (>10 µg/ml), the effect of co-coated glycosphingolipids was negligible. However, when the adhesion protein concentration decreased, the role of glycosphingolipids in the adhesion was very important showing increased adhesion from absence of glycosphingolipids to presence of LacCer or Gg3. Moreover, even when adhesion protein concentration is high, if the plastic surface is coated with

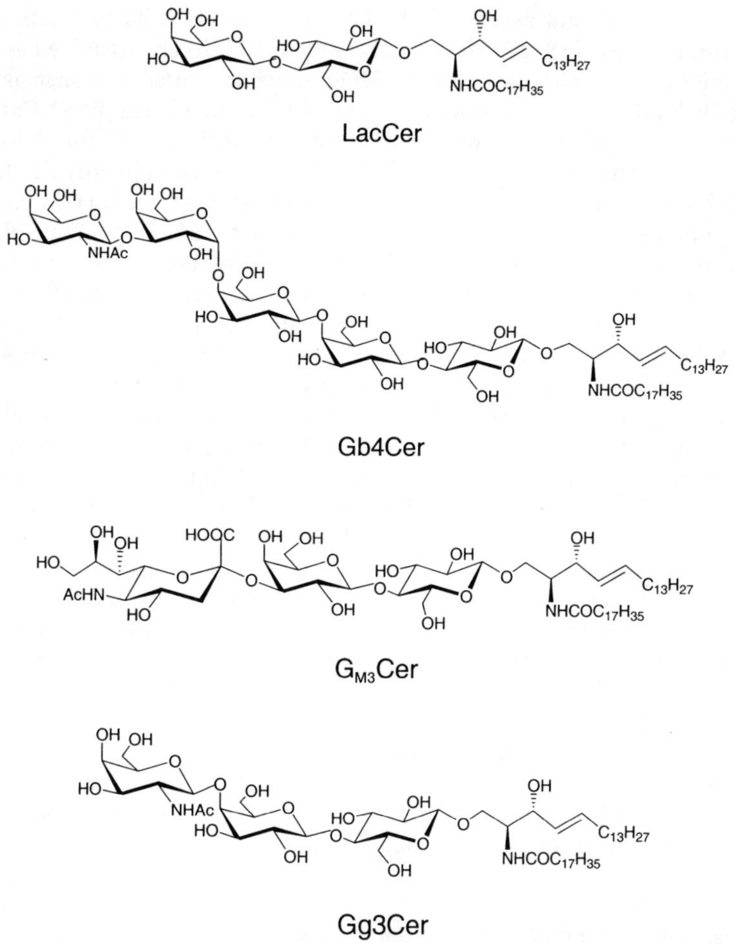

LacCer

Gb4Cer

G_{M3}Cer

Gg3Cer

Fig. 5. Structure of G_{M3}, Gg3, lactose, and Gb4 glycosphingolipids

G_{M3} a clear inhibition of B16 cell adhesion was observed; in other words, specific glycosphingolipid repellent interaction (G_{M3}-G_{M3}) was able to abolish the adhesion. Another important point was the time-dependent adhesion. At short times, the interaction of B16 cells with plates coated with Gg3 or LacCer was stronger than with plates coated with adhesion protein alone. The adhesion was faster if the plates were coated with both adhesion protein and the glycosphingolipids, showing a clear synergistic effect. The inhibition of G_{M3}-Gg3 or G_{M3}-LacCer interactions in the adhesion of melanoma cells to endothelial cells may represent an anti-adhesion therapy against tumor metastasis. Hakomori et al. have shown that spontaneous metastasis from subcutaneously-grown tumors was reduced if G_{M3}- or Gg3-liposomes were injected during tumor progression. In contrast, treatment with other paraglobosides had not effect on the reduction lung colonization [54].

A further important aspect of GSL-GSL interactions is its association with signal transduction [55–57]. GSL-clusters in animal cells are organized as membrane microdomains associated with signal transducer molecules such as c-Src, Src family kinases, small G proteins, and focal adhesion kinase (FAK). Carbohydrate-carbohydrate interaction through complementary GSL-microdomains causes activation of the signal transducers, leading to cell phenotypic changes [56]. The association of G_{M3} with c-Src and Rho has been found in G_{M3}-enriched microdomains at B16 murine melanoma cell surfaces [58]. In human embryonal carcinoma cells the adhesion of globosides, via carbohydrate-carbohydrate interaction, initiates signal transduction and enhances the activity of transcription factor AP1 (a gene regulatory protein) and CREB (cyclic AMP response element binding protein) [59]. These and other results suggest the presence of GSL-microdomains involved in signal transduction, which was termed "glycosignaling domains" (GSD) [60]. These domains seem to be structurally and functionally independent from caveolae. GSL-clustering is necessary to maintain initiation of signaling through gangliosides. If GSL-clustering is inhibited by addition of gangliosides analogues, this function could be blocked [61]. In order to assess the essential components that makes GSD functional, membranes with properties similar to those of GSD were reconstituted using G_{M3}, sphingomyelin, and c-Src. The reconstituted membranes displayed G_{M3}-dependent adhesion to plate coated with Gg3 or anti-G_{M3} antibody, resulting in enhanced c-Src phosphorylation. This response was not observed in the absence of G_{M3} or after replacement of G_{M3} by other gangliosides (GM1, GD1a, or LacCer) in the reconstituted membrane. Analogues of G_{M3} such as lyso-G_{M3} also blocked adhesion and signaling, probably due to the disruption of G_{M3}-G_{M3} *cis*-interaction and subsequent loss of G_{M3} clustering [62]. It seems that association between clustered GSLs interaction and signal transducers play a central role in cellular interactions coupled with signal transduction.

3.3
Multistep Cell Recognition and Adhesion Models

Based on all this strong experimental evidence in favor of carbohydrate-carbohydrate interaction, Hakomori et al. proposed a model mechanism where this interaction participated as the initial step in specific recognition between two cells (Fig. 6) [48].

This initial step consists of a highly specific low affinity multivalent interaction between the carbohydrate moieties of glycolipids at cell surface. This interaction was dependent on bivalent cations. The multivalency could be obtained through patches of glycosphingolipids found at the cell surface as it was demonstrated by freeze-etch electron microscopic studies [16, 17]. All the experiments performed to study this interaction based on liposomes, coated plastic surface, etc. reviewed above, tried to reproduce these glycolipids cluster to get multivalency.

The initial step is followed by a stronger but non-specific protein-protein interaction between pericellular adhesive proteins (fibronectin, laminin, thrombospondin, cadherin) and the corresponding integrin receptor. At this point,

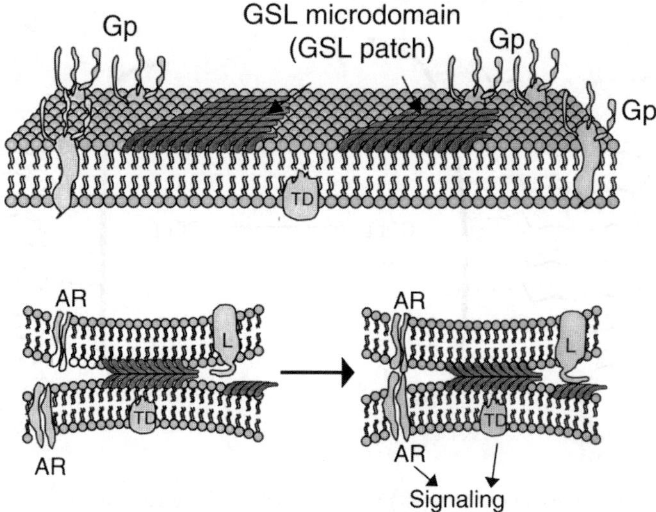

Fig. 6. Cartoon of the adhesion model via carbohydrate-carbohydrate interaction proposed by Hakomori. Modified from [55]. Gp = glycoprotein; GSL = glycosphingolipid; TD = transducer; AR = adhesion receptor; L = lectin

interactions between sugar-binding protein and their glycoconjugate substrates can also take place. Finally, the last step established intercellular junctions through plasma membrane junction proteins. This interaction allowed the formation of communication channels between cells.

Spillmann has compared the interaction between carbohydrates in the presence of Ca^{2+} to a velcro pad or a zipper, a useful model system based on polyvalent arrays of carbohydrates. The stumps of the zipper (the carbohydrate moieties) have to be complementary to give enough strength to the binding, while Ca^{2+} may be the driving force to approach the carbohydrates moieties and maintain the zipper closed (Fig. 7). The complementary carbohydrates should be arranged on a scaffold which should not participate in the interaction but should allow the correct orientation of the groups involved in the interaction [63].

The cell adhesion process takes place normally between the same sort of cells through a highly specific and homotypic association. Due to the lack of specificity of protein-protein interaction and because carbohydrate-carbohydrate interaction is specific and occurs prior to adhesion protein interactions as described above, the former interactions have been suggested as an explanation for this mechanism process. It seems reasonable that the first step of the adhesion process should be a rapid and a weak interaction allowing the contact to be broken easily in case the contact were wrong. When the contact is the right one, other stronger forces would take place to re-enforce the interaction (protein-carbohydrate and/or protein-protein).

Whitesides has proposed a model where polyvalent low affinity interactions produce a *conformal* change in the cell which causes a better surface-surface contact to help other stronger interactions to take place (Fig. 8). Applying this

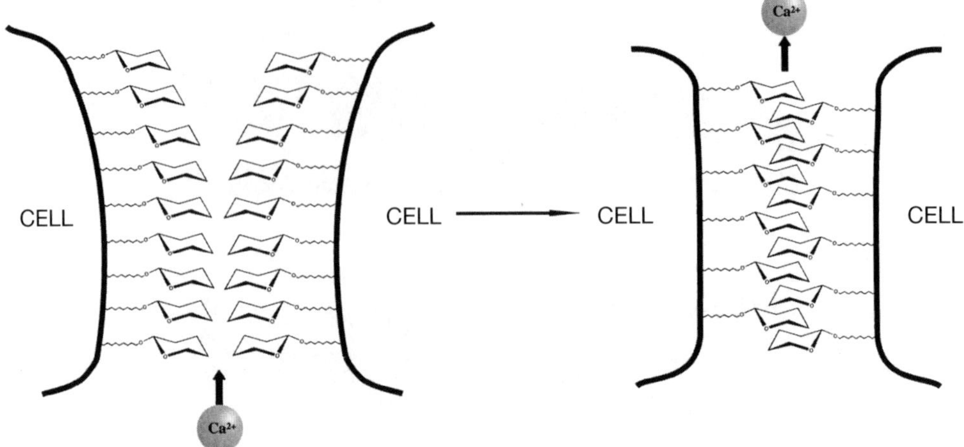

Fig. 7. Cartoon of the zipper adhesion model via carbohydrate-carbohydrate interaction proposed by Spillmann

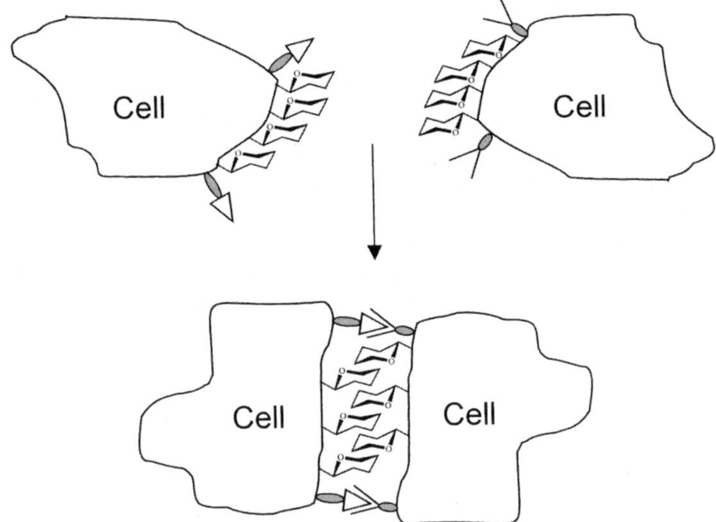

Fig. 8. Implication of carbohydrate-carbohydrate interaction on conformal change (Whitesides model [18]) in cell adhesion process

model to the carbohydrate-carbohydrate interaction, carbohydrate patches at the cell surface of two cells could interact and this interaction could induce a *conformal* change on both cells, allowing the participation of other stronger interactions such as protein-carbohydrate and protein-protein [18].

Very recently, a more complicated system has been described where neutral glycosphingolipids are implicated in binding of soluble form of fibronectin to

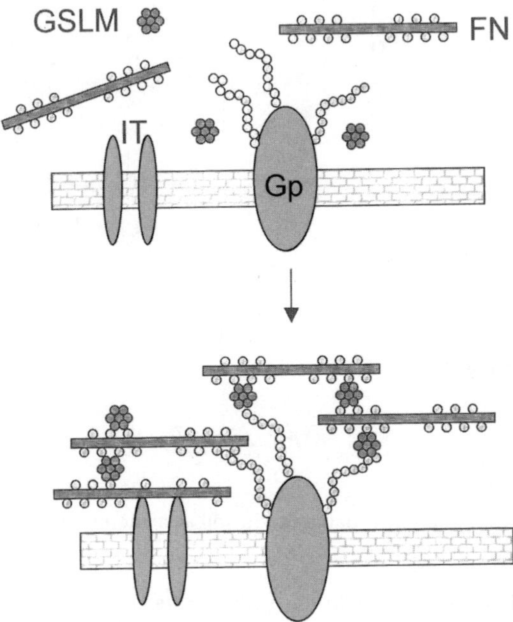

Fig. 9. Cartoon of the carbohydrate-carbohydrate interactions among soluble pericellular fibronectin (FN), glycosphingolipid micelles (GSLM), and glycoproteins (Gp)

cell surface and also in the promotion of pericellular matrix formation [64]. Glycosphingolipids are organized in microdomains at cell membranes but they are also organized extracellularly in micellar form. Hakomori has proposed that neutral glycosphingolipids such as LacCer, Gb4, or Gg3, in this micellar form, may be responsible for organizing a pericellular matrix through interaction with a carbohydrate epitope expressed on a soluble form of fibronectin identified as disialyl-I and with carbohydrates of glycoproteins of the cell membranes as represented in Fig. 9.

4
Carbohydrate-Carbohydrate Interactions as Possible DNA Binding Motif

Carbohydrates in biological systems are often associated with specific recognition and signaling processes through their interaction with proteins. This information transfer between the oligosaccharide and the protein normally leads to important biological functions and diseases.

A very important aspect of carbohydrate recognition, although much less studied, is the interaction between oligosaccharides and DNA. As a matter of fact, DNA-binding drugs that contain carbohydrates have been known for over 40 years. The role of the carbohydrate portion of these DNA-binding conjugates was not clear for a long time. First studies on the importance of the carbohydrate in DNA binders for their biological activity were carried out with the family of anthracycline antibiotics (Fig. 10). It has been shown that the number and the

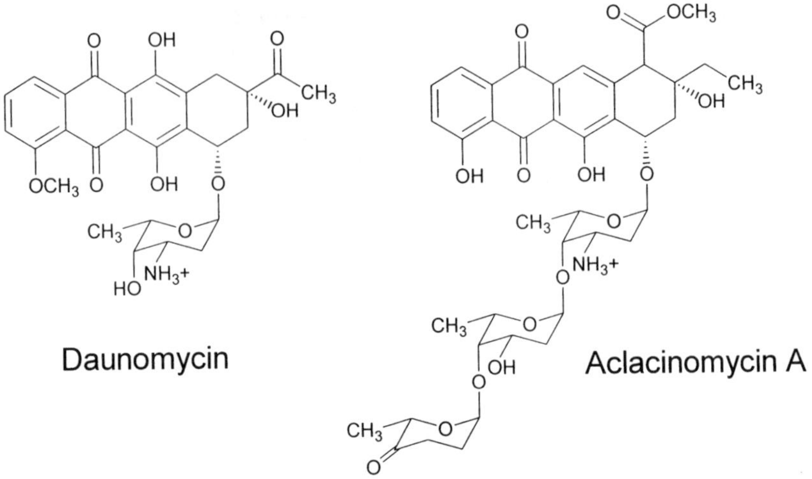

Daunomycin

Aclacinomycin A

Fig. 10. Structure of the antibiotics daunomycin and aclacinomycin A

substitution pattern of the carbohydrate chains of anthracyclines have a profound effect on biological activity, and removal of the sugar side chain results in biological inactivity [65, 66]. At the same time, early structural evidence pointed to the possible role of the carbohydrate moiety during the binding of DNA. Structures of complexes of DNA with daunomycin [67, 68], adriamycin [68], nogalamycin [69–71], MAR-70 [72], and aclacinomycin [73], antibiotics of the family of anthracyclines, have been elucidated by X-ray crystallography and NMR studies. They show that the carbohydrate makes van der Waals contacts with the minor groove of the DNA.

Further evidence of the relevance of the carbohydrate during the DNA binding process came from comparison of the association constants of three members of the anthracycline family, doxorubicin, betaclamycin A, and ditrisarubicin B, which contain a monosaccharide, a trisaccharide, and two trisaccharides, respectively, attached to the chromophore [74]. A much higher affinity constant was found for ditrisarubicin compared with doxorubicin and betaclamycin A, proving the additional stabilization due to the second trisaccharide.

There is enough evidence to support the importance of the sugar side chains of the anthracyclines for the binding of DNA, but this is not as clear for other DNA-binding drugs that contain carbohydrates. For example, metallobleomycins, a group of potent antitumor drugs that cleave DNA, posses a disaccharide moiety consisting of gulose and mannose sugars connected to the metal binding domain (Fig. 11). Recent studies carried out by Boger and colleagues [75] with analogs lacking one or both saccharide units have shown that the terminal α-D-mannopyranoside residue has a small effect upon cleaving activity, whereas the α-L-gulopyranoside residue assists in proper association of the drug to the DNA. Actually, the DNA binding of Co·deglycobleomycin is about 35 times less than for its parent. This has been related to loss of nonspecific van

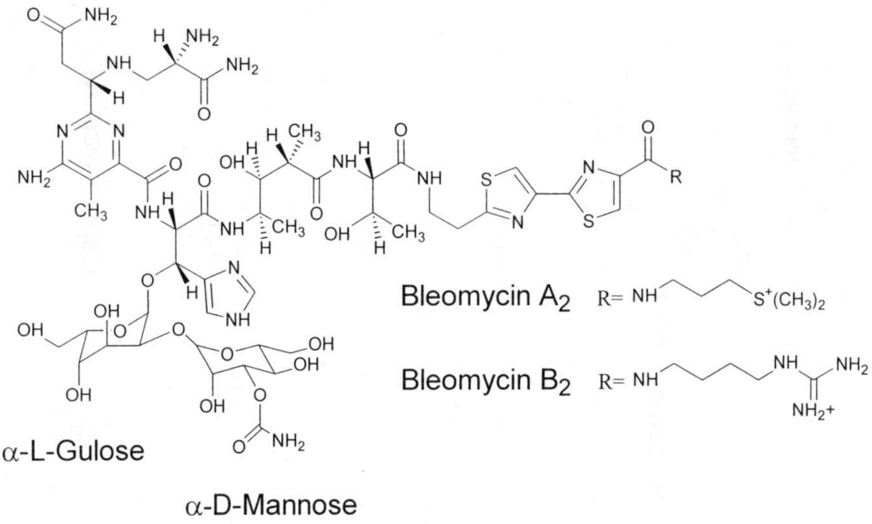

α-L-Gulose

α-D-Mannose

Bleomycin A₂   R= NH ... S⁺(CH₃)₂

Bleomycin B₂   R= NH ...

Fig. 11. Structure of the antitumor drugs bleomycin A_2 and bleomycin B_2

der Waals contacts between the sugar domain and the DNA strand that have been observed in the NMR structure of the complex bleomycin-DNA [76, 77].

Another crucial aspect of these carbohydrate-containing drugs is the role of the sugar in DNA sequence specificity. For example, DNA footprinting studies of several anthracycline drugs with different number of carbohydrates attached to the chromophore showed similar binding sites [78, 79]. These results suggest that the sugars do not play a significant role in DNA sequence specificity. While this statement seems to be true for the anthracycline family, a new view of the carbohydrates as DNA specific binders started with the discovery of the enediyne antitumor antibiotic calicheamicin.

Calicheamicin, a member of the enediyne family, is a drug with a fascinating structure (Fig. 12). It consists of a bicyclic enediyne moiety and a tetrasaccharide fragment. The oligosaccharide tail also includes an arene ring which also plays a role in DNA binding. This potent antitumor agent interacts in vitro with DNA in the minor groove and causes site-specific double-stranded cleavage. The mechanism involves a rearrangement of the enediyne moiety into a 1,4-aryl diradical under reducing conditions, which abstracts hydrogen atoms from the DNA backbone initiating DNA strand scission [80]. Early reports by Ellestad et al. [81] indicated a remarkable sequence specificity of the cleavage preferentially at 5′-TCCT.AGGA and 5′-CTCT.AGAG sites within duplex DNA. Later studies showed that the cleavage-site preferences for cytosine-containing oligopyrimidine tracts could be extended to a variety of sites containing three or more pyrimidines in a row, such as 5′-TTTT.AAAA [82]. This rather confusing sequence specificity where very different functionality can be selected during DNA binding cannot be explained by direct hydrogen bonds between the drug and the minor groove of DNA like in the case of Dervan's derivatives of netropsin and distamycin [83, 84].

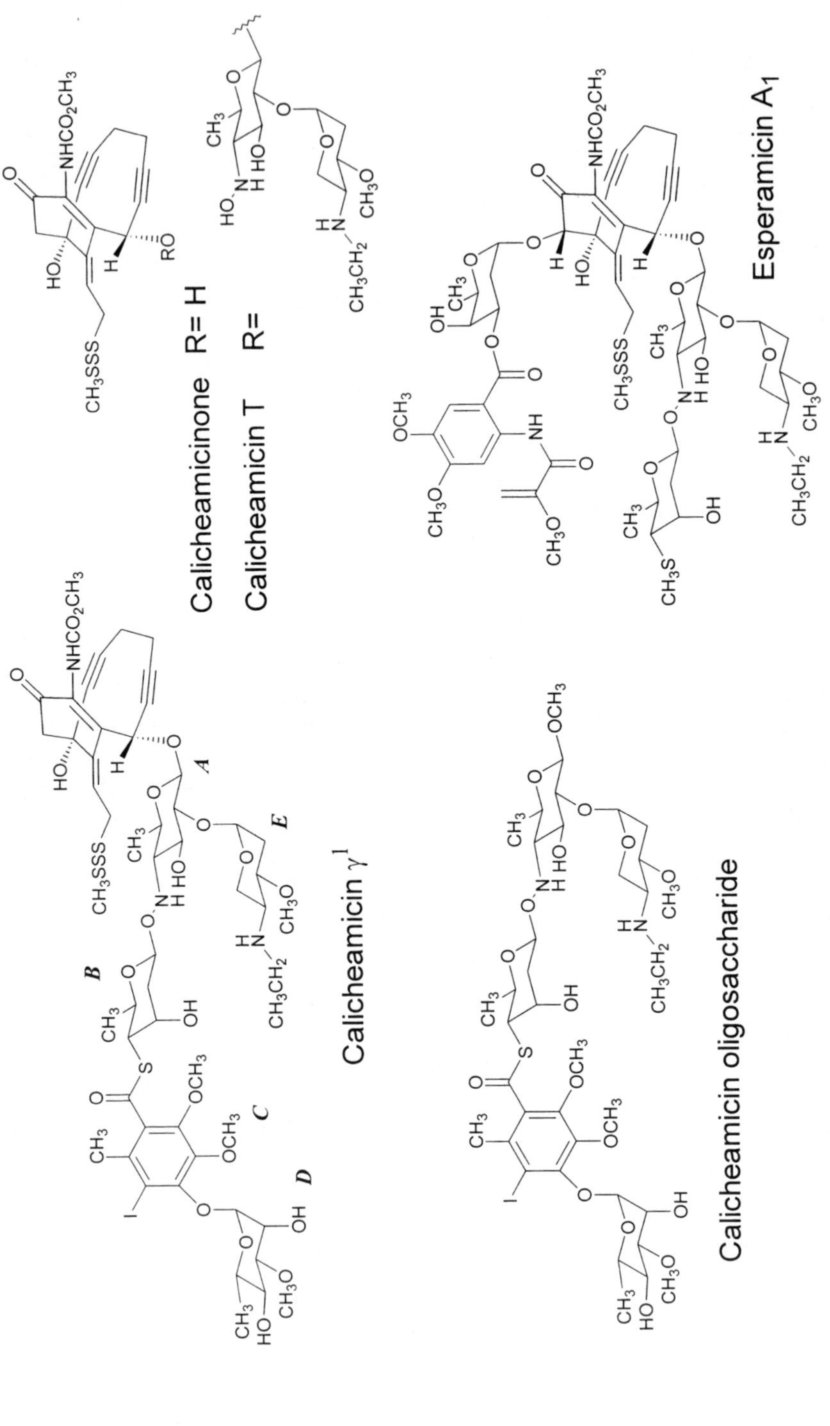

Calicheamicinone R= H

Calicheamicin T R=

Esperamicin A$_1$

Calicheamicin γ_1^I

Calicheamicin oligosaccharide

Fig. 12. Some structures of the calicheamicin family

Moreover, DNA cleaving studies carried out with calicheamicin derivatives lacking two or more carbohydrate units, such calicheamicin T and calicheamicinone (Fig. 12), exhibited less efficient and non-selective cleavage [82–85]. This clearly indicated that the oligosaccharide portion of calicheamicin is the principal DNA-binding element and is largely responsible for the oligopyrimidine selectivity. Furthermore, calichaemicin is the first and only example to date in which the carbohydrate portion itself binds site-selectively to DNA [86, 87]. Then, which is the mechanism used by the carbohydrate for achieving binding selectivity?

Kahne and coworkers [82, 88] proposed an *induced-fit mechanism* for the binding selectivity of enediyne drugs, such as calicheamicin γ and esperamicin A, based on an analysis of different pyrimidine-rich binding sites. Structural evidences that support this hypothesis came from circular dichroism (CD) measurements by the Sugiura group [89] and by the Ellestad group [90] and from NMR studies by the Kahne [91, 92], Nicolaou [93], and Patel groups [94–96]. NMR experiments showed that calicheamicin binds to the DNA minor groove with its aryltetrasaccharide segment in an extended conformation spanning the recognition sequence of the duplex. The oligosaccharide conformation seems to fit perfectly in the minor groove following the natural curvature of DNA. Moreover, the position of the carbohydrate portion of calicheamicin is basically identical in all the structures obtained as shown by NMR [97], independently if the sequence recognized was ACCT, TTTT, or TCCT.

However, again, what makes these oligopyrimidine sequences recognized by calicheamicin when they posses so different H-bonding arrays in their minor groove? The answer may be associated with the ability of oligopyrimidine-oligopurine tracts to distort in order to accommodate relatively rigid DNA minor groove binders. Actually, Kahne and coworkers [91, 92] detected conformational changes upon binding in their NMR studies, including increase in groove width and changes in the conformation of some of the ribose sugars. In contrast, a more refined structure of the complex solved by Patel and colleagues [96] exhibited essentially unperturbed minor groove width at the aryltetrasaccharide binding-site and widening of the groove at the adjacent enediyne binding-site. However, they also observed local conformational perturbations in the DNA helix on complex formation, such as altered stacking between the third and forth base pairs and changes in some ribose puckers within the recognition site.

CD studies on complexes of esperamicin A [89] and calicheamicin γ [90] with DNA also support the induced fit mechanism for the binding selectivity of these enediyne drugs. The CD titration experiments of DNA with the aromatized version of esperamicin A_1 (Fig. 12) showed a reorganization of the host DNA due to the association with the drug. This conformational change of the DNA appears to be associated with a decrease in hydration of the minor groove due to the binding of the hydrophobic esperamicin A. In fact, hydrophobic interaction has been shown to be an important factor in the enediyne/DNA association [98, 99]. These conclusions were presented on the basis of the effect of various inorganic salts on the cleavage of DNA. It is important to note that the hydrophobicity of calicheamicin or esperamicin is due to both the enediyne unit and the carbohydrate [100]. Hydrophobic effects may also be involved in carbohydrate-carbohydrate interactions. In contrast to typical highly hydroxylated oligosaccharides, many of the sugars in carbohydrate-containing drugs such as chromomycin [101], an aureolic acid, or hedamycin [102], a pluramycin, are 2,6-dideoxysugars, which makes them unusually hydrophobic and possess a very small number of hydrogen-bonding partners. Nevertheless, these remaining hydroxy groups make hydrogen bonds with minor groove acceptor atoms and phosphates (note that all these groups are in a very similar position for all DNA sequences) as observed in the refined structure of the calicheamicin γ-DNA complex [96]. Even the positively charged ethylamino group at sugar E interacts with the backbone phosphate. However, the importance of the electrostatic interaction in the binding affinity has been questioned since salt-dependence studies of binding energetics suggest that the association is depending on other forces rather than electrostatic forces [99]. These results again indicate the relevance of the hydrophobicity as the main driving force during the binding process, as in carbohydrate-carbohydrate interaction.

An important question about binding selectivity still remains to be answered: how calicheamicin detects the oligopyrimidine-oligopurine tracts, that supposedly contains more flexible DNA, in order to take advantage of an induced fit recognition? NMR studies of calicheamicin E, the rearrangement product of calicheamicin γ, carried out in different solvents indicate that this enediyne drug is conformationally quite rigid, and therefore is substantially preorganized [103]. Not only are six-member rings quite rigid in their chair conformation but, at the same time, many of the glycosidic linkages are conformationally restricted as well. This rigidity and its extended conformation is probably used by calicheamicin to search the more flexible sequences of DNA and bind them selectively through an induced fit mechanism.

So what does it take to prepare a good carbohydrate minor groove binder? Should hydrophobic 2,6-dideoxysugars and rigid interglycosidic linkages be enough elements to design oligosaccharides that will bind DNA selectively? Actually, a disaccharide and a trisaccharide, both similar to the oligosaccharides found in two anthracycline drugs, were screened as DNA binders. DNA footprinting studies did not show significant DNA binding for any of them [79]. It is possible that the conformation of these oligosaccharides does not fit well in the minor groove of DNA or that they are not rigid enough. Other elements found in calicheamicin could also be important to consider since the carbohydrate portion of calicheamicin by itself is still the only example that can bind selectively DNA [86, 87]. What are the key differences between these two examples?

An important feature found in calicheamicin but not in other oligosaccharide portion of these DNA-binding conjugates is the hydroxylamine glycosidic linkage. In fact, Kahne and coworkers [103] showed that the unusual N-O bond, with its distinctive torsion angles, organizes the two halves of calicheamicin into a shape that complements the shape of the minor groove.

Another critical element only found in calicheamicin is the arene ring and especially the iodine substitution on it. Nicolaou and collaborators [104] replaced this iodine with bromine, chlorine, fluorine, methyl, and hydrogen in the calicheamicin oligosaccharide and observed progressive reduction in binding affinity up to 2.3 kcal mol^{-1} for the hydrogen analogue. It has been suggested that this iodine moiety interacts with the exocyclic C2-NH2 of both guanine residues that lie within the oligopurine strand of the target DNA [105]. Recent NMR data from two different groups [91, 96] suggest that iodine is in a position to interact with the first (5′-most) but not the second guanine C2-NH2. It is possible that these two, seemingly secondary elements, the iodine-NH2 interaction and the unusual N-O bond, are crucial to increase DNA binding affinity and to obtain the rigidity needed to achieve DNA selectivity through an induced fit mechanism. Nevertheless, the Khane group compared the iodine-NH2 interaction with SMe-NH2, a functionality of similar size using capillary electrophoresis [106]. They found that the iodine-NH2 interaction is worth only about 0.6 kcal mol^{-1}, demonstrating that the iodine in pyrimidine selectivity seems to be largely steric, and that electronic contributions are not so important.

4.1
Carbohydrate-Containing Model Systems as Minor Groove Binders

Site-selective binding of both enediyne and aureolic acid antibiotics seems to be achieved largely by an induced process, so a design strategy has been suggested, focused on exploiting sequence-dependent DNA flexibility [88]. There are no examples to date using such an approach probably due to the poorly understood relationship between DNA sequence and its particular ability to get deformed.

Nevertheless, new minor groove binders that posses carbohydrates have been designed based on the knowledge acquired studying calicheamicin and other carbohydrate-containing DNA binders. A straightforward approach is to maintain the calicheamicin carbohydrate as the main recognition element and to add other recognition units to it. Nicolaou and coworkers synthesized two dimeric forms of the calicheamicin oligosaccharide, the head-to-head (h-h) dimer [107], and the head-to-tail (h-t) dimer [108] (Fig. 13). While the calicheamicin binds selectively to 5′-TCCT DNA binding sites, the two dimers were designed by computer modeling to bind into the minor groove along the sequences 5′-AGGAXXTCCT (where X is any natural base) and 5′-TCCTTCCT, respectively.

Inhibition of the cleavage activity of calicheamicin γ by the h-h and h-t dimers confirmed their affinity to the targeted DNA sequences, and showed an increment of 100-fold and 1000-fold, respectively, compared to the affinity of the monomeric oligosaccharide [109]. Furthermore, cross-inhibition experiments demonstrated that each dimer prefers its own binding-site by a factor of 100 or more. NMR studies on both complexes of the dimers with duplex DNA [110, 111] have shown that the binding mode of each oligosaccharide unit of each dimer in the minor groove seems to be very close to that observed in the case of the monomeric calicheamicin oligosaccharide bound to its corresponding TCCT recognition site. It is worth noting that a comparative analysis of the carbohydrate-DNA interactions at two different binding sites was possible since the h-t dimer was complexed to the duplex DNA through the sequence 5′-ACCTTCCT that contains two of four specific sites for calicheamicin [111]. It was observed

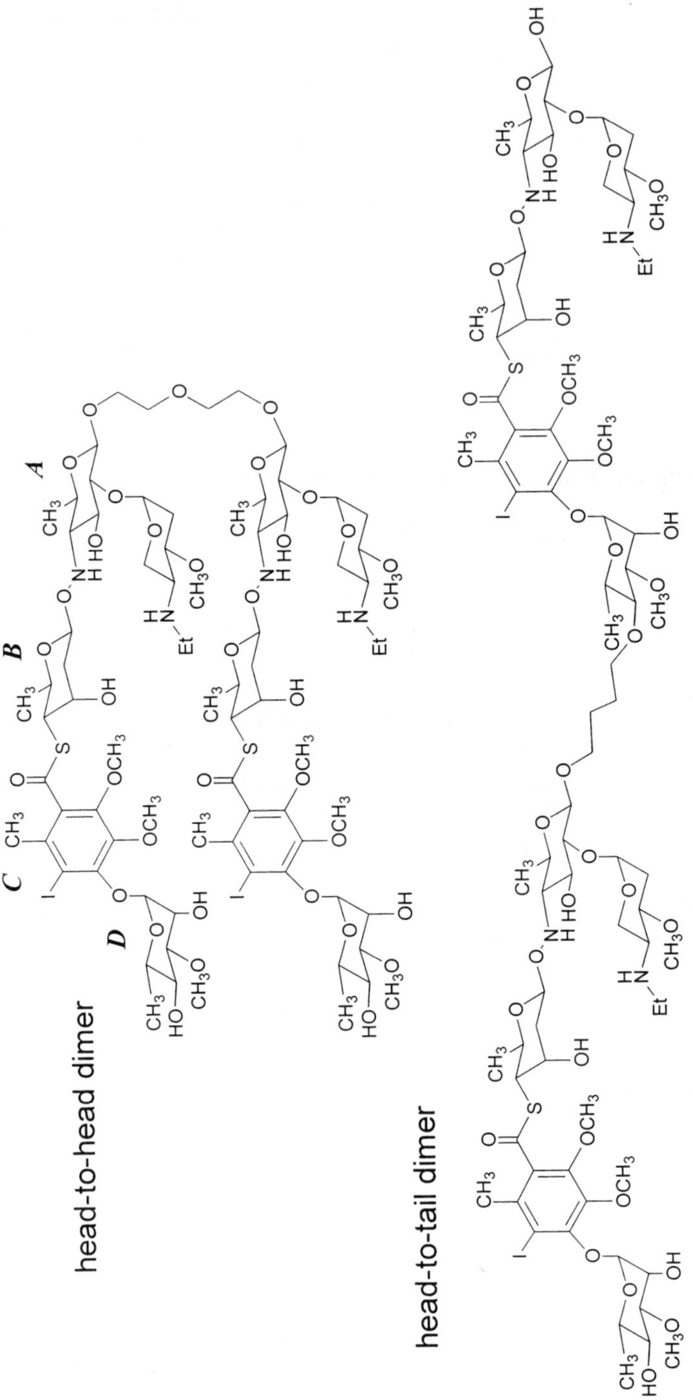

Fig. 13. Structure of the synthetic dimers of the calicheamicin carbohydrate moiety

that the molecular interactions are very similar even for the position where the sequences are different. Here, a hydrogen bond is formed from the carbohydrate B-3OH of each subunit, either to thymine O2 in the ACCT site or the adenine N3 in the TCCT site. Thus, both AT and TA base pairs can be recognized interchangeably. The great similarity in the intermolecular interactions in the two binding sites of the complex of the h-t dimer with duplex DNA and the little distortion found for the B-DNA geometry led Bifulco et al. [111] to propose that sequence selectivity might be due to some degree of induced-fit of the DNA but also due to specific interactions with the bound sequence.

Another example where the calicheamicin oligosaccharide is utilized as the minor groove recognition element are the hybrids with intercalating agents. Depew et al. [112] designed DNA binders where the calicheamicin oligosaccharide is joined to the anthracycline arene moiety, daunorubicinone (Fig. 14). Whereas the directly linked hybrid would seem to be incapable of intercalating dsDNA, the hybrid linked through a bis(ethylene glycol) chain binds DNA with specificity (Fig. 14a). Curiously, the DNA binding sequences are CTTC and TTTC, both of them different from the binding sites of calicheamicin. Unfortunately, there is no structural data for a complex between this calichearubicin hybrid and duplex DNA that could give some insight into the molecular interactions and explain the different specificity between this molecule and its parent calicheamicin.

A similar approach to the calichearubicin hybrid was developed by Toshima and coworkers [113]. They designed a hybrid molecule with another intercalating molecule, an anthraquinone, and a minimum unit of carbohydrate as the minor groove binder (Fig. 14b). The carbohydrate sources were 2,6-dideoxysugars, due to their hydrophobicity. Using DNA footprinting experiments they found DNA binding with some specificity for regions with the sequence 5′-TGC for the hybrids containing an aminosugar but none for the hybrids with the neutral sugars. Recently, the Toshima group [114] has synthesized 2-phenylquinoline hybrids as a new family of light-activateble DNA-cleaving agents (Fig. 14c). The hybrids showed significant cleavage of DNA upon photoirradiation with long-wavelength UV light, whereas neither 2-phenylquinoline nor the sugar moieties show any cleavage.

Another strategy utilized by Kahne and colleagues [115] involves the use of an oligosaccharide as scaffold and attachment of positively charged guanidinium chains to it. 2-Deoxyfucose was chosen as the monomer building block due to its hydrophobicity (Fig. 14d). The oligosaccharide binds to duplex DNA in the micromolar range and NMR studies showed that it does it through the minor groove of the DNA.

In Sect. 6 some quantitative evaluation of the contribution of the carbohydrate moiety to the energetics of DNA-carbohydrate interactions will be presented.

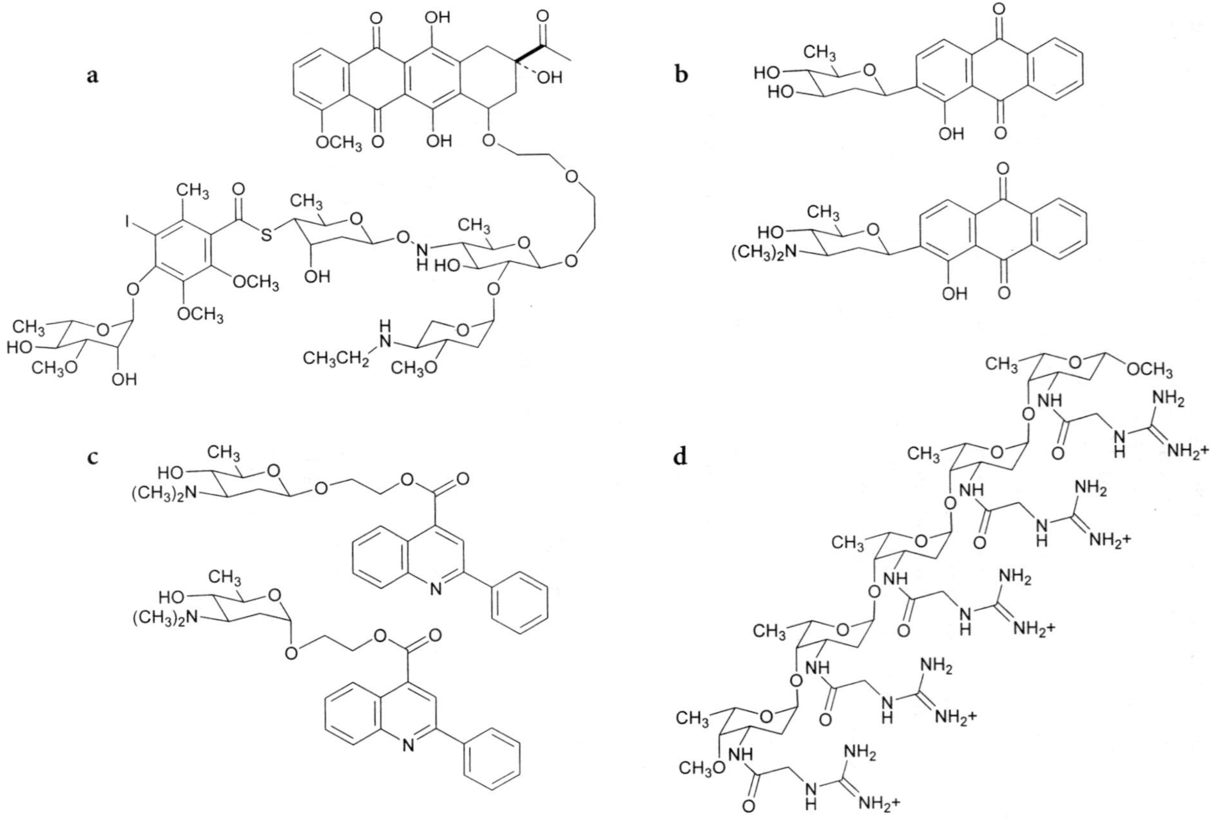

Fig. 14 a–d. Some structures of synthetic DNA minor groove binders: **a** hybrid of calicheamicin oligosaccharide and daunorubicinone; **b** hybrids of anthraquinone and simple monosaccharides; **c** hybrids of 2-phenyl quinoline and simple monosaccharides; **d** 2-deoxyfucose oligomer substituted with guanidinium groups

5
Studying Carbohydrate-Carbohydrate Interactions with Polyvalent Model Systems

The interaction among carbohydrates is defined as low affinity and calcium dependent interaction. These characteristics have made a real challenge of investigating and evaluating carbohydrate-carbohydrate interactions. In fact, there is only qualitative demonstration of this interaction. Thermodynamic or kinetic data of this interaction in biological systems do not exist and only a few examples are described, using model systems, with energetic information on carbohydrate interactions. An understanding of the molecular basis of this interaction demands structural and energetic information.

In the biological context, carbohydrate-carbohydrate interactions co-exist with carbohydrate-protein and protein-protein interactions. This complexity makes it difficult to dissect and analyze the individual contributions of each interacting system. As in the study of other biological interactions, model systems can help to overcome this problem.

The advantage of having available a variety of different model systems lies in the ability to design the components for a well defined chemical system. This advantage is obscured by the disadvantage of using systems far away from the real situation where carbohydrate moieties are not in their native cellular environment (a similar situation found when comparing in vivo or in vitro experiments). Nevertheless, the use of model systems has been a very productive approach in the understanding of many biological mechanisms.

The use of model systems in studying carbohydrate-carbohydrate interactions goes from the very simple monovalent to polyvalent carbohydrate presentation and even to whole cells.

5.1
Cells and Polyvalent Surfaces

The first studies of carbohydrate-carbohydrate interactions with model systems started with the work of Misevic and Burger [9] on the cell aggregation in marine sponges and by Hakomori [10] on the GSL-GSL interaction in embryonic and tumor cells. They have used whole cells and poly-functionalized surfaces (coated plates, liposomes, affinity supports, beads) as interaction partners. The studies with cells have been already described in Sect. 3.

Isolated sponge cells alone or in conjunction with agarose beads have been used as a model system to study the species-specific self-aggregation of marine sponges. The binding of *Microciona* aggregation factor (AF) has been characterized using cells from different sponge species [116]. A model system based on functionalized beads was utilized to investigate whether Ca^{2+} promoted association between AF molecules. Sepharose beads covalently coupled to purified AF aggregate spontaneously under standard aggregation assay conditions in the presence of Ca^{2+} and aggregation was enhanced by addition of exogenous AF.

In an attempt to identify the epitopes responsible for aggregation, *Microciona prolifera* aggregation factor (AF) was dissociated and the smallest functionally intact fragments were tested for interaction with AF-conjugated agarose beads in the presence or absence of Ca^{2+} ions. The fragments were unable to bind to the beads. They were able, however, to inhibit AF-promoted cell aggregation [117].

Proteoglycans isolated from three different marine sponge species, *Microciona prolifera*, *Halicondria panicea*, and *Cliona celata*, were attached to latex-amidine beads of different colors (AF of *Microciona prolifera* to pink, AF of *Halicondria panicea* to yellow, and AF of *Cliona celata* to white beads) and mixed the three bead types in sea water. Within 5–15 min, species-specific bead aggregation occurred in the presence of calcium [118].

In a different research line related to the interaction between glycosaminoglycans, hyaluronic acid-derivatized beads were also used as model systems to demonstrate the specific binding, via a carbohydrate-carbohydrate interaction, of chondroitin sulfate and hyaluronic acid [33]. This result indicates that interactions between glycosaminoglycans located at the cell surface may be involved in cellular adhesive and recognition phenomena.

Cell lines that express preferentially a GSL have been extensively used by Hakomori et al. [119] as model systems in the study of Le^x-Le^x mediated morula compaction and metastasis. F9 teratocarcinoma cells, which express Le^x at the cell surface have been used as model systems in conjunction with Le^x functionalized liposomes, Le^x-coated plastic surface, and Le^x-functionalized octyl sepharose column to study the homotypic Le^x-Le^x interactions. The results suggest that the homotypic Le^x-Le^x interaction on opposing cells provide a basic mechanism for cell recognition during early development.

Le^x-coated plastic beads have also been used as a model to show the selective self-aggregation of Le^x to a series of paraglobosides. The results indicate that Le^x-Le^x interaction can be based on both Le^x GSL and Le^x-bearing glycans of glycoprotein [49]. More examples and references on model systems using cell lines and different polyvalent ligands are also given in Sect. 3.2.1.

Similar model systems as for Le^x-Le^x interactions have been used to demonstrate the heterotypic interaction between the glycosphingolipid Gg3 and G_{M3} expressed in lymphoma and melanoma cell lines, respectively [52]. Liposomes containing different glycosphingolipids were prepared and their interaction with Gg3 coated plastic surface was analyzed. Only when liposomes contained G_{M3} as glycosphingolipids were attractive interactions observed [51]. This interaction was strongly enhanced in the presence of Ca^{2+} and inhibited with EDTA. Moreover, if G_{M3} was used to coat the plastic surface the interaction with G_{M3} liposomes was repulsive, showing clearly that the repulsive interactions may also determine specificity.

Theoretical calculations with G_{M3} and Gg3 were carried out in the same way as that for Le^x [49]. Minimum energy conformations were obtained based on HSEA (Hard Sphere Exo-anomeric) calculations and on glycosidic torsion angles. It was found that, in the minimum energy conformation, the carbohydrate moiety of this glycosphingolipids presented two faces, one more hydrophobic and one more hydrophilic. CPK models were built and it was found that the hydrophobic surface of each molecule fitted closely together head-to-head with a 40–60° angle. Bivalent cation might stabilize this interaction joining the hydrophilic surfaces through the oxygen of the hydroxylic groups. This model is very similar to that proposed for the interaction of Le^x-Le^x and implies the interdigitated interaction of these molecules from two cells surfaces approaching from opposite sides. More details on these models and references have been already given in Sect. 3.2.2.

5.2
Liposomes

Liposomes represent one of the best model systems to mimic glycosphingolipid clustering at the plasma membrane. Liposomes can integrate natural glycosphingolipids as well as synthetic neoglycolipids opening the way for the construction of a broad range of model systems. In decreasing order of complexity, besides cells and polyvalent surfaces, the interaction between fuctionalized liposomes has been investigated as model system to mimic the interaction between cells.

Brewer and Matinyan [120] have used large spherical artificial membranes made from phospholipids and fluorescence derivatives of gangliosides to test the hypothesis that gangliosides can congregate to form an adhesive junction between two membranes. Two types of experiments were carried out. First, the polarized fluorescence from the adhesive junction between two membranes containing the fluorescence gangliosides was compared to that in non-adhesive regions. Increasing fluorescence in the junction suggested congregation of gangliosides. Independent evidences for congregation of charge-bearing gangliosides was found with the conductance probe nonactin. The experiments suggested a structural rearrangement of gangliosides in an adhesive junction. Based on these results an interesting molecular model (Fig. 15) for ganglioside-mediated contact sensation in biological membranes is proposed.

A heterotypic carbohydrate-carbohydrate interaction between galactoceramide (GalCer) and cerebroside sulfate (CBS) – the major glycolipid found in myelin – has been demonstrated in liposomes model systems by Boggs and colleagues [121] (Fig. 16). Small unilamellar phosphatidylcholine/cholesterol liposomes containing GalCer have been prepared and the interaction with similar liposomes containing 3-O-sulfate-GalCer (CBS) was monitored by measuring the increase in optical density at 450 nm when aggregation occurs. Heterotypic aggregation occurred only in the presence of Ca^{2+}. Homotypic aggregation also occurred in the presence of divalent cations but to a lesser extent. No aggregation of vesicles with liposomes lacking glycolipids was found. The influence of the ceramide composition on the aggregation was tested. Increasing the fatty acid chain length of either GalCer or CBS resulted in increasing aggregation. Also, the influence of hydroxyl groups in the lipid chain was tested. These findings indicate that GalCer-CBS interactions mediated by a divalent cation could play a role in the formation of compact multilamellar myelin.

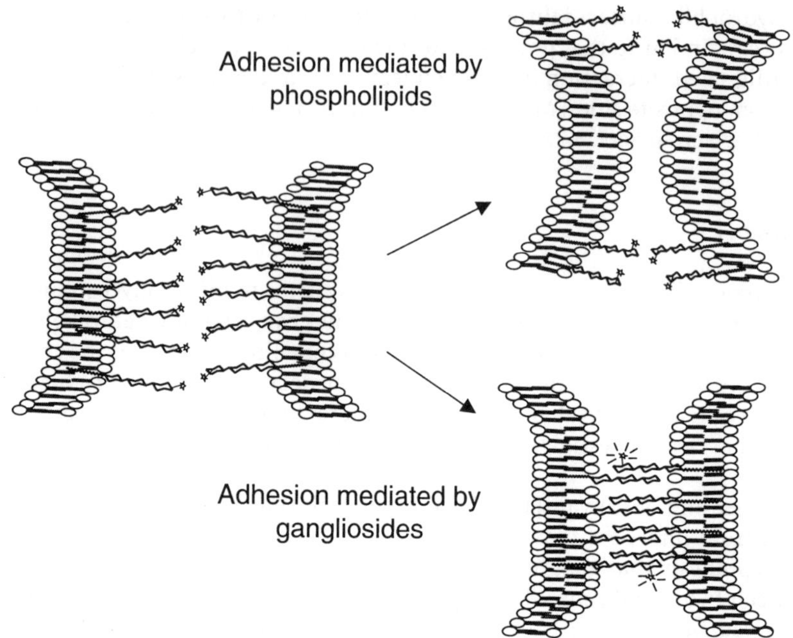

Fig. 15. Two possible ways for the junction between model membranes: adhesion mediated by phospholipid interactions and adhesion mediated by carbohydrate-carbohydrate interactions

Fig. 16. Structure of the galactosyl ceramide and the 3-*O*-sulfate galactosyl ceramide (GBS)

5.3
Affinity Supports

Polyvalent models based on affinity supports have also helped to understand molecular selectivity of carbohydrate-carbohydrate interactions.

Hakomori [10] has studied the interaction of Le^x with itself by affinity chromatography using a modified C-18 column. Le^x-LysLys and lactosyl-LysLys conjugates were applied to an Le^x glycolipid modified and unmodified C-18 silica columns. Only in the case of Le^x conjugates and modified C-18 column retention was observed indicating an association of Le^x solute and Le^x at support.

Affinity chromatography was also used to probe that association between Gg3 and G_{M3} is based on carbohydrate-carbohydrate interaction. LysLys conjugated oligosaccharides and functionalized C-18 columns with glycolipids were prepared. Sialosyllactose (the carbohydrate moiety of G_{M3}) LysLys conjugate was absorbed on a Gg3-bound C-18 column but monovalent sialosyllactose was only weakly absorbed. However, when the C-18 column was modified with G_{M3} or phophatidylcholine, no absorption was observed indicating a specific interaction between the carbohydrate moieties of G_{M3} and Gg3 [122].

Fransson and colleagues [123] have developed an affinity chromatography procedure to measure binding of heparan sulfate species to agarose gels substituted with different heparan sulfates. Modifications of heparan sulfate have been carried out to assess the specificity of these carbohydrate-carbohydrate interactions. The procedure used in this study was based on binding of free chain to immobilized chains. Although the method is only qualitative, the results strongly suggest that heparan sulfate self-association is specific and may participate in specific cell-cell adhesion phenomena

A more complex system formed by proteoheparan sulfate – a cell surface constituent of adherent mammalian cells – and heparan sulfate-agaroses has also been studied by affinity chromatography. Evidence for carbohydrate-carbohydrate interaction was obtained from selective experiments that show (a) interaction of intact proteoglycan with heparan sulfate functionalized column, (b) interaction of heparan sulfate side chain released from proteoglycan with the column, and (c) no interaction of the core protein obtained after heparan sulfate-lyase digestion of the proteoglycan [124].

5.4
Gold Glyconanoparticles

The approaches described above to identify carbohydrate-carbohydrate interaction are based mainly on polyvalent model systems derived from biological products whose composition and structure are not always well defined and cannot be manipulated at all. The utility of these models to evaluate qualitatively carbohydrate-carbohydrate interaction is enormous. However, they are not appropriate for defining geometry, structural requirements, and energetics of these interactions.

The unambiguous characterization of carbohydrate-carbohydrate interactions requires the preparation of polyvalent model systems with well-defined chemical structure that will enable to control their density and presentation.

In contrast to the plethora of polyvalent carbohydrate models developed to study carbohydrate-protein interactions (for reviews see Chaps. 1 and 7 of this book and references therein), few model systems have been designed and used for the study of carbohydrate-carbohydrate interactions. This will change as soon as the scientific community is aware of the importance of clarifying the mechanism of this interaction to understand fully cell adhesion and recognition processes.

The approaches for designing synthetic model systems to study carbohydrate-carbohydrate interaction requires one to take into account the special characteristics of the biological system to mimic. The structural basis of GSL-GSL interaction is expected to differ from that of proteoglycan-proteoglycan interaction and this from that of carbohydrate-protein interaction. The low affinity of carbohydrate-carbohydrate related to other interactions demands a careful manipulation of polyvalence to be effective. Also important is the technique used to study the interactions. The zipper model proposed by Spillmann [63] for carbohydrate-carbohydrate interaction may be considered a two-dimensional polyvalent array of glycoconjugates. To investigate carbohydrate-protein interactions several groups have recently prepared, using self-assembled monolayers (SAM) on gold, 2D polyvalent arrays of carbohydrates (see Chap. 1 in this volume). We have also used this approach to quantify adhesion forces between carbohydrate antigens by atomic force microscopy (AFM) (see Sect. 6.7). In addition to the 2D polyvalent models, spherical arrays of carbohydrates, as represented by liposomes and functional beads, has been preferred to investigate carbohydrate-carbohydrate interactions. Spherical arrays of carbohydrates may offer an efficient alternative to interfere with cell-cell recognition processes. Some of these models are not soluble in aqueous solution and their study has been carried out in suspension.

To study interactions between carbohydrates we have designed model systems that range from synthetic small receptors to biomimetic surfaces with 2D- and 3D-multivalent display of carbohydrates. We have devised two simple approaches based on self-assembled monolayers, by which neoglycoconjugates of the natural epitopes involved in carbohydrate-carbohydrate interactions are attached to 2D- and 3D-gold surfaces, creating in this way multivalent arrays of carbohydrates (Fig. 17). The strategy for tailoring highly polyvalent 3D-carbohydrate surfaces with globular shapes to investigate in solution carbohydrate interactions is based on sugar modified gold nanoclusters [125]. This strategy provides a versatile way to prepare, using a simple synthetic step, a great variety of water-soluble highly polyvalent carbohydrate arrays. Any thiol derivatized neoglycoconjugate of biologically significant oligosaccharides can be attached in situ to gold nanoparticles. The *glyconanoparticles* so prepared are stable, water soluble, and can be manipulated as a water-soluble biological macromolecule.

This approach opens the way to tailor glyconanoparticles with differing ligand density, providing an under-control model for investigating the influence of ligand density on their interactions with specific receptors [126, 127]. Hybrid glyconanoparticles built up of carbohydrates and other molecules (biotin, fluorescent probe, peptides, etc.) can also be prepared by formation of the gold clusters in the presence of the equivalent mixture of the components.

The gold nanoparticles functionalized with self-assembled monolayer (SAM) of glycoconjugates are our 3D-model to mimic glycosphingolipid clustering at the cell surface and to investigate the existence of carbohydrate-carbohydrate interactions in aqueous solutions. Gold nanoparticles functionalized with the disaccharide lactose or the trisaccharide Lewis X determinant have been used to demonstrate the selective self-recognition of the Lewis X via carbohydrate-carbohydrate interactions [125].

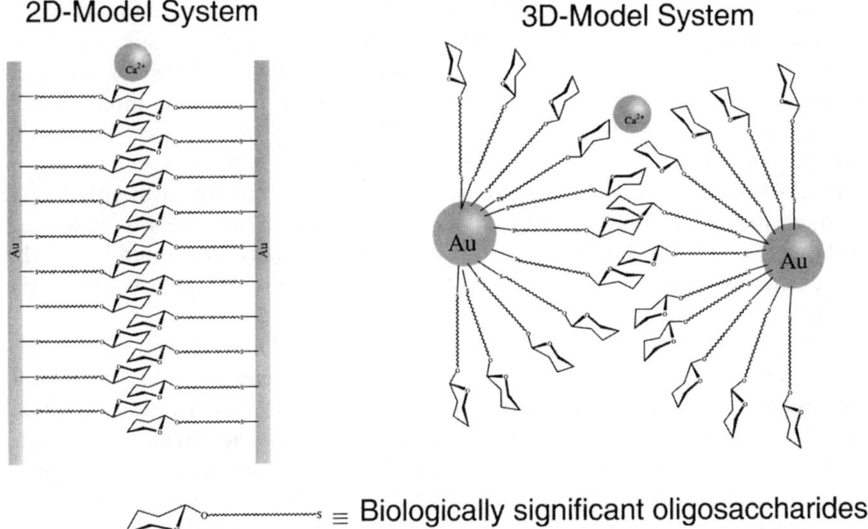

Fig. 17. 2D- and 3D-polyvalent carbohydrate arrays on gold surfaces to study carbohydrate-carbohydrate interactions

The well-defined composition of the glyconanoparticles and their water solubility allow us to carry out experiments (NMR, calorimetry, biosensors) to quantify carbohydrate-carbohydrate interaction. Gold glyconanoparticles functionalized with carbohydrate antigens are currently being applied in cell adhesion processes in vitro as well as in vivo.

6
Model Systems to Study and Quantify the Energetics of Carbohydrate-Carbohydrate Interactions

Establishing the structural origin of affinity and selectivity as well as the connection between structure and thermodynamics in carbohydrate interactions remains a major challenge. Until very recently no energetics data were known in carbohydrate-carbohydrate interactions. In this section we collect the few examples where thermodynamic data are obtained using different approaches. They have been classified based on the model system used.

6.1
Monovalent Ligands

First attempts to identify and quantify carbohydrate-carbohydrate interactions in solution with monomeric ligands were unsuccessful. The homotypic interaction and metal binding properties of the Le^x pentasaccharide (Galβ1–4 [Fucα1–3]GlcNAcβ1–3Galβ1–4Glcβ1, LNFP III) has been investigated by NMR spectroscopy in aqueous solution in the presence and absence of Ca^{2+} ions [128].

No evidence for metal binding or homotypic self-association were found. Molecular modeling of the Le^x-Le^x interaction in the presence of Ca^{2+} suggested that preferred interaction of two Le^x molecules involves contact between two nonpolar patches and Ca^{2+} could subsequently cross-link the two Le^x molecules by being coordinated to four oxygens [49]. The affinity of Ca^{2+} cations for the methyl glycoside of Le^x trisaccharide is not detectable by NMR spectroscopy in water. When two molecules of Le^x trisaccharides, however, are covalently linked through the O-6 of the glucosamine, the affinity for calcium ions in water was just detectable and an affinity constant around 5 M^{-1} was determined [129]. The use of methanol as solvent has allowed one to determine a higher value for the binding constant [129, 130].

Ion Spray Mass Spectrometry (IS-MS) has been used to investigate the molecular basis of carbohydrate associations. Siuzdak and colleagues [131] have used Electrospray Ionization Mass Spectrometry (ESI-MS) to detect cation-mediated glycolipid oligomers of Le^x trisaccharide and Le^x pentasaccharide glycosphingolipid. Le^x trisaccharide as well as Le^x glycolipid form homodimers in the presence of divalent cations (Ca^{2+}, Mn^{2+}, Mg^{2+}). In addition, Le^x glycosphingolipid was found to undergo calcium-dependent heteroassociation with LacCer, GlcCer, and even ceramide alone with decreasing binding affinities in this order. In the presence of monovalent ions only monomers of these molecules were detected.

The tandem Ion Spray Mass Spectrometry (IS-MS) with Collision Induced Decomposition (CID) revealed more details about the structure of the Ca^{2+} dimers. The first conclusion was that Ca^{2+} bound the sugar moiety and neither Cer nor GlcCer moieties participated in the stabilization of the dimer. In summary, results in IS-MS suggested the formation of Le^x homodimers in the presence of Ca^{2+} [131], although this formation is generated in the vacuum phase and could not represent the real species in solution. Based on these data, the authors proposed a mechanism of dimerization starting with a single Ca^{2+} bound to a Le^x molecule and then, in a second step, a conformational change could take place providing an adequate surface for the formation of a homodimer with another Le^x molecule. This model is in contradiction with a different approach which has been proposed by Hakomori to explain this Le^x homotypic interaction [49].

In order to determine if association occurs between the monomeric galactoseceramide (GalCer) and cerebroside molecules (CBS), Koshy and Boggs [132] have studied the non-covalent complexes formed on addition of calcium chloride to GalCer and CBS in methanol by ESI-MS. In this study the authors demonstrated the existence of a specific interaction between GalCer and CBS anion mediated by Ca^{2+}. Strong differences were, however, observed in the aggregation behavior of the glycosphingolipids in ESI-MS conditions in comparison with that observed in the liposome experimental conditions. This is not surprising taking into account the very different experimental conditions for MS and the solution liposome experiments.

A similar study was carried out by ESI-MS in methanol using synthetic lipids with uniform molecular mass in order to assign unambiguously the most relevant peaks. It seems that carbohydrate-carbohydrate interaction can only be demonstrated in solution between multivalent ligands fixed to carrier molecules [133].

6.2
Glycophanes and Cyclodextrins

The first thermodynamic data on carbohydrate-carbohydrate were obtained in our laboratory with a very simple system constituted by a new type of synthetic, water soluble receptors, named glycophanes, and a series of 4-nitrophenyl glycosides (PNPGly). Glycophanes are amphiphilic receptors constituted by disaccharides and aromatic moieties [134]. They may be considered as cyclodextrin-cyclophane hybrids because carbohydrate molecules, lipophilic cavities, and aromatic moieties are involved. The amphiphilic character of these glycophanes, with a hydrophobic cavity surrounded by a hydrophilic area, makes them appropriate models to mimic the interaction of glycoconjugates with their natural receptors in water (Fig. 18).

We have synthesized a series of glycophanes incorporating either α,α-trehalose (a disaccharide with C2 symmetry) or maltose (the constituent disaccharide of cyclodextrins) and naphthalene or (4-hydroxymethyl) benzoic acid as aromatic segments. In the complexes formed by these glycophanes and a series of 4-nitrophenyl glycosides of different stereochemistry, the favorable interaction between the electron rich-electron poor aromatic moieties is a driving force to approach both host and guest carbohydrate moieties in water. As a comparative system we considered the complex formed by cyclodextrin and the same ligands. The free energy of binding, ΔG, determined by NMR titrations, for the interaction between the glycophanes and the PNPGly of D-gluco-, D-galacto-, D-manno-, D-xylo-, L-arabino- and L-fuco- with axial and equatorial configuration at the anomeric center, showed that both glycophanes and α-cyclodextrin form complexes with all glycosides, the free energy of binding ranging from 1.7 to 3.3 kcal mol^{-1} depending on the sugar stereochemistry. As reference compounds the aglycone p-nitrophenol (PNP) and a p-nitrophenyl glycerol derivative (PNPG) were also studied [135, 136].

The most interesting result of this study was that in the glycophane systems a stabilizing contribution to the binding due to the presence of the pyranose ring was observed. This stabilization was evaluated from the difference in the binding energy values of the glycosides and those obtained for the PNP or 1-(4-nitrophenyl) glycerol (PNPG). The highest contribution was observed for the complex formed between α,α-trehalose glycophane and α-L-Fuc derivative with 1.8 kcal mol^{-1} stabilization related to the PNP aglycon. This stabilization is due to carbohydrate-carbohydrate interaction between non-polar patches of both host and guest carbohydrate moieties. In contrast, no stabilizing contribution was found in the α-cyclodextrin system. We showed that these differences in complexing properties are due to the different geometry of both glycophane and cyclodextrin complexes (Fig. 19). In the calculated geometry for the glycophane-PNPGly complexes, the carbohydrate moiety is embedded in the receptor, making van der Waals contacts between non-polar carbohydrate patches of both host and guest possible, while in the cyclodextrin complexes the carbohydrate moiety of the guests remains mainly in contact with the bulk water. Differences in van der Waals contacts between axial and equatorial complexes could be the origin of the observed α/β selectivity shown by glycophanes and absent in cyclodextrin.

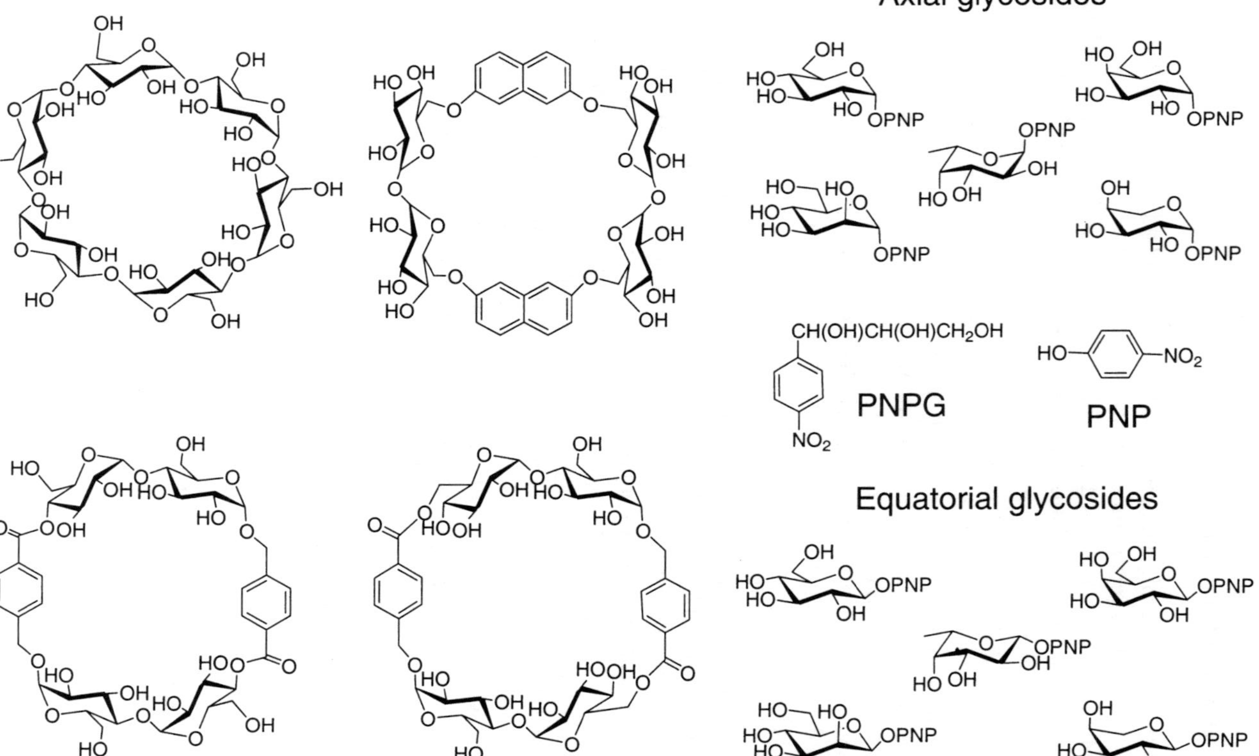

Fig. 18. *Left side:* structure of α-cyclodextrin and glycophane receptors based on α,α-trehalose and maltose disaccharides. *Right side:* structure of the *p*-nitrophenyl glycosides (PNPGly) ligands and the reference compounds PNPG and PNP

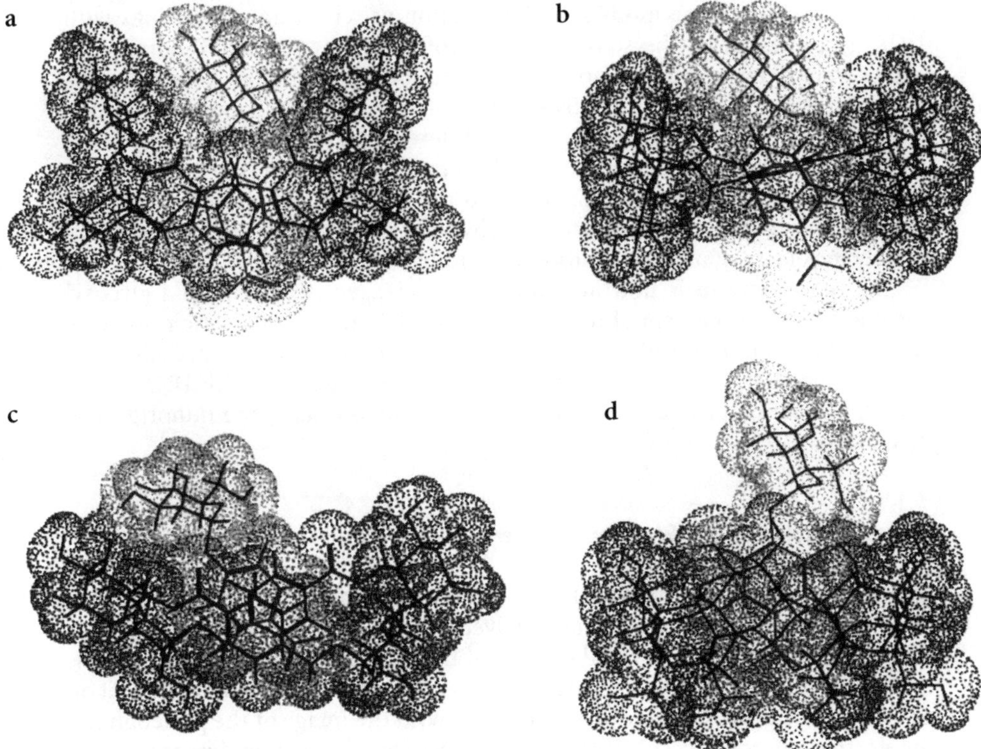

Fig. 19a–d. Calculated geometries of the complex form between: **a** α,α-trehalose glycophane; **b** 4,4′-maltose glycophane; **c** 6,6′-maltose glycophane; **d** α-cyclodextrin; and the p-nitrophenyl α-D-mannopyranoside ligand

In order to understand better the origin of the additional stabilization found in glycophanes we have also tried to get accurate values of enthalpy and entropy of binding of these systems. With these values we have also expected to have insight into the role of water in the mechanism of carbohydrate associations. Enthalpy and entropy values, determined either by microcalorimetry [137] or by van't Hoff analysis [135], indicate a favorable enthalpy and a small unfavorable entropy of binding for all complexes studied. This thermodynamic pattern is similar to that found for monosaccharide association to lectins [32], but also for many other associations in water of amphiphilic receptors with apolar ligands. This is noteworthy taking into account that the only structural homology between these simple systems and biological systems (as for example lectins) is the solvent water. We interpret these results as water being the main driving force in this interaction [15].

Resemblance can be established between our glycophane system and the system formed by DNA and most of the intercalating antitumoral drugs. Carbohydrate moieties can also stabilize the interaction of rebeccamycin – a glycosilated indolocarbazole antitumoral drug – with duplex DNA [138]. In a similar way to that of our glycophane models, the interaction of rebeccamycin analogs with DNA is controlled by the stereochemistry of the sugar residue. Binding constants were determined by fluorescence-titration experiments using the aglycone and the α and β anomers. The binding free energy contribution from the carbohydrate moiety, estimated from the difference in the binding energy of the β glycoside and the aglycone, was -1.6 kcal mol^{-1}. This value shows clearly that the β anomer contributes significantly to the binding free energy. It was not elucidated whether carbohydrate-carbohydrate interactions are involved, but clearly the electrostatic contribution to binding was negligible.

The rebeccamycin is not the only anticancer agent that require a glycosyl residue for its mechanism of interaction (Sect. 4). Nuclear magnetic resonance and crystal structural studies with daunomycin show that the sugar residue participates not only in specific hydrogen-bonding interactions with DNA bases, but also in numerous van der Waals contacts with atoms in the minor groove [68, 139].

6.3
Adhesion Factor (AF) of *Microciona prolifera*

Atomic force microscopy (AFM) has been recently applied to quantify binding forces between the proteoglycans involved in specific cell adhesion in marine sponges, giving an idea about the range of the forces implicated in this process [140]. First of all, an image of proteoglycan was obtained by AFM absorbing it on freshly cleaved mica surface. In comparison with the image of the proteoglycan by electron microscopy (EM), AFM allows one to obtain a similar image but with 3D information about the ring "sunburst" proteoglycan.

For the adhesion forces measurement, a cantilever tip and a flat surface were functionalized with the proteoglycan and approach and retract cycles were collected at a physiological Ca^{2+} concentration (10 mmol l^{-1}) in seawater. Several jump-offs, which indicated a polyvalent binding, were observed and adhesion forces of 40 ± 15 pN were calculated for the interaction between two molecules. When the monoclonal antibody Block 2 was used to block part of the proteoglycan on the surface, the adhesion force between the tip and the surface was reduced significantly; however, when a different antibody (a control) was used no changes were observed, and therefore the forces found were due to specific interactions. It is important to notice that the complete chemical structure of AF is unknown and it is not possible to say how to represent these measurements in chemical terms. Chemical characterization of this proteoglycan has to be performed to define what epitopes are responsible for interaction. One point that still remains to be resolved is whether the interaction is homologous or heterologous involving the association of one epitope with itself or with other epitopes. Many of these questions could be answered with a clearer idea about the structure of the glycan and with the unmistakable determination of the real epitopes. This study is the only one reported to date on carbohydrate-carbohydrate interactions by AFM (see the last section of this chapter and Chap. 4 of this volume for recent AFM application to measure carbohydrate-carbohydrate interaction forces).

Model systems built up with synthetic well-defined structures of the epitopes involved in the interaction have been foreseen as an useful tool for gaining insight into the molecular mechanism of adhesion. In the next pages the few examples on quantification of binding between carbohydrates head groups reported to date will be presented. The surface force apparatus, NMR spectroscopy, biosensor, and AFM have been used as detection techniques. As model systems, mixed bilayer membranes, liposomes, Langmuir monolayers, and SAM have been used respectively.

6.4
Mixed Bilayer Membranes

Leckband et al. [141] have used the surface forces apparatus to determine the forces governing the adhesion between mixed bilayer membranes comprising lactosyl ceramide (LacCer) and di-tridecanoyl-phosphatidyl choline. Forces between membranes were quantified as a function of the glycolipid surface densities. Adhesion forces increased with increasing amounts of glycosphingolipid in the membrane. The energy increased from -0.5 to -3.5 mN/m when the glycolipid density increased from 0 to 30%.

Authors assumed that the values found should be considered as semiquantitative because some approximations had been made with respect to the thickness of the layer and the refractive indices. It was difficult to determine what was the real contribution of carbohydrates to the increment of adhesion forces between the layers. Also, when the adhesion forces were measured during the retract cycle, it might be possible that some LacCer molecules were pulled out of the bilayer and complicate more the interpretation of these results. However, based on the combined thickness of the headgroup of the glycosphingolipids, it was concluded that the carbohydrate layers interacted in an interdigitated way.

6.5
Functionalized Liposomes

A different approach to measure binding between carbohydrates has been recently published using a functionalized liposome and transfer NOEs technique. Le^x neoglycolipid was inserted in dimyristoylphosphatidylcholine/dihexanoylphosphatidylcholine liposomes and the interaction between these liposomes (in the presence and absence of Ca cations) and the methyl glycoside of Le^x was evaluated by differences in the observed transfer NOEs. Due to the huge molecular weight of the Le^x conjugated liposome compare to the Le^x molecule, transient interactions between them could be detected by transfer NOE. The monomer Le^x in the presence of Ca^{2+} showed negative cross-signals in the NOESY spectrum. The addition of EDTA hijacked Ca^{2+} from the complex and therefore broke the interaction decreasing the NOE intensity. Using this method, an affinity constant of 2–3 M^{-1} was found for Le^x-Le^x interaction [142].

6.6
Langmuir Monolayers and Polymers

A quantitative estimation of the binding between G_{M3} and Gg3 have been carried out by Surface Plasmon Resonance (SPR) [143]. The association constants between Langmuir monolayers deposited on a hydrophobized gold-deposited glass plate of glycosphingolipid G_{M3}, lactosylceramide, glucosylceramide, and ceramide, and glycoconjugate polystyrenes of Gg3, lactose, and cellobiose were determined by SPR. Polystyrene conjugate of Gg3 was bound to the G_{M3} monolayer with an apparent affinity constant Ka $\cong 10^{-6}$ M^{-1}, whereas lactose and cellobiose conjugate had little selectivity to the glycolipid monolayers. The affinity constant of the glycoconjugate Gg3 to the lipids decreased in the order $G_{M3} >$ LacCer > GlcCer $\gg$ Ceramide. The decrease order was the same as those qualitatively estimated by Kojima and Hakomori [51] for carbohydrate-carbohydrate interaction using liposomes and cells.

A review on the application of SPR on carbohydrate-protein interactions and on carbohydrate-carbohydrate interaction is given in Chap. 3 of this volume.

6.7
Self-Assembled Monolayers (SAMs)

Self-assembled monolayers (SAMs) on gold have successfully been used to create polyvalent surfaces and to study carbohydrate-protein interactions using biosensors and other detection techniques. We are using SAMs of neoglycoconjugates on gold in an attempt to quantify carbohydrate-carbohydrate interactions by different techniques. The thiol derivatized neoglycoconjugates of the di- and trisaccharide epitopes of the *Microciona prolifera* sponge and lactose and Le^x antigen have been prepared to attach them to gold surfaces. These neoglycoconjugates have allowed us to built up 2D-(SAM) model systems to determine adhesion forces between carbohydrates by atomic force microscopy (AFM) [144].

AFM is a form of scanning probe microscopy where a very sharp tip is attached to the end of a cantilever. The optical monitoring of the cantilever deflection when the tip scans the sample surface produces either topographic images (imaging mode) or force-distance curves (force spectroscopy mode). From the force-distance curves the individual adhesion force between molecules can be calculated (for more details and references see Chap. 4 of this volume). The microscope can incorporate a liquid cell to perform experiments in solution. To measure adhesion forces between carbohydrate molecules, first a thin gold layer is deposited on the tip and on a freshly cleaved mica surface. The tip and the sample are immersed in a solution of the thiol derivatized oligosaccharide to form the self-assembled monolayer (Fig. 20).

We have measured the adhesion forces between the trisaccharide Le^x antigen and compared them with those between lactose disaccharide. The organization of the monolayers on the tip and on the sample provides a polyvalent surface mimicking the GSL presentation of these antigens at the outer cell membrane. The thiol neoglycoconjugates 1 and 2, used to prepared the gold glyconanopar-

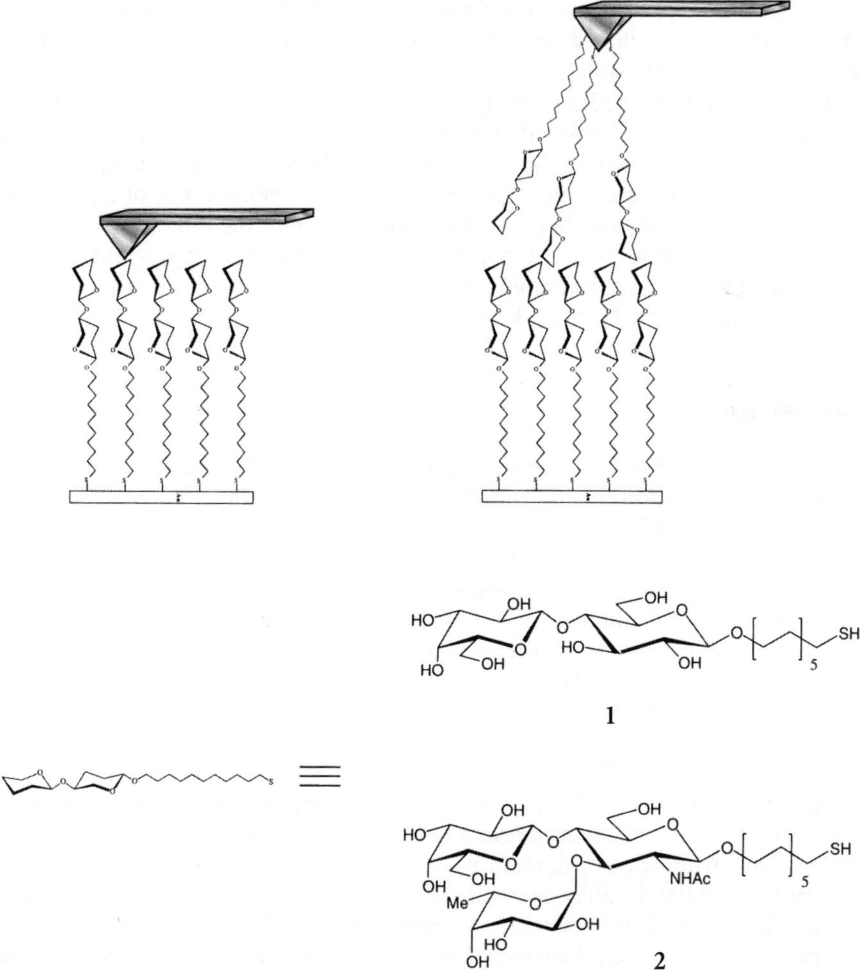

Fig. 20. Schematic representation of atomic force microscopy (AFM) application to measure the adhesion forces between carbohydrate antigen neoglycoconjugate 1 and 2

ticles (Sect. 5.4) have been attached to the gold surfaces of the AFM cantilever and sample. The force-distance curves obtained with the Le^x molecules showed a stepwise profile characteristic of molecular interactions. Statistical analysis of the step high reveals that the measured unbinding forces are entire multiples of 20 ± 4 pN. The strong periodicity of the autocorrelation function is proof that the interaction observed is due exclusively to adhesion forces between Le^x molecules. The 20 ± 4 pN was assigned to the binding between two individual Le^x determinant molecules.

In control experiments between lactose functionalized tip and sample, no interaction was observed. Additionally, crossed-experiments between Le^x functionalized tip and lactose functionalized sample and vice-versa gave the same results as the lactose-lactose experiments. No interaction was observed between these surfaces. The 20 ± 4 pN value is significant taking into account that the force of a single adhesion binding event of the proteoglycan in marine sponges [140] is in the same order of magnitude. Using the approach of Bell [145] an energy of − 5.7 kJ mol^{-1} was calculated for the 20 ± 4 pN force. According to the equation $E = F \times d$, where E stands for energy and F for maximal force, the interaction distance d corresponds approximately to 0.5 nm, the size of a pyranose ring of the Le^x trisaccharide [144]. The contribution of the interaction between the pyranose moieties to the stabilization of the complex formed by the p-nitrophenyl glycosides and glycophanes [135, 136] is also in the same order of magnitude, ranging from 2.0 to 7.5 kJ mol^{-1}.

7
Perspectives

Carbohydrate-carbohydrate interaction may represent a novel and versatile form of cellular adhesion and recognition. The glycosphingolipid microdomain is being considered as a functional portal for bacterial and virus infection, tumor cell adhesion, metastasis, and many others normal and pathological processes. If carbohydrate-carbohydrate interactions are involved in these processes, the understanding of the molecular basis of this mechanism should lead to the benefit of a clearer understanding of the role of surface glycoconjugates in the initial steps of cell recognition.

In this chapter we have reviewed the current status of research in this topic and we have tried to glimpse the potential work that can be done in this field. The synergy among cell biologists, biochemists, and carbohydrate chemists is of major importance in establishing unambiguously the involvement of these interactions in biological processes and in discovering new processes where carbohydrate-carbohydrate interactions are involved.

The weak and polyvalent character of carbohydrate-carbohydrate association makes the study of this interaction a big challenge. The application and development of new analytical surface techniques to measure weak affinity binding together with the preparation of new polyvalent model systems will be decisive in unraveling the basis of this interaction.

The carbohydrate-carbohydrate interaction landscape is almost unexplored. There are presently not many ways to carry out this exploration and new ways have to be opened.

Acknowledgement. We would like to dedicate this chapter to the memory of the late Prof. R. U. Lemieux, who always encouraged us to investigate the role of carbohydrates in biological molecular recognition and to Prof. M. Martín-Lomas, who introduced us to the field and has been supporting and advising us for many years. We thank África G. Barrientos, Jesús M. de la Fuente and Juan Antonio Gallardo for experimental work and for their help in the preparation of the figures.

8
References

1. Sler M, Desplat-Jego S, Vacher B, Ponsonnet L, Fraterno M, Bongrad P, Martin J-M, Foa C (1998) FEBS Lett 429:89
2. Varki A (1993) Glycobiology 3:97
3. Dwek RA (1996) Chem Rev 96:683
4. Lee YC, Lee RT (1995) Acc Chem Res 28:321
5. Bourne Y, van Tilbeurgh H, Cambillau C (1993) Curr Opin Struct Biol 3:681
6. Spillmann D, Burger MM (2000) Carbohydrate-carbohydrate interactions. In: Ernst B, Hart GW, Sinaÿ P (eds) Carbohydrates in chemistry and biology. Wiley-VCH, Weinheim, vol 2, p 1061
7. Bovin NV (1997) Carbohydrate-carbohydrate interactions. In: Gabius HJ, Gabius S (eds) Glycoscience: status and perspectives. Chapman and Hall, Weinheim, p 277
8. Spillmann D, Burger MM (1996) J Cell Biochem 61:562
9. Misevic GN, Burger MM (1993) J Biol Chem 268:4922
10. Hakomori S-I (1991) Pure Appl Chem 63:473
11. Moser HE (2000) Carbohydrate-nucleic acid interactions. In: Ernst B, Hart GW, Sinaÿ P (eds) Carbohydrates in chemistry and biology. Wiley-VCH, Weinheim, vol 2, p 1095
12. Sucheck SJ, Wong C-H (2000) Curr Opin Chem Biol 4:678
13. Walter F, Vicens Q, Westhof E (1999) Curr Opin Chem Biol 3:694
14. Balasubramanian D, Raman B, Sundari CS (1993) J Am Chem Soc 115:74
15. Lemieux RU (1996) Acc Chem Res 29:373
16. Tillack TW, Allietta M, Moran RE, Young WW Jr (1983) Biochim Biophys Acta 733:15
17. Rock P, Allietta M, Young WW Jr, Thompson TE, Tillack TW (1990) Biochemistry 29:8484
18. Mammen M, Choi S-K, Whitesides GM (1998) Angew Chem Int Ed Eng 37:2754
19. Schnaar RL (1991) Glycobiology 1:477
20. Borman S (2000) Chem Eng News 48
21. Sharon N, Lis H (1989) Science 246:227
22. Lasky LA (1992) Science 258:964
23. Varki A (1994) Proc Natl Acad Sci USA 91:7390
24. Quiocho FA (1989) Pure Appl Chem 61:1293
25. Quiocho FA (1986) Ann Rev Biochem 55:287
26. Lemieux RU, Boullanger PH, Bundle DR, Baker DA, Nagpurkar A, Venot A (1978) Nouv J Chim 2:321
27. Rand RP (1992) Science 256:618
28. Ernst JA, Clubb RT, Zhou H-X, Gronenborn AM, Clore GM (1995) Science 267:1813
29. Ladbury JE (1996) Chem Biol 3:973
30. Israelachvili J, Wennerström H (1996) Nature 379:219
31. Cheng Y-K, Rossky PJ (1998) Nature 392:696
32. Burkhalter NF, Dimick SM, Toone EJ (2000) Protein-carbohydrate interaction: fundamental considerations. In: Ernst B, Hart GW, Sinaÿ P (eds) Carbohydrates in chemistry and biology. Wiley-VCH, Weinheim, vol 2, p 863
33. Turley EA, Roth S (1980) Nature 283:268
34. Wilson HV (1907) J Exp Zool 5:245
35. Humphreys T (1963) Dev Biol 8:27
36. Henkart P, Humphreys S, Humphreys T (1973) Biochemistry 12:3045
37. Cauldwell CB, Henkart P, Humphreys T (1973) Biochemistry 12:3051
38. Burkart W, Burger MM (1981) J Supramol Struct Cell Biochem 16:170
39. Misevic GN, Burger MM (1986) J Biol Chem 261:2853
40. Misevic GN, Burger MM (1993) J Biol Chem 268:4922
41. Misevic GN, Finne J, Burger MM (1987) J Biol Chem 262:5870
42. Spillmann D, Hard K, Thomas-Oates J, Vliegenthart JFG, Misevic G, Burger MM, Finne J (1993) J Biol Chem 268:13,378

43. Spillmann D, Hard K, Thomas-Oates JE, Van Kuik JA, Vliegenthart JFG, Misevic G, Burger MM, Finne J (1995) J Biol Chem 270:5089
44. Yoshida-Noro C, Heasman J, Goldstone K, Vickers L, Wylie C (1999) Glycobiology 9:1323 and references cited therein
45. Fenderson BA, Eddy EM, Hakomori S-I (1990) BioEssays 12:173
46. Toyokuni T, Hakomori S-I (1994) Methods Enzymol 247:325
47. Fenderson BA, Zehavi U, Hakomori S-I (1984) J Exp Med 160:1591
48. Eggens I, Fenderson B, Toyokuni T, Dean B, Stroud M, Hakomori S-I (1989) J Biol Chem 264:9476
49. Kojima N, Fenderson BA, Stroud MR, Goldberg IR, Habermann R, Toyokuni T, Hakomori S-I (1994) Glycoconj J 11:238
50. Boubelík M, Floryk D, Bohata J, Dráberová L, Macák J, Smíd F, Dráber P (1998) Glycobiology 8:139
51. Kojima N, Hakomori S-I (1989) J Biol Chem 264:20,159
52. Kojima N, Shiota M, Sadahira Y, Handa K, Hakomori S-I (1992) J Biol Chem 267:17,264
53. Kojima N, Hakomori S-I (1991) Glycobiology 1:623
54. Otsuji E, Park YS, Tashiro K, Kojima N, Toyokuni T, Hakomori S-I (1995) Int J Oncol 6:319
55. Hakomori S-I, Handa K (2000) Glycosphingolipid microdomain in signal transduction, cancer, and development. In: Ernst B, Hart GW, Sinaÿ P (eds) Carbohydrates in chemistry and biology. Wiley-VCH, Weinheim, vol 4, p 771
56. Hakomori S-I (2000) Glycoconj J 17:143
57. Kasahara K, Sanai Y (2000) Glycoconj J 17:153
58. Yamamura S, Handa K, Hakomori S-I (1997) Biochem Biophys Res Commun 236:218
59. Song Y, Withers DA, Hakomori S-I (1998) J Biol Chem 273:2517
60. Iwabuchi K, Handa K, Hakomori S-I (1998) J Biol Chem 273:33,766
61. Zhang Y, Iwabuchi K, Nunomura S, Hakomori S-I (2000) Biochemistry 39:2459
62. Iwabuchi K, Zhang Y, Handa K, Withers DA, Sinaÿ P, Hakomori S-I (2000) J Biol Chem 275:15,174
63. Spillmann D (1994) Glycoconj J 11:169
64. Zheng M, Hakomori S-I (2000) Arch Biochem Biophys 374:93
65. Lown JW (1988) Anthracycline and anthracenedione-based anticancer agents. Elsevier, New York
66. Lown JW (1993) Chem Soc Rev 93:165
67. Quigley GJ, Wang AH-J, Ughetto G, van der Marel GA, van Boom JH, Rich A (1980) Proc Natl Acad Sci USA 77:7204
68. Frederick CA, Williams LD, Ughetto G, van der Marel GA, van Boom JH, Rich A, Wang AH-J (1990) Biochemistry 29:2538
69. Williams LD, Egli M, Gao Q, Bash P, van der Marel GA, van Boom JH, Rich A, Frederick CA (1990) Proc Natl Acad Sci USA 87:2225
70. Smith CK, Davies GJ, Dodson EJ, Moore MH (1995) Biochemistry 34:415
71. Caceres-Cortes J, Wang AH-J (1996) Biochemistry 35:616
72. Gao Y-G, Liaw Y-C, Li Y-K, van der Marel GA, Boom JH, Wang AH-J (1991) Proc Natl Acad Sci USA 88:4845
73. Yang D, Wang AH-J (1994) Biochemistry 33:6595
74. Kunimoto S, Takahoshi Y, Uchida T, Takeuchi T, Umezawa H (1988) J Antibiot 41:655
75. Boger DL, Teramoto S, Zhou J (1995) J Am Chem Soc 117:7344
76. Wu W, Vanderwall DE, Lui SM, Tang X-J, Turner CJ, Kozarich JW, Stubbe J (1996) J Am Chem Soc 118:1268
77. Wu W, Vanderwall DE, Teramoto S, Lui SM, Hoehn ST, Tang X-J, Turner CJ, Boger DL, Kozarich JW, Stubbe J (1998) J Am Chem Soc 120:2239
78. Fox KR (1988) Anti-Cancer Drug Des 3:157
79. Shelton CJ, Harding MM, Prakash AS (1996) Biochemistry 35:7974
80. Zein N, Sinha AM, McGahren WJ, Ellestad GA (1988) Science 240:1198
81. Zein N, Poncin M, Nilakantan RWJ, Ellestad GA (1989) Science 244:697

82. Walker S, Landovitz R, Ding W-D, Ellestad GA, Kahne D (1992) Proc Natl Acad Sci USA 89:4608
83. White S, Szewczyk JW, Turner JM, Baird EE, Dervan PB (1998) Nature 391:468
84. Wemmer D (1998) Nat Struc Biol 5:169
85. Drak J, Iwasawa N, Danishefsky SJ, Crothers DM (1991) Proc Natl Acad Sci USA 88:7464
86. Aiyar J, Danishefsky SJ, Crothers DM (1992) J Am Chem Soc 114:7552
87. Nicolaou KC, Tsay S-C, Suzuki G, Joyce GF (1992) J Am Chem Soc 114:7555
88. Kahne D (1995) Chem Biol 2:7
89. Uesugi M, Sugiura Y (1993) Biochemistry 32:4622
90. Krishnamurthy G, Ding W-D, O'Brien L, Ellestad GA (1994) Tetrahedron 50:1341
91. Walker S, Murnick J, Kahne D (1993) J Am Chem Soc 115:7954
92. Walker S, Andreotti AH, Kahne D (1994) Tetrahedron 50:1351
93. Gomez Paloma L, Smith JA, Chazin WJ, Nicolaou KC (1994) J Am Chem Soc 116:3697
94. Ikemoto N, Ajay Kumar R, Dedon PC, Danishefsky SJ, Patel DJ (1994) J Am Chem Soc 116:9387
95. Ikemoto N, Ajay Kumar R, Ling T-T, Ellestad GA, Danishefsky SJ, Patel DJ (1995) Proc Natl Acad Sci USA 92:10,506
96. Ajay Kumar R, Ikemoto N, Patel DJ (1997) J Mol Biol 265:187
97. Kalben A, Pal S, Andreotti AM, Walker S, Gange D, Biswas K, Kahne D (2000) J Am Chem Soc 122:8403
98. Ding W-D, Ellestad GA (1991) J Am Chem Soc 113:6617
99. Krishnamurthy G, Brenowitz MD, Ellestad GA (1995) Biochemistry 34:1001
100. Walker S, Valentine KG, Kahne D (1990) J Am Chem Soc 112:6428
101. Silva DJ, Kahne D (1993) J Am Chem Soc 115:7962
102. Hansen MR, Hurley LH (1996) Acc Chem Res 29:249
103. Walker S, Gange D, Gupta V, Kahne D (1994) J Am Chem Soc 116:3197
104. Li T, Zeng Z, Estevez VA, Baldenius KU, Nicolaou KC, Joyce GF (1994) J Am Chem Soc 116:3709
105. Hawley RC, Kiessling LL, Schreiber SL (1989) Proc Natl Acad Sci USA 86:1105
106. Biswas K, Pal S, Carbeck JD, Kahne D (2000) J Am Chem Soc 122:8413
107. Nicolaou KC, Ajito K, Komatsu H, Smith BM, Li T, Egan MG, Gomez Paloma L (1995) Angew Chem Int Ed Eng 34:576
108. Nicolaou KC, Ajito K, Komatsu H, Smith BM, Bertinato P, Gomez Paloma L (1996) J Chem Soc Chem Comm 96:1495
109. Nicolaou KC, Smith BM, Ajito K, Komatsu H, Gomez Paloma L, Tor Y (1996) J Am Chem Soc 118:2303
110. Bifulco G, Galeone A, Gomez Paloma L, Nicolaou KC, Chazin WJ (1996) J Am Chem Soc 118:8817
111. Bifulco G, Galeone A, Nicolaou KC, Chazin WJ, Gomez Paloma L (1998) J Am Chem Soc 120:7183
112. Depew KM, Zeman SM, Boyer SH, Denhart DJ, Ikemoto N, Danishefsky SJ, Crothers DM (1996) Angew Chem Int Ed Eng 35:2797
113. Toshima K, Ouchi H, Okazaki Y, Kano T, Moriguchi M, Asai A, Matsumura S (1997) Angew Chem Int Ed Eng 36:2748
114. Toshima K, Takano R, Maeda Y, Suzuki M, Asai A, Matsumura S (1999) Angew Chem Int Ed Eng 38:3733
115. Xuereb H, Maletic M, Gildersleeve J, Peltzer I, Kahne D (2000) J Am Chem Soc 122:1883
116. Jumblatt JE, Schlup V, Burger MM (1980) Biochemistry 19:1038
117. Misevic GN, Jumblatt JE, Burger MM (1982) J Biol Chem 257:6931
118. Popescu O, Misevic GN (1997) Nature 386:231
119. Eggens I, Fenderson BA, Toyokuni T, Hakomori S-I (1989) Biochem Biophys Res Commun 158:913
120. Brewer GJ, Matinyan N (1992) Biochemistry 31:1816
121. Stewart RJ, Boggs JM (1993) Biochemistry 32:10,666
122. Kojima N, Hakomori S-I (1991) J Biol Chem 266:17,552

123. Fransson L-A, Havsmark B, Sheehan JK (1981) J Biol Chem 256:13,039
124. Fransson L-A, Carlstedt I, Cöster L, Malmström A (1983) J Biol Chem 258:14,342
125. De la Fuente JM, Barrientos AG, Rojas TC, Rojo J, Cañada J, Fernández A, Penadés S (2001) Angew Chem Int Ed 40:7758
126. Houseman BT, Mrksich M (1999) Angew Chem Int Ed Eng 38:782
127. Horan N, Yan L, Isobe H, Whitesides GM, Kahne D (1999) Proc Natl Acad Sci USA 96:11,782
128. Wormald MR, Edge CJ, Dwek RA (1991) Biochem Biophys Res Commun 180:1214
129. Geyer A, Gege C, Schmidt RR (2000) Angew Chem Int Ed 39:3246
130. Benoit H, Hervé D, Pristchepa M, Berthault P, Zhang Y-M, Mallet J-M, Esnault J, Sinaÿ P (1999) Carbohydrate Res 315:48
131. Siuzdak G, Ichikawa Y, Caulfield TJ, Munoz B, Wong C-H, Nicolaou KC (1993) J Am Chem Soc 115:2877
132. Koshy KM, Boggs JM (1996) J Biol Chem 271:3496
133. Koshy KM, Wang J, Boggs JM (1999) Biophys J 77:306
134. Coterón JM, Vicent C, Bosso C, Penadés S (1993) J Am Chem Chem 115:10,066
135. Jiménez-Barbero J, Junquera E, Martín-Pastor M, Sharma S, Vicent C, Penadés S (1995) J Am Chem Soc 117:11,198
136. Morales JC, Zurita D, Penadés S (1998) J Org Chem 63:9212
137. Junquera E, Laynez J, Menéndez M, Sharma S, Penadés S (1996) J Org Chem 61:6790
138. Bailly C, Qu X, Graves DE, Prudhomme M, Chaires JB (1999) Chem Biol 6:277
139. Wang AHJ, Ughetto G, Quigley GJ, Rich A (1987) Biochemistry 26:1152
140. Dammer U, Popescu O, Wagner P, Anselmetti D, Güntherodt H-J, Misevic GN (1995) Science 267:1173
141. Yu ZW, Calvert TL, Leckband D (1998) Biochemistry 37:1540
142. Geyer A, Gege C, Schmidt RR (1999) Angew Chem Int Ed Eng 38:1466
143. Matsuura K, Kitakouji H, Sawada N, Ishida H, Kiso M, Kitajima K, Kobayashi K (2000) J Am Chem Soc 122:7406
144. Tromas C, Rojo J, de la Fuente JM, Barrientos AG, García R, Penadés S (2001) Angew Chem Int Ed (in press)
145. Bell GI (1978) Science 200:618

Unravelling Carbohydrate Interactions with Biosensors Using Surface Plasmon Resonance (SPR) Detection

Simon R. Haseley, Johannis P. Kamerling, Johannes F. G. Vliegenthart

Department of Bio-Organic Chemistry, Bijvoet Center, Utrecht University, P. O. Box 80.075, 3508TB Utrecht, The Netherlands
E-mail: simon@boc.chem.uu.nl

Surface plasmon resonance (SPR) is an optical phenomenon occurring at a metal coated interface between two media of different refractive index, e.g., water and glass. Exploitation of this phenomenon for investigating biomolecular interactions has occurred since biomolecules can be attached to a chemically modified gold surface and that an interaction between the surface and other biomolecules in solution will affect the SPR of the system. A great deal of the literature on this subject has involved an investigation of protein-ligand interactions, but this chapter will review the use of this technique for investigating carbohydrate-ligand interactions.

Keywords. Surface plasmon resonance, Carbohydrate interactions, Kinetics

Topics in Current Chemistry, Vol. 218
© Springer-Verlag Berlin Heidelberg 2002

List of Abbreviations

BSA	bovine serum albumin
ConA	*Concanavalin A* agglutinin
DMPC	dimyristoylphosphatidylcholine
DSA	*Datura stramonium* agglutinin
GAG	glycosaminoglycan
Gal	galactose
GalNAc	*N*-acetylgalactosamine
GlcNAc	*N*-acetylglucosamine
LPL	lipoprotein lipase
LPS	lipopolysaccharide
MAA	*Maackia amurensis* agglutinin
MM	molecular mass
MP	*Microciona prolifera*
Neu5Ac	sialic acid/*N*-acetyl neuraminic acid
PSA	polysialic acid
RCA	*Ricinus communis* agglutinin
RI	refractive index
RU	response unit
SAM	self-assembled monolayer
SNA	*Sambucus nigra* agglutinin
SPR	surface plasmon resonance

1
Introduction

Carbohydrate recognition is a subject of increasing importance with the fine specificity of these molecules now being increasingly recognized and exploited. Carbohydrates are molecules carrying an enormous amount of information and have been implicated in a number of biochemical events including cellular adhesion and mediation of protein folding [1–8]. Techniques for studying these interactions vary depending on the interactions of interest.

Biosensors using surface plasmon resonance (SPR) as detection were introduced to the scientific community in the early 1990s [9–13]. These biosensors are used to derive kinetic and affinity data for ligand-analyte interactions in real-time [14, 15]. SPR is an optical phenomenon occurring at a metal coated interface between two media of different refractive index (RI). Under conditions of total internal reflection of a plane polarized light source, an evanescent wave will penetrate into the medium of lower RI, causing free electrons in the metal layer to oscillate – the generation of surface plasmon waves. This phenomenon is induced by one specific angle of the incident light, and can be monitored in the reflected light since, at that angle, a reduction in intensity occurs.

The SPR of the system is dependent, amongst other things, on the RI of the two media. If the RI of one medium changes then the SPR of the system will also change, indicated by reduction in intensity of a different angle of the reflected

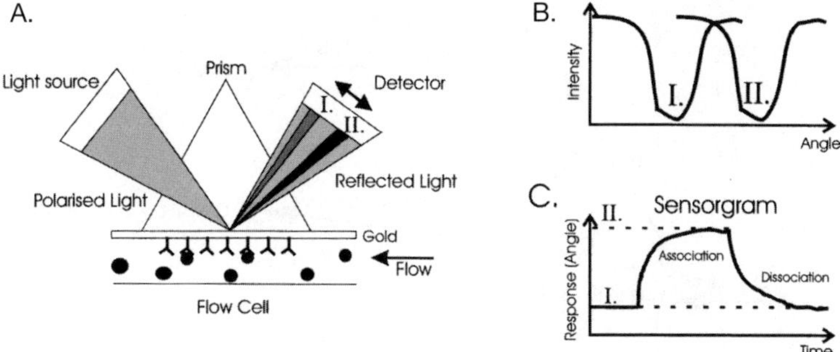

Fig. 1. **A** Diagram of an SPR system containing a prism as one medium and a solution of analyte (●), in the flow cell, as the other. At the interface is a gold surface containing substrate (Y). A light source is internally reflected at the interface and monitored at the detector. **B** The angle of the incoming light at which SPR occurs is indicated by reduction in intensity of the same angle in the reflected light (**I**; see A and B). As more analyte binds to substrate the refractive index of the solution close to the surface will change, with the result that the angle of light at which SPR is induced is changed (**II**; see A and B). **C.** The speed and extent of this gradual change is monitored in real-time in a plot known as a sensorgram

light source (Fig. 1 A, B). The difference between the angles is indicative of the amount of substrate bound to the surface, whereas the rate at which the change occurs depends on the kinetics of the interaction. Biosensors based on this technique contain a prism as one medium and usually an aqueous solution as the other, with a thin layer of gold at the interface (Fig. 1 A). One biomolecule (substrate) of interest is attached to the surface of the gold in contact with the solution, whilst a second biomolecule (analyte) is dissolved in the aqueous solution. If an interaction between these two biomolecules occurs the local refractive index at the surface will change, thus affecting the SPR of the system. A large amount of data involving protein-ligand interactions has been derived from this type of system [16], whereas that of carbohydrate-related interactions has produced less, and until now has not been evaluated and reviewed. In this review we will consider some of the common problems encountered when analyzing carbohydrate interactions, such as multivalency, weak affinity interactions, and handling low molecular mass (MM) compounds. Studies illustrating the scope and limits of SPR detection systems for investigating carbohydrate interactions will be especially highlighted.

1.1
Experimental Set-Up

Biomolecules can be attached to the gold surface of the sensor in one of several ways. One of the most common means of attachment (or immobilization) is to couple the molecules to a carboxymethylated dextran-coated gold surface by covalent attachment to the carboxyl groups of the dextran [17]. This can be achieved in several ways (Fig. 2).

Other immobilization procedures are direct coupling of thiol-containing molecules to a gold surface and incorporation of molecules into immobilized

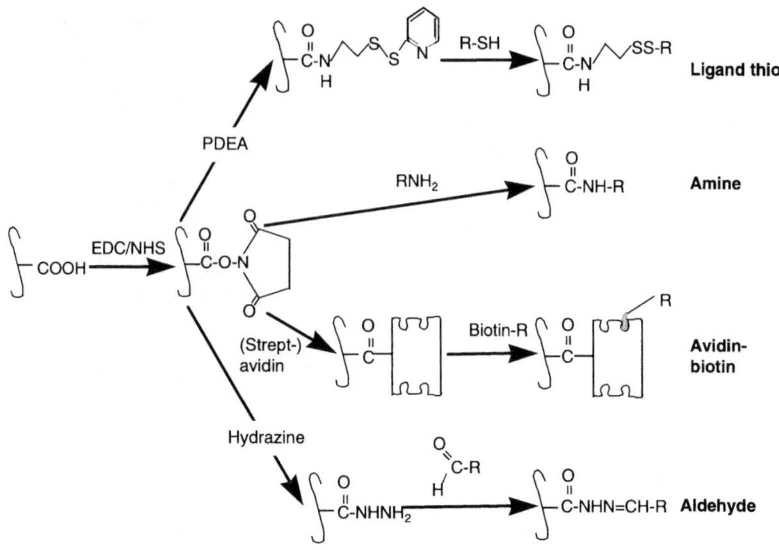

Fig. 2. Chemical approaches for attaching biomolecules to a carboxymethylated dextran coated surface. NHS, *N*-hydroxysuccinimide; EDC, *N*-ethyl-*N'*-(dimethylaminopropyl)carbodiimide; PDEA, 2-(2-pyridinyldithio)ethaneamine hydrochloride

lipid membranes [18]. Careful choice of experimental conditions and immobilization reactions is a prerequisite. It is essential that reference surfaces are used so that non-specific binding effects are taken into account, avidity binding is avoided or at least taken into account, mass transport and aggregation are minimized, regeneration of the surface is complete, and data processing is exact [19, 20]. The majority of experiments in this review have been performed on systems in which the analyte is continuously flowed across the gold surface, as opposed to the cuvette system, which is static. The sensorgram (Fig. 1C) is a graphical representation of the change in SPR angle, or response, with time as the interaction between substrate and analyte proceeding in a continuously flowing system. The response, measured in number of response units (RU), is a representation of the amount of surface bound substance: $1000\,RU = 1\,nmol\,l^{-1}\,mm^{-2}$ of protein. The kinetics of the interaction can be derived from the change in response with time recorded in the sensorgram.

The kinetics [20–22] are governed by the real-time on- and off-rates which are described by the association rate constant (k_A; $M^{-1}\,s^{-1}$) and the dissociation rate constant (k_D; s^{-1}). The overall affinity of the system is expressed as the equilibrium association or dissociation constant, K_A (M^{-1}) or K_D (M), respectively, in which the one is the reciprocal of the other. Equilibrium kinetic constants can be calculated either from the level of binding recorded at equilibrium, and so can be derived from results obtained with static SPR biosensors as well as continuously flowing SPR systems, or can be obtained by dividing the real-time rate constants k_A and k_D by each other. Good agreement between K_A (or K_D) derived in both ways is an indication that the experimental set-up is of high quality.

Within this review equilibrium constants will be quoted in the form that they appear in the relevant publication. This will mean that both K_A and K_D will be interchanged throughout the text. K_D may be referred to as a power of ten (e.g., 10^{-3}, 10^{-6}, 10^{-9} mol l^{-1}), or as the abbreviated unit (e.g., mmol l^{-1}, µmol l^{-1}, nmol l^{-1}).

2
Carbohydrate-Protein Interactions

The most commonly studied carbohydrate interaction is that with protein, a recognized and abundant recognition process throughout nature. The specificity and intensity of these interactions have been studied by numerous methods including microcalorimetry [23], fluorescence anisotropy [24, 25], NMR spectroscopy [26, 27], atomic force microscopy [28], mass spectrometry [29, 30], and molecular modeling techniques [31], of which all have their relative merits. The attraction of being able to monitor binding events in real-time, without the use of labels, has facilitated the introduction of SPR biosensors into the field of carbohydrate-protein interactions.

The affinity (K_D) of this type of interaction ranges from mmol l^{-1} to µmol l^{-1} for lectin-carbohydrate interactions, mmol l^{-1} to nmol l^{-1} for antibody-carbohydrate interactions, and even lower values for the interaction of glycosaminoglycans with protein, largely because these interactions tend to involve multiple, and co-operative, binding events. Within this section data will be discussed that has been acquired using this technique, including comparisons with other techniques when data are available.

2.1
Carbohydrate-Lectin

Lectins are carbohydrate binding proteins of non-immuno origin, occurring in both animals and plants, involved in cellular recognition [8, 32, 33]. By using SPR detection the specificity, affinity, and stoichiometry of several lectin-based interactions have been investigated, as well as exploitation of this technique for characterizing glycans of unknown structure with lectins of known specificity. Exploring the possibilities of developing a system for determining oligosaccharide structures began in 1994. Hutchinson [34] investigated the interaction of a number of lectins with the glycoprotein bovine fetuin, and subsequent glycosidase treated bovine fetuin. The lectins, of known specificity, were shown to bind to the correct epitopes, for example the sialic acid (Neu5Ac) specific lectin from *Maackia amurensis* (MAA) bound to intact fetuin but not to sialidase treated fetuin. The level of binding was calculated as a percentage of the maximum change in SPR response, the kinetics of the interactions were not calculated or taken into account. In a similar study, Shinohara et al. investigated the interaction of surface-bound sialidase treated fetuin with a group of lectins [35]. Both the galactose (Gal) binding lectin from *Ricinus communis* (RCA) and the lectin from *Datura stramonium* (DSA), which recognizes *N*-acetylglucosamine (GlcNAc) but also polylactosamine, bound but as expected not the Neu5Ac binding lectins. RCA had a k_A of 3.4×10^5 M^{-1} s^{-1} and a k_D of 2.1×10^{-3} s^{-1}, which gave a K_A of 1.6×10^8 M^{-1}. DSA had a k_A of 5.7×10^5 M^{-1} s^{-1}, a k_D of 1.3×10^{-3} s^{-1}, and a K_A of 4.3×10^8 M^{-1}. These K_A values are higher than expected for common lectin-saccharide interactions, and values 100-fold less would be more probable. One possible explanation for these high values, particularly in the case of RCA with two Gal binding sites, is that the interaction may have been stabilized by a second interaction. Co-operative binding due to one extra binding site can increase the affinity by a factor of 100 (see throughout the review). However, it could be as high as 10,000-fold, as was shown for the interaction of wheat germ lectin with chitooligosaccharides, the difference in affinity being derived from experiments in which both lectin and oligosaccharide were immobilized to a gold surface, enabling the comparison of the affinities and calculation of the effect of co-operativity [36]. Co-operative binding will be seen throughout this review to play both positive and negative roles in the investigation of carbohydrate interactions.

Since it was clear that lectin-carbohydrate interactions could be measured using SPR, it became important to evaluate and assess the different possibilities for recording accurate and reproducible data. Kalinin et al. revealed *Concanavalin A* (ConA) lectin to bind (K_A 2.5×10^5 M^{-1}) the carboxymethylated dextran matrix commonly used on sensor surfaces [37]. They also stressed that this interaction did not follow first order kinetics, and that care should be taken when using multivalent solutes: many lectins are dimers or tetramers, and often contain more than one carbohydrate binding site (e.g., RCA and ConA). This point was further emphasized in work utilizing carbohydrate-derivatized self-assembled monolayers (SAMs) [38]. Not only did multivalent binding take place, but as the surface density of the carbohydrate ligands increased, the binding selectivity of *Bauhinia purpureas* lectin was shown to switch from one carbohydrate ligand to another. At low surface densities [mole fraction of sugar (X_{sugar}) ~ 0.1] the lectin bound more strongly to mixed SAMs containing 1 than to those containing 2 or 3 (Fig. 3). As the surface density increased (X_{sugar} ~ 0.6), the avidity of the lectin for monolayers containing 1 decreased and that containing 2 increased. It would appear that secondary interactions play a significant role on the binding of the lectin at high surface densities, and that this mechanism may have a real influence on biological recognition processes.

In order to avoid 'non-specific' binding to the carboxymethylated dextran surface, Mann et al. generated synthetic glycolipid surfaces (see Sect. 5), which avoided the use of the dextran, in their development of a system for rapidly evaluating the ability of various inhibitors to block ConA binding [39]. The affinity of ConA tetramer, containing four possible carbohydrate binding sites, to a lipid membrane layer containing 10% glycolipid (containing terminal α-mannose) had a K_A of 2.7×10^4 M^{-1}. This value appears to be lower than would be expected for a carbohydrate-lectin interaction involving co-operative binding, although multivalent binding was assumed to occur. A reduced k_A value could possibly be due to either performing experiments at a lipid layer or hindered presentation of the glycolipid at the surface (see Sect. 2.5). The interaction could be inhibited by methyl α-mannoside or methyl α-glucoside, for which the calculated K_D values were 92 and 290 μmol l^{-1}, respectively. These values were

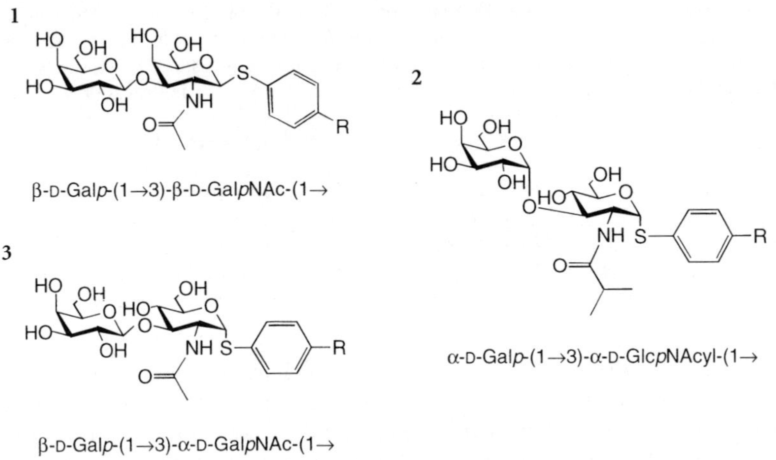

Fig. 3. structures of ligands 1, 2, 3

β-D-Gal*p*-(1→3)-β-D-Gal*p*NAc-(1→

α-D-Gal*p*-(1→3)-α-D-Glc*p*NAcyl-(1→

β-D-Gal*p*-(1→3)-α-D-Gal*p*NAc-(1→

Fig. 3. The structure of ligands **1**, **2**, and **3** used in the study of the binding selectivity of *Bauhinia purpureas* lectin at carbohydrate-derivatized self-assembled monolayers. R denotes a hydrophobic spacer

only marginally different to those derived from microcalorimetry (132 and 416 µmol l^{-1}, respectively) [40], and the ratio of the two was identical in both cases. These conclusions would appear to support a lack of co-operative binding, at least in the sense of multivalent binding as opposed to that in terms of communication between binding sites [41, 42], since microcalorimetry should only facilitate elucidation of the monovalent interaction. The results were also very similar to those derived from fluorescence anisotropy (181 and 588 µM, respectively) [40]. The 'multivalent properties' of this system were illustrated by the increased inhibitory potency (up to 40-fold higher than that for methyl *α*-mannoside) of mannose polymers containing more than 15 monosaccharide residues: the proposed minimum length required to span two sub-units of the lectin.

In order to develop a system for accurately calculating the monovalent interaction kinetics of lectin-carbohydrate binding, Haseley et al. have devised and evaluated a method in which the lectin and the respective denatured lectin, established as being the most suitable blank surface, are immobilized to the gold surface [43]. They also investigated buffer composition and both immobilization and regeneration conditions. The definitive system proved very successful, with determination of kinetic data for the binding of ten pure oligosaccharides to the Neu5Ac binding lectins from *Sambucus nigra* (SNA) and MAA, in which certain recurring changes in oligosaccharide structure brought about similar subtle changes in the kinetic parameters. This effect is nicely illustrated by the values in Table 1 recorded for four of the oligosaccharides (Fig. 4). The lectin SNA, which recognizes Neu5Ac linked to the 6-position of Gal, had a similar affinity for all four oligosaccharides. This was as expected since all oligosaccharides contained this Neu5Ac epitope in very similar chemical environments. The lectin MAA, however, which recognizes Neu5Ac linked to the 3-position of

Table 1. Equilibrium association rate constants (K_A) for the binding of oligosaccharides 1–4 to *Sambucus nigra* agglutinin (SNA) and *Maackia amurensis* agglutinin (MAA). Average values from five experiments, error between 10–20% for each result. The closeness of fit for each estimated parameter is indicated by the statistical value of χ^2 in brackets

Oligosaccharide	SNA K_A (M^{-1})	MAA K_A (M^{-1})
1	1.36×10^6 [0.054]	2.92×10^5 [0.16]
2	1.58×10^6 [0.41]	3.83×10^5 [2.94]
3	7.88×10^5 [0.74]	1.42×10^6 [0.71]
4	1.86×10^6 [0.34]	2.40×10^6 [2.46]

Gal, was able to differentiate between oligosaccharides containing this Neu5Ac epitope on both the upper (as indicating in Fig. 4) and lower branches of the glycan and those only containing the epitope on the lower branch. The 5- to 10-fold increase in K_A brought about by the addition of Neu5Ac to the upper branch may be the result of increased availability, or reduced steric hindrance, of the epitope at this position, which in turn could cause an increase in the overall affinity of these oligosaccharides for MAA.

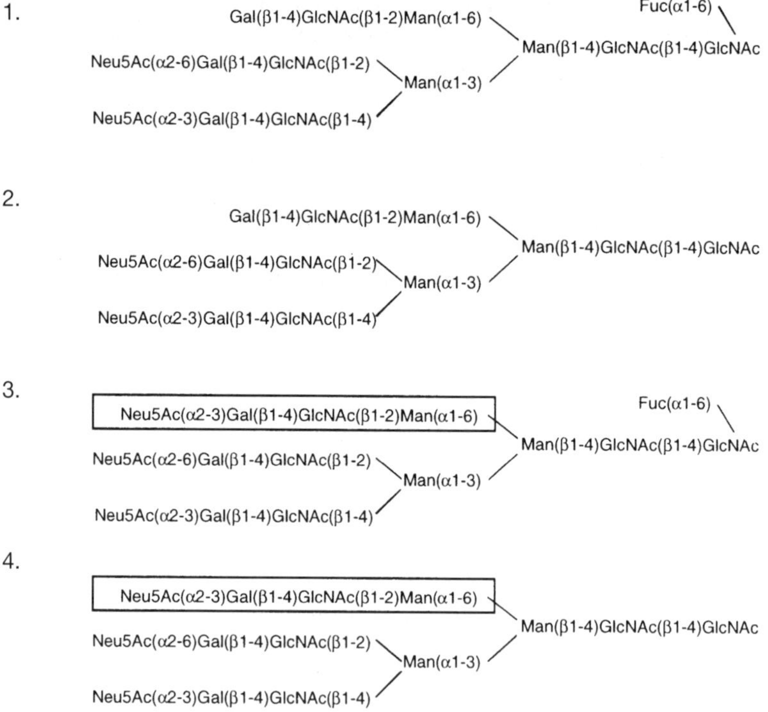

Fig. 4. The structure of oligosaccharides 1–4 used to study the specificity of lectins from *Sambucus nigra* and *Maackia amurensis*. The *circled epitopes* denote structural motifs of particular interest (see text)

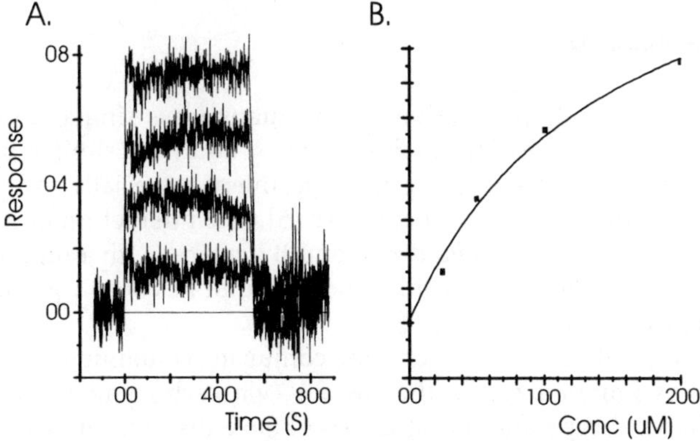

Fig. 5. **A** Overlaid sensorgrams produced by the interaction of different concentrations (0, 25, 50, 100 and 200 µmol l^{-1}) of the Sd^a-determinant tetrasaccharide with immobilized *Dolichos biflorus* lectin. **B** A plot of equilibrium response against concentration (steady-state analysis). The interaction K$_D$ was derived from this plot using known mathematical equations [71]

This system has now been employed for the analysis of several lectins, including that from *Dolichos biflorus* [44] for which the affinity of the Sd^a-determinant, a tetrasaccharide, has been determined via steady-state analysis (Fig. 5 B), i.e., from a plot of the level of binding at equilibrium (Fig. 5 A) against concentration. The interaction had a K$_D$ of 1.44 × 10^{-4} mol l^{-1}. Biosensors based on SPR usually require the analyte to have a molecular mass larger than 1000 Da (SPR response is related to the size of the molecule). The sensitivity of this system has enabled determination of the affinity of a trisaccharide mimic of the Sd^a-determinant (MM 600 Da) for *Dolichos biflorus* [44]. Regeneration of lectin surfaces can, in the majority of cases, be successfully achieved by using the methyl glycoside based on the monosaccharide epitope recognized.

Other studies involving lectin-carbohydrate interactions are as follows.

A mouse C-type macrophage lectin was revealed [45] to bind glycopeptides and oligosaccharides containing terminal N-acetylgalactosamine (GalNAc) residues with relatively high affinity (K$_A$ 6.2 × 10^7 M^{-1}).

An investigation of P-selectin binding to immobilized P-selectin glycoprotein ligand-1 (PSGL-1) has revealed [46] that this Ca^{2+}-dependent interaction has both a rapid on- (k$_A$ 4.4 × 10^6 M^{-1}s^{-1}) and off-rate (k$_D$ 1.4 s^{-1}). These features are in agreement with the fact that leukocytes use PSGL-1 to tether and roll on P-selectin on activated endothelial cells. The affinity of this interaction (K$_D$ 320 nmol l^{-1}) was relatively high for a lectin-carbohydrate interaction.

Hamster galectin-3 binds to laminin with a k$_A$ of 1 × 10^4 M^{-1} s^{-1}, and a k$_D$ of 0.2 s^{-1} resulting in a K$_A$ of 1 × 10^5 M^{-1} [47], whereas calreticulin, a molecular chaperone thought to possess lectin-like properties, bound to laminin with a K$_A$ of 2.1 × 10^6 M^{-1} [48]: it did not show any binding to deglycosylated laminin.

2.2
Carbohydrate-Antibody

Carbohydrate-antibody interactions are in many ways similar to interactions with lectins. The analyses follow similar rules, and have similar properties to take into account when designing an experiment, especially the problems of multivalency, which lead to co-operativity [5]. The interaction of an O-specific polysaccharide from *Salmonella* serogroup B bacteria with a monoclonal antibody specific for the tetrasaccharide repeating unit has been rigorously studied by MacKenzie et al. [49].

In this study a BSA-O-polysaccharide conjugate, containing up to 25 repeating units, was immobilized to the surface (via carboxymethylated dextran). Whereas the wild type antibody (SE155–4) gave distinct biphasic association and dissociation data, monomeric fractions of the antibody showed little evidence of biphasic behavior with both fast on- and off-rates (Table 2), with overall K_D values of between 4.1×10^{-6} and 7.5×10^{-6} mol l^{-1}. Dimeric forms of the antibody showed off-rates approximately 20-fold slower and association rates 5-fold faster than the monomers, and therefore at least a 100-fold increase in the affinity (Table 2). These values were slightly higher than that of the wild type antibody (K_D 2.3×10^{-7} mol l^{-1}).

The dissociation stage of all experiments was performed in the presence of the free trisaccharide epitope to minimize rebinding of the antibody (dissociation is commonly performed only in the presence of the system buffer). The effect of this was to increase the rate of release, or alternatively minimize rebinding, of the antibody, although whether dissociation in the presence of free trisaccharide resembles more the natural dissociation situation is debatable. It is important for this phenomenon to be further investigated in order to develop a standard protocol to secure comparable SPR biosensor results. Equilibrium binding analyses provided K_D values (Table 2) in good agreement with those derived from active association and dissociation rate constants, thus confirming the quality of the data and supporting the models proposed. In addition, the

Table 2. Comparison of the kinetic values calculated for the active rate constants k_A and k_D, K_D derived from k_A and k_D, and K_D calculated from Scatchard analysis of the responses at equilibrium (final column) for wild-type and monomeric and dimeric fractions of antibody SE155–4 specific for the *Salmonella* serogroup B O-polysaccharide

Antibody fractions		k_A (M^{-1}s^{-1})	k_D (s^{-1})	k_D/k_A (M)	K_D (M)
Monomeric	SK4	3.0×10^4	2.0×10^{-1}	4.1×10^{-6}	6.8×10^{-6}
	3B1	3.8×10^4	2.4×10^{-1}	7.5×10^{-6}	7.6×10^{-6}
	B5–6	4.8×10^4	2.9×10^{-1}	6.5×10^{-6}	6.1×10^{-6}
	B3–20	4.3×10^4	2.5×10^{-1}	5.3×10^{-6}	5.8×10^{-6}
Dimeric	SLA-1	1.5×10^5	1.1×10^{-2}	4.5×10^{-8}	7.2×10^{-8}
	B5–1	3.2×10^5	1.3×10^{-2}	8.1×10^{-8}	4.1×10^{-8}
	B4–3	1.6×10^5	7.8×10^{-3}	5.6×10^{-8}	5.0×10^{-8}
Wild-type IgG	SE155–4	8.7×10^4	1.2×10^{-2}	2.3×10^{-7}	1.4×10^{-7}

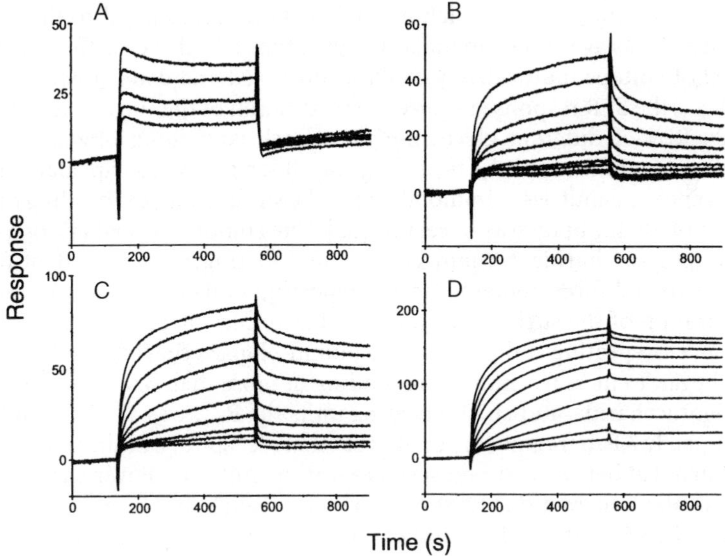

Fig. 6A–D. Sensorgrams produced by the interaction of different concentrations of PSA with antibody: **A** Neu5Ac$_{16}$; 50, 25, 12.5, 6.25, and 3.125 µmol l⁻¹; **B** Neu5Ac$_{30}$; 200, 100, 50, 25, 12.5, 6.25, 3.12, 1.56, 0.78, and 0.39 nmol l⁻¹; **C** Neu5Ac$_{100}$; **D** Neu5Ac$_{200}$; 100, 50, 25, 12.5, 6.25, 3.12, 1.56, 0.78, 0.39, and 0.195 nmol l⁻¹

binding affinities for the monovalent antibodies were in good agreement with data acquired from titration microcalorimetry [50]. This SPR investigation was evaluated using the BIAevaluation software supplied by BIAcore and is a good example of the data that can be produced.

In an analysis of the interaction between (α2–8)-linked polysialic acid (PSA) with a monoclonal antibody [51] in which the antibody was bound to the surface, both fast on- and off-rates (average K$_D$ of 8 µmol l⁻¹) were obtained for molecules containing up to sixteen residues of Neu5Ac (Fig. 6A). Molecules containing a higher number of Neu5Ac residues however, appeared to possess some biphasic binding characteristics (Fig. 6B, C), whereas molecules with ~ 200 residues appeared to follow a triphasic interaction mechanism with tight binding (average K$_D$ of 900 pmol l⁻¹) and negligible dissociation of the complex (Fig. 6D).

This phenomenon is due to the presence of co-operativity and illustrates the care that should be taken when performing experiments, and how quickly co-operative binding can enhance the affinity, or avidity, of the system. Since 1 ng protein per mm² on the sensor surface results in an SPR response of ~1000 RU, it was possible to calculate the amount of surface bound antibody corresponding to the increase in SPR response (6000 RU) obtained following immobilization of the antibody and therefore the average distance between single molecules on the surface. The calculated average of one antibody molecule every 100 Å denoted that at least 16 residues of Neu5Ac were needed to span two antibodies, and thus needed to initiate a biphasic binding mechanism, a value in good agreement with the experimental data.

In a similar study, but in which glycoconjugates were bound to the surface, multivalency was again seen to make an appearance. In this case the emergence and level of epitope multivalency could be observed and determined by monitoring R_{max}, the SPR response required for saturation of the surface. A fixed density (200 RU) of different glycoconjugates, each containing oligosaccharides with differing numbers of the type III group B *streptococcus* capsular polysaccharide repeating unit, were immobilized to the surface, and the binding of a Fab fragment of an antibody was observed [52]. The conformational epitope of the capsular polysaccharide recognized by the Fab fragment was identified as between two and three pentasaccharide repeating units (K_D 4.6 × 10⁻⁶ mol l⁻¹), and saturation of the surface was identified as occurring at an R_{max} of 148 RU. Above seven repeating units of surface bound oligosaccharide the R_{max} of the system began to increase, indicating binding of more than one Fab fragment to each oligosaccharide, and at 20 repeating units the R_{max} was 486 RU, binding of approximately 3 Fab fragments per oligosaccharide. Epitope multivalency therefore occurred at between six and seven repeating units. Moreover, the affinity of the monovalent interaction was shown to increase threefold from 7 to 20 repeating units (K_D 6.5 × 10⁻⁷ mol l⁻¹), indicating optimization and further stability of the conformational epitope.

Evaluation of antibody recognition of synthetic oligosaccharide fragments of the gut-associated circulating anodic antigen excreted by the *Schistosoma* worm has led to the development of a rapid and efficient system, based on SPR detection, for evaluating titers of antibodies from humans [53]. This system has been proposed as a means of diagnosing *Schistosoma mansoni* infection.

2.3
Carbohydrate-Enzyme

Only one study exists to date for the analysis of the interaction between carbohydrate and enzyme. The main reason for this is that enzyme interactions are not simple monovalent interactions, since once the reaction has been catalyzed the product vacates the enzyme. Laroy et al. [54] have avoided this problem by cloning mutants of a sialyltransferase in which the catalytic domain, but not the binding site, has been deactivated. However, they do report that binding of wild type sialyltransferase to sialidase treated fetuin does occur at low salt concentrations (<25 mmol l⁻¹ NaCl) and that binding can be inhibited with CMP-Neu5Ac, the natural donor substrate. It should be noted, however, that SPR experiments are generally performed in buffer containing 150 mmol l⁻¹ NaCl to avoid non-specific ionic interactions with the surface. One of the mutants bound at similar salt concentrations to those required for the wild type enzyme but, since CMP-Neu5Ac did not inhibit the binding, the sialyltransferase activity had most likely been lost. A further two mutants bound at physiological salt concentrations. As SPR biosensors become more refined it is likely that the analysis of enzyme catalyzed reactions, such as the rather elegant analysis of a DNA-polymerase [55], which may also involve carbohydrate interactions, will become more common.

2.4
Glycosaminoglycan-Protein

Glycosaminoglycans (GAGs) are highly heterogeneous, highly sulfated glycans coating animal cells and tissues. Included within this group are heparin and heparan sulfate, molecules highly sulfated by N-deacetylation and N-sulfation of N-acetylglucosamine (GlcNAc), and various O-sulfation patterns. This coat can be modified during differentiation, development, and during disease, and so it is important to investigate the functional role of these molecules. There has been a considerable amount of research using SPR into the interactions of heparin and other glycosaminoglycans [56–69]. One particularly interesting study, which highlights many of the advantages and disadvantages of this technique, is that covered in two publications by Lookene et al. [56, 57]. They initially investigated the interaction of lipoprotein lipase (LPL) with heparin and heparan sulfate [56]. Attached to the vascular endothelium, LPL binds circulating lipoproteins and hydrolyses their triglycerides. Attachment to the endothelium most likely occurs via interaction with heparin or heparan sulfate. To prepare a heparin or heparan sulfate coated surface the GAG was biotinylated via free amino groups [70], and bound to previously immobilized streptavidin. The interaction between LPL and heparan sulfate was found to be a fast exchange process with a k_A of 1.7×10^8 M^{-1} s^{-1}, and a k_D of 5×10^{-2} s^{-1}, resulting in a K_D of 3×10^{-10} mol l^{-1}. These results were obtained at 150 mmol l^{-1} NaCl. In order to calculate the electrostatic contribution of the interaction, the kinetics were determined at NaCl concentrations between 150 mmol l^{-1} and 900 mmol l^{-1} (the concentration at which binding to the GAG surface was shown to be identical to that to the blank surface). A plot of log K_D against log [NaCl] indicated a linear relationship. Using known equations the electrostatic contribution of the interaction of heparan sulfate with LPL was calculated to be 44%, while that to heparan sulfate was 49%. Other interesting and subtle points included the observation that the dissociation rate followed a biexponential decay in buffer containing less than 300 mmol l^{-1} NaCl, but that at higher concentrations it followed a single exponential decay process. Of all the possible explanations for this phenomenon, the most likely were either heterogeneity of the heparan sulfate chains, resulting in heterogeneous binding along the polymer, or rebinding of LPL during dissociation. To determine whether rebinding occurred, free heparin fragments were injected into the system during dissociation (see [49]). This increased the dissociation rate by a factor of 100; therefore the biexponential decay process was due to rebinding of LPL. To determine the affinity of size-fractionated heparin fragments (tetra-, hexa-, octa- and decasaccharides) and low molecular mass heparin for LPL, a solution-affinity experiment [71] was performed, whereby the concentration of LPL was kept constant, and the concentration of the heparin fragments in solution, varied. In this way the affinity of these fragments for LPL could be determined by inhibiting the interaction of LPL to heparan sulfate. The change in K_D from tetra- to decasaccharide was large, while that between decasaccharide and low molecular mass heparin (mean length of 24 monomers) was significantly less: the decasaccharide had a K_D only 4.5-fold lower than that of the low molecular mass heparin, which had a

K_D of 0.2 nmol l^{-1}. LPL is a dimer in rapid and reversible equilibrium with its inactive monomer, and it is known that heparin protects LPL from inactivation, i.e., from returning to the monomeric form. Since both octa- and decasaccharides blocked LPL binding to heparin at a molar ratio of 1:1 with LPL, the authors concluded that an octasaccharide was the minimal length heparin fragment sufficient for maximum stabilization of the dimer. This conclusion was also reflected in the calculated K_D values. This led to the conclusion that an octasaccharide possessed the minimum length required to span both monomers at the same time, therefore stabilizing the dimer. Monomeric LPL had an association rate constant for heparan sulfate 1000 times lower than dimeric LPL. To summarize, the interaction of LPL dimer with a heparan sulfate surface is a rapid process with a K_D of 0.3 nmol l^{-1}. It is partially electrostatically driven, and leads to accumulation of the enzyme close to the surface, i.e., it rebinds to the surface rather than dissociating. Heparin fragments longer than octasaccharide span the binding sites of both subunits of the dimer. Previously reported values for the interaction of low molecular mass heparin to LPL were 13 nmol l^{-1} [72] and 43 nmol l^{-1} [73]. The differences between these two results and the values obtained by SPR detection (K_D 0.2 nmol l^{-1}), are factors 40- and 100-fold lower, respectively. These may be due to heparin binding to immobilized LPL in the first experiment [72], or perhaps partial inactivation of LPL during both of the previous studies, resulting in an average binding K_D between inactive monomer (K_D 1.2 µmol l^{-1}) and active dimer.

In their second publication the interaction between lipoprotein and heparan sulfate, in both the presence and absence of LPL, was investigated [57]. In short, they identified that lipoprotein binds to heparan sulfate, but that the binding is amplified in the presence of LPL, and that heparin has a higher binding capacity for lipoproteins than does heparan sulfate. They also discovered that two to four lipase dimers per lipoprotein were necessary for efficient binding to the heparan sulfate surface, indicating a role for co-operativity in the interaction. The association rate constants were about tenfold higher for the interaction between lipoprotein and LPL bound to heparan sulfate than for lipoprotein to immobilized LPL, demonstrating a contribution of heparan sulfate to the interaction. The role of LPL in the interaction of lipoprotein with heparan sulfate, was not to increase the affinity of the interaction, but actually to increase the number of binding sites at the surface. These conclusions were all derived from SPR biosensor experiments and further illustrate the power of this technique.

Among the many SPR studies investigating glycosaminoglycan interactions the following are of particular interest.

The binding of human acidic fibroblast growth factor (aFGF) to heparin was found to have an affinity K_D of 50–140 nmol l^{-1} [58], which was in agreement with results obtained from affinity electrophoresis experiments [74].

The interaction of platelet-derived growth factor with heparin had an affinity K_A of 1.7×10^8 M^{-1}, and was shown to involve two heparin binding sites of which the carboxy-terminus played an important role [59].

Shuvaev et al. showed that hyperglycation of apolipoprotein E, as in diabetic patients, impairs lipoprotein-cell interactions, which are mediated via heparan sulfate proteoglycans [60, 61].

The interaction between heparin and heparin binding growth-associated molecule had a K_D (4 nmol l^{-1}) 100-fold lower than that determined by isothermal titration calorimetry [62]. This tighter binding was suggested to have been the result of immobilizing the growth-associated molecule to the surface, as opposed to the interaction in solution.

2.5
Glycolipid-Protein

Investigating glycolipid interactions with SPR biosensors is an almost completely different and separate subject to other carbohydrate-related interactions as these studies usually employ membranes or vesicles attached to the gold surface. The presence of a lipid monolayer in the analysis can not only produce quite different results to those using molecules attached to carboxymethylated dextran, but may also produce data more closely resembling the in vivo situation. Using a liposome-capture method, MacKenzie et al. have analyzed in great depth the interaction of cholera toxin with a selection of gangliosides [75]. These glycolipid receptors were incorporated into artificial liposomes that also contained a small amount of lipopolysaccharide (LPS) to allow capture of these liposomes by an immobilized anti-LPS monoclonal antibody (immobilized to a carboxymethylated dextran surface). Cholera toxin demonstrated an absolute requirement for terminal galactose and internal sialic acid residues (as in G_{M1}; Fig. 7) with tolerance for substitution with a second internal sialic acid (as in G_{D1b}, Fig. 7).

One of their major findings was that the percentage of glycolipid incorporated into the liposomes affected the affinity of the toxin interaction. Both the association and dissociation rates decreased as the percentage of glycolipid changed from 2% to 4%, the K_D values for G_{M1} being 1.7 nmol l^{-1} and 6.8 nmol l^{-1}, respectively. The reason for this was that two distinct on-rates existed at the surface containing 4% G_{M1}, but only the faster on-rate was observed at G_{M1} concentrations of 2%, resulting in a slight increase in the calculated affinity. Compared to conventional assays, the liposome-capture method showed highly restricted specificities for bacterial toxin binding to glycolipids, and the data were in good

G_{M1}

β-D-Gal*p*-(1→3)-β-D-Gal*p*NAc-(1→4)-β-D-Gal*p*-(1→4)-β-D-Glc*p*-(1→
　　　　　　　　　　　　　　3
　　　　　　　　　　　　　　↑
　　　　　　　　　　　　　　2
　　　　　　　　　　　α-Neu*p*5Ac

G_{D1b}

β-D-Gal*p*-(1→3)-β-D-Gal*p*NAc-(1→4)-β-D-Gal*p*-(1→4)-β-D-Glc*p*-(1→
　　　　　　　　　　　　　　3
　　　　　　　　　　　　　　↑
　　　　　　　　　　　　　　2
　　　　α-Neu*p*5Ac-(2→8)-α-Neu*p*5Ac

Fig. 7. Structures of the glycolipids G_{M1} and G_{D1b}

agreement with the majority of previous studies [76–81]. The results were inconsistent with an earlier SPR study [82] which observed the order of binding strength $G_{M1} > G_{M2} > G_{D1a} > G_{M3} > G_{T1b} > G_{D1b} >$ asialo-G_{M1}, with the interaction of G_{M1} possessing a K_D (4.61×10^{-12} mol l^{-1}) 1000-fold lower than that reported by MacKenzie et al. (1.7×10^{-9} mol l^{-1}). The binding of G_{D1a}, G_{M3}, and asialo-G_{M1} contradicted binding specificities by other means, particularly the high affinity reported for G_{D1a} (K_D 31.8 pmol l^{-1}). This ligand was previously used as a negative control in titration experiments of cholera toxin with gangliosides [78]. MacKenzie et al. suggested that the specificity differences may have been the result of different presentation of the two lipid bilayer environments. The glycolipid surfaces used in the latter experiment [82] were prepared by immobilizing glycolipid: palmitoyloleoylglycero-3-phosphocholine (5:95 mol%) to an alkylthiol monolayer on the gold surface to form hybrid bilayer membranes. There is considerable evidence that glycolipid carbohydrate can be modulated by the bilayer microenvironment, although non-specific binding, which is rife in analyses using exposed alkane-thiol groups, may also have contributed [82, 83].

In a study characterizing the binding of an IgM antibody to glycolipid asialo-G_{M1} Harrison et al. compared the affinity recorded using liposomes fused to an alkane-thiol layer (formation of a hybrid bilayer) and liposomes captured to an antibody previously immobilized to a carboxymethylated dextran surface [84]. The sensorgrams indicated that the hybrid bilayers had approximately ten times the capacity for binding, lower association rates (factor of ~ 5) and lower dissociation rates (factor of ~ 10), resulting in derivation of similar K_Ds by both methods. By injecting asialo-G_{M1} tetrasaccharide during dissociation the K_D value decreased 60-fold from 3.0×10^{-8} mol l^{-1} to 5.2×10^{-10} mol l^{-1}. It is believed that this unusual effect (the dissociation rate is usually increased [49, 56]) is due to selective inhibition of lower valency binding, which promotes higher affinity antibody-carbohydrate interactions.

The most interesting result from the interaction of murine R24 IgG with G_{D3}-bearing liposome surfaces was that saturation of the surface was never attained [85]. This was the result of additional homophilic binding of the antibody, as had been previously reported for the interaction of this antibody to human melanoma cell lines expressing G_{D3} [86]. The binding kinetics were complex at all concentrations of antibody. Data collected at low IgG concentrations fitted relatively well to a one-to-one interaction model with a k_A of 1.3×10^5 M^{-1} s^{-1}, a k_D of 2.3×10^{-3} s^{-1}, and a K_D of 18 n. Calculation of the interaction kinetics for the homophilic binding was not performed.

The interaction of the lectin from *Ulex europaeus* (UEA) with H-fucolipid embedded in liposomes attached to immobilized ricin [87] had a k_A of 2.1×10^4 M^{-1} s^{-1} and a k_D of 0.09 s^{-1}, and therefore a K_A (2.3×10^{-5} M^{-1}) in accordance with the expected affinity for carbohydrate-lectin interactions. The K_A was shown to decrease from 2.42×10^{-5} M^{-1} to 1.72×10^{-5} M^{-1} by increasing the temperature of the system from 15 °C to 30 °C. The activation energy for the association phase was calculated, with the aid of Arrhenius plots, to be 29.1 kJ/mol, indicating that a large amount of energy needed to be expended in order for binding to occur. Thus, thermodynamic data may also be derived from SPR experiments, although this is the only example for a carbohydrate-related interaction.

The interaction of an IgA antibody with a lipid monolayer containing LPS from *Vibrio cholerae* (Inaba strain) was determined [88] to have a K_A of 1.0×10^8 M^{-1}. The interaction of detergent solubilized LPS had a K_A of 6.6×10^5 M^{-1}, thus indicating the care that should be taken when designing experiments to investigate the interactions of glycolipids.

3
Carbohydrate-Carbohydrate Interactions

The study of carbohydrate-carbohydrate interactions is a subject of immense interest in cellular recognition as it is a relatively unexplored field even though some groups have postulated that this type of interaction may play an important role in a number of cellular binding events [7, 89–94]. Since carbohydrate-carbohydrate interactions are of low affinity, careful choice of the techniques and experimental methods used is essential. The majority of investigations performed have focused on an analysis of these weak interactions in the context of multivalent models [7, 89–91] since co-operative binding will in general result in an increased avidity of interaction [5]. Monovalent interactions have also been investigated, for example for the self-interaction of Lewis X epitopes [95, 96], although the affinity was difficult to evaluate and has been reported to have a K_A as low as 2 M^{-1} [95]. Another example of carbohydrate self-interaction, which has been implied but never proven, is that of defined carbohydrate epitopes at the surface of marine sponge cells [7, 92–94]. For the red beard sponge, *Microciona prolifera*, two carbohydrate epitopes, a sulfated disaccharide (Fig. 8) [97] and a pyruvylated trisaccharide [98], have been implicated in Ca^{2+}-dependent cellular adhesion. In order to investigate this phenomenon a model system, using SPR detection, is being developed [99].

In this system a BSA-conjugate (MP-conjugate), containing on average 7.8 moieties of the sulfated disaccharide [100] has been immobilized to a carboxymethylated dextran coated gold surface, along with BSA and underivatized carboxymethylated dextran as control surfaces. The interaction of these three substrates with the analytes BSA and MP-conjugate will then be monitored both in the presence and absence of 10 mmol l^{-1} Ca^{2+} ions. The system buffer consists of 20 mmol l^{-1} tris.HCl (pH 7.4) and 500 mmol l^{-1} NaCl, the concentration of salt commonly found in sea-water. The presentation of multiple epitopes of disaccharide per conjugate means that the specific interaction of carbohydrate to car-

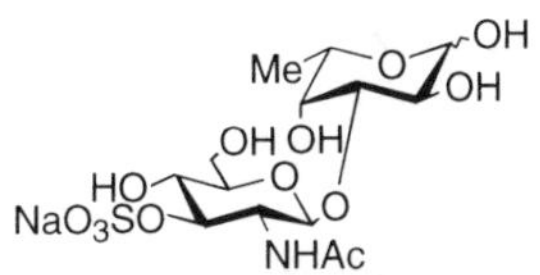

β-D-Glc*p*NAc3*S*-(1→3)-L-Fuc*p*

Fig. 8. Structure of the sulfated disaccharide implicated in the Ca^{2+} dependent cellular adhesion of the marine sponge *Microciona prolifera*

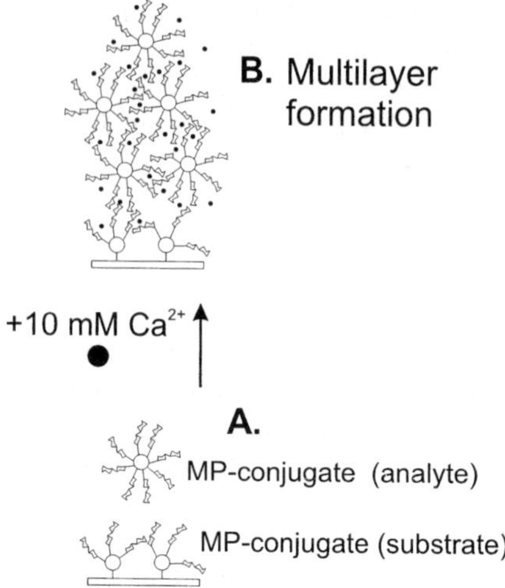

Fig. 9A, B. Illustration of the polyvalent multilayer formation of MP-conjugate: **A** before addition of Ca^{2+} ions; **B** upon addition of Ca^{2+} ions

bohydrate will result in polyvalent multilayer formation of MP-conjugate at the surface (Fig. 9): the interaction of two conjugates will result in the presentation of further carbohydrate epitopes at the surface. In addition the binding profile of the sensorgram should have characteristics very different to the standard profile as saturation of the surface cannot be attained. If successful, this system may provide a powerful means of investigating other types of carbohydrate interactions, such as the self-interaction of Lewis X.

4
Other Carbohydrate-Related Interactions

In addition to the subjects mentioned above, SPR and carbohydrate-related interactions have also been used for the development of detection assays, such as the detection of molecules in human serum [101], and detection of oligosaccharides by continuous sampling [102], which exploits weak-affinity interactions (immediate on- and off-rates) between oligosaccharides and antibodies [103, 104]. Within our group the recombination of two sub-units of a glycoprotein has been performed at the surface of a SPR biosensor, in which the role of the glycans and the question of whether a correctly folded protein can be formed, have been investigated [105].

5
Preparation of Carbohydrate Surfaces

The surface of the gold can be coated by carbohydrate in several ways. The most common method is to attach neo-glycoconjugates to a carboxymethylated dextran surface, often via amine functions on the surface of the protein (Fig. 2).

Alternatively, biotinylated carbohydrate can be attached to a streptavidin coated surface, previously attached via the amine coupling procedure. Heparin has been biotinylated in several ways, these include labeling of amine functions using NHS-LC-biotin [61], or sulfo-NHS-biotin [56, 58], or carbonyl functions, after periodate oxidation of *cis*-diol groups, using biotin hydrazide [58] or amino biotin [17]. A comparison of the k_As for the interaction of acidic fibroblast growth factor with heparin biotinylated either at amine or carboxyl functions, derived from periodate oxidation, has shown a marked (tenfold) decrease due to carboxyl labeling [58]. The k_A for the *cis*-diol biotinylated product had a value of 1.9×10^5 M^{-1} s^{-1}, as opposed to 1.1×10^6 M^{-1} s^{-1} for the amino biotinylated heparin. Chondroitin sulfate has been attached to a carboxymethylated surface by activating the surface with a 1:1 mixture of *N*-hydroxysuccinimide (NHS) and *N*-ethyl-*N'*-(dimethylaminopropyl)carbodiimide (EDC), followed by attachment of adipic acid dihydrazide (ADHZ) to the activated carboxyl functions, and finally reaction of formyl groups at the reducing end of the carbohydrate to form Schiff bases (as indicated in Fig. 2: aldehyde). The Schiff bases were then reduced, by reductive amination using $NaBH_3CN$ to give stable alkylamine bonds [106]. Alternatively, carbohydrate may be directly coupled to a dextran surface via a carboxyl or amine containing spacer molecule.

There are several ways of preparing glycolipid surfaces. One way is to capture liposomes, via an independent molecule, to the carboxymethylated dextran surface. An example of this is the attachment of glycolipid/DMPC liposomes containing a low percentage of *Salmonella* group B LPS to an antibody specific for the LPS [75, 84, 85]. In an alternative approach liposomes are fused to an alkane thiol monolayer present on a gold sensor chip [39, 82, 84]. This results in a hybrid bilayer in which the ligand of interest is presented to the analyte. As already mentioned, different immobilization protocols for similar problems can result in contrasting results [83, 84].

6
Conclusions and Outlook

Biosensors using surface plasmon resonance detection have been employed to investigate and solve some interesting carbohydrate-related problems. It is clear that the reliability of the results depends on the type of experiment performed, and the care taken in designing the experiment, e.g., the use of blank surfaces in the analysis. This technique would at present appear to be a suitable alternative to other techniques, but perhaps as the instruments available are improved it may become even more powerful. Possible improvements could involve the design of gold-bound lipid membrane bilayers more closely resembling cell surfaces, and also fine-tuning of the sensitivity of the system, in order to analyze low molecular mass compounds. One major downside of this type of system is that one biomolecule or the other is likely to be changed or modified in some way during the experiment, e.g., by attachment to the surface. However, the use of lipid membranes may also, in the future, avoid this problem. In conclusion, these systems obviously offer something more to the scientific community than was already available, which is clear from the thousands of articles that have appeared over the last decade [16]. It can be anticipated that they will be utilized to their full extent for characterizing interactions involving or mediated by carbohydrates.

7
References

1. Fukuda M, Vliegenthart JFG (eds) (1999) Structure and biosynthesis of glycoproteins. Biochim Biophys Acta 1473 (and references cited therein)
2. Stevens FJ, Argon Y (1999) Semin Cell Dev Biol 10:43
3. Imperiali B, O'Connor SE (1999) Curr Opin Chem Biol 6:643
4. Seydel U, Schromm AB, Blunk R, Brandenburg K (2000) Chem Immunol 74:5
5. Mammen M, Choi S-K, Whitesides GM (1998) Angew Chem Int Ed 37:2754
6. Hakomori S-I, Igarishi Y (1995) Biochem J 118:1091
7. Spillmann D, Burger MM (1996) J Cell Biochem 61:562
8. Sharon N (1998) Protein Sci 7:2042
9. Fägerstam LG, Frostell A, Karlsson R, Kullman M, Larsson A, Malmqvist M, Butt H (1990) J Mol Recog 3:208
10. Jönsson U, Malmqvist M (1992) Adv Biosensors 2:291
11. Stevenson R (1991) Am Biotechnol Lab 9:36
12. Attridge JW, Daniels PB, Deacon JK, Robinson GA, Davidson GP (1991) Biosens Bioelectron 6:201
13. Johnsson B, Lofas S, Lindquist G (1991) Anal Biochem 198:268
14. Rich RL, Myszka DG (2000) Curr Opin Biotechnol 11:54
15. Green RJ, Frazier RA, Shakesheff KM, Davies MC, Roberts CJ, Tendler SJB (2000) Biomaterials 21:1823
16. See www.biacore.com
17. BIAcore AB (1994) BIAapplications Handbook. Pharmacia Biosensor AB, Uppsala, Sweden
18. Cooper MA, Try AC, Carroll J, Ellar DJ, Williams DH (1998) Biochim Biophys Acta 1373:101
19. Myszka DG (1999) J Mol Recog 12:1
20. Karlsson R, Fält A (1997) J Immunol Methods 200:121
21. Davies JS (1992) Methods Enzymol 210:374
22. O'Shannessy DJ, Brigham-Burke M, Soneson KK, Hensley P, Brooks I (1994) Methods Enzymol 240:323
23. Dam TK, Roy R, Das SK, Oscarson S, Brewer CF (2000) J Biol Chem 275:14,223
24. Albani JR, Sillen A, Coddeville B, Plancke YD, Engelborghs Y (1999) Carbohydr Res 322:87
25. Iida S, Yamamoto K, Irimura T (1999) J Biol Chem 274:10,697
26. Homans SW, Field RA, Milton MJ, Probert M, Richardson JM (1998) Adv Exp Med Biol 435:29
27. Jiménez-Barbero J, Asensio JL, Cañada FJ, Poveda A (1999) Curr Opin Struct Biol 9:549
28. Misevic GN (2000) Methods Mol Biol 139:111
29. Bundy J, Fenselau C (1999) Anal Chem 71:1460
30. Pramanik BN, Bartner PL, Mirza UA, Liu YH, Ganguly AK (1998) J Mass Spectrom 33:911
31. Imberty A, Casset F, Gegg CV, Etzler ME, Perez S (1994) Glycoconjugate J 11:400
32. Van Zundert M (ed) (1998) Carbohydr Eur 23
33. Goldstein IJ, Winter HC, Poretz RD (1997) In: Montreuil J, Vliegenthart JFG, Schachter H (eds) Glycoproteins II. Elsevier, Amsterdam, chap 12
34. Hutchinson AM (1994) Anal Biochem 220:303
35. Shinohara Y, Kim F, Shimizu M, Goto M, Tosu M, Hasegawa Y (1994) Eur J Biochem 223:189

36. Shinohara Y, Hasegawa Y, Kaku H, Shibuya N (1997) Glycobiology 7:1201
37. Kalinin NL, Ward LD, Winzor DJ (1995) Anal Biochem 228:238
38. Horan N, Yan L, Isobe H, Whitesides GM, Kahne D (1999) Proc Natl Acad Sci USA 96:11,782
39. Mann DA, Kanai M, Maly DJ, Kiessling LL (1998) J Am Chem Soc 120:10,575
40. Weathermann RW, Mortell KH, Chervenak M, Kiessling LL, Toone EJ (1996) Biochemistry 35:3619
41. Uchiumi T, Sato N, Wada A, Hachimori A (1999) J Biol Chem 274:681
42. Ketis MV, Grant CW (1982) Biochim Biophys Acta 689:194
43. Haseley SR, Talaga P, Kamerling JP, Vliegenthart JFG (1999) Anal Biochem 274:203
44. Jiménez Blanco JL, Haseley SR, Kamerling JP, Vliegenthart JFG (2000) Unpublished results
45. Yamamoto K, Ishida C, Shinohara Y, Hasegawa Y, Konami Y, Osawa T, Irimura T (1994) Biochemistry 33:8159
46. Mehta P, Cummings RD, McEver RP (1998) J Biol Chem 273:32,506
47. Barboni EAM, Bawumia S, Hughes RC (1999) Glycoconjugate J 16:365
48. McDonnell JM, Jones GE, White TK, Tanzer ML (1996) J Biol Chem 271:7891
49. MacKenzie CR, Hirama T, Deng S-J, Bundle DR, Narang SA, Young NM (1996) J Biol Chem 271:1527
50. Deng S-J, MacKenzie CR, Sadowska J, Michniewicz J, Young NM, Bundle DR, Narang SA (1994) J Biol Chem 269:9533
51. Häyrinen J, Talaga P, Haseley SR, Mühlenhoff M, Finne J, Vliegenthart JFG (manuscript in preparation)
52. Zou W, MacKenzie R, Thérien L, Tomoko H, Yang Q, Gidney MA, Jennings HJ (1999) J Immunol 163:820
53. Vermeer HJ (2000) PhD thesis, Utrecht University, The Netherlands
54. Laroy W, Ameloot P, Contreras R (2000) Personnal communication
55. Tsoi PY, Yang M (1999) BIAsymposium 1999, San Fransisco, USA
56. Lookene A, Chevreuil O, Østergaard P, Olivecrona G (1996) Biochemistry 35:12,155
57. Lookene A, Savonen R, Olivecrona G (1997) Biochemistry 36:5267
58. Mach H, Volkin DB, Burke CJ, Russel Middaugh C, Linhardt RJ, Fromm JR, Loganathan D (1993) Biochemistry 32:5480
59. Lustig F, Hoebeke J, Östergren-Lundén G, Velge-Roussel F, Bondjers G, Olsson U, Rüetschi U, Fager G (1996) Biochemistry 35:12,077
60. Shuvaev VV, Laffont I, Siest G (1999) FEBS Lett 459:353
61. Shuvaev VV, Fujii J, Kawasaki Y, Itoh H, Hamaoka R, Barbier A, Ziegler O, Siest G, Taniguchi N (1999) Biochim Biophys Acta 1454:296
62. Fath M, VanderNoot V, Kilpeläinen I, Kinnunen T, Rauvala H, Linhardt RJ (1999) FEBS Lett 454:105
63. Lookene A, Stenlund P, Tibell LAE (2000) Biochemistry 39:230
64. Amara A, Lorthioir O, Valenzuela A, Magerus A, Thelen M, Montes M, Virelizier J-L, Delepierre M, Baleux F, Lortat-Jacob H, Arenzana-Seisdedos F (1999) J Biol Chem 274:2396
65. Capila I, VanderNoot VA, Mealy TR, Seaton BA, Linhardt RJ (1999) FEBS Lett 446:327
66. Cotman SL, Halfter W, Cole GJ (1999) Exp Cell Res 249:54
67. Persson E, Ezban M, Shymko RM (1995) Biochemistry 34:12,775
68. Munakata H, Takagaki K, Majima M, Endo M (1999) Glycobiology 9:1023
69. Gaus K, Hall EAH (1999) J Coll Int Sci 217:111
70. Lee WT, Conrad DH (1984) J Exp Med 159:1790
71. BIAtechnology Handbook (1994) Pharmacia Biosensor AB, Uppsala, Sweden
72. Larnkjaer A, Nykjaer A, Olivecrona G, Thogersen H, Østergaard PB (1995) Biochem J 273:747
73. Clarke AR, Luscombe M, Holbrook JJ (1983) Biochim Biophys Acta 747:130
74. Lee MK, Lander AD (1991) Proc Natl Acad Sci USA 88:2768
75. MacKenzie CR, Hirama T, Lee KK, Altmann E, Young NM (1997) J Biol Chem 272:5533

76. Teneberg S, Hirst TR, Ångström J, Karlsson K-A (1994) Glycoconjugate J 11:533
77. Schön A, Freire E (1989) Biochemistry 28:5019
78. Masserini M, Freire E, Palestini P, Calappi E, Tettamanti G (1992) Biochemistry 31:2422
79. Holmgren J, Elwing H, Fredman P, Svennerholm L (1980) Eur J Biochem 106:371
80. Fukuta S, Magnani JL, Twiddy ED, Holmes RK, Ginsburg V (1988) Infect Immun 56:1748
81. Ångström J, Teneberg S, Karlsson K-A (1994) Proc Natl Acad Sci USA 91:11,859
82. Kuziemko GM, Stroh M, Stevens RC (1996) Biochemistry 35:6375
83. Evans SV, MacKenzie CR (1999) J Mol Recog. 12:155
84. Harrison BA, MacKenzie R, Hirama T, Lee KK, Altman E (1998) J Immunol Meth 212:29
85. Kaminski MJ, MacKenzie CR, Mooibroek MJ, Dahms TES, Hirama T, Houghton AN, Chapman PB, Evans SV (1999) J Biol Chem 274:5597
86. Chapman PB, Yuasa H, Houghton AN (1990) J Immunol 145:891
87. Thomas CJ, Surolia A (2000) Arch Biochem Biophys 374:8
88. Lüllau E, Heyse S, Vogel H, Marison I, von Stockar U, Kraehenbuhl J-P, Corthésy B (1996) J Biol Chem 271:16,300
89. Eggens I, Fenderson B, Toyokuni T, Dean B, Stroud M, Hakomori S-I (1989) J Biol Chem 264:9476
90. Kojima N, Fenderson BA, Stroud MR, Goldberg RI, Habermann R, Toyokuni T, Hakomori S-I (1994) Glycoconjugate J 11:238
91. Bovin NV (1997) In: Gabius H-J, Gabius S (eds) Glycosciences. Chapmann and Hall, Weinheim, chap 15
92. Popescu O, Misevic GN (1997) Nature 386:231
93. Dammer U, Popescu P, Wagner D, Anselmetti D, Güntherodt H-J, Misevic GN (1995) Science 267:1173
94. Coombe DR, Jakobsen KB, Parish CR (1987) Exp Cell Res 170:381
95. Geyer A, Gege C, Schmidt RR (1999) Angew Chem Int Ed 38:1466
96. Henry B, Desvaux H, Pristchepa M, Berthault P, Zhang Y-M, Mallet J-M, Esnault J, Sinaÿ P (1999) Carbohydr Res 315:48
97. Spillmann D, Thomas-Oates JE, van Kuik JA, Vliegenthart JFG, Misevic G, Burger MM, Finne J (1995) J Biol Chem 270:5089
98. Spillmann D, Hård K, Thomas-Oates J, Vliegenthart JFG, Misevic G, Burger MM, Finne J (1993) J Biol Chem 268:13,378
99. Haseley SR, Vermeer HJ, Kamerling JP, Vliegenthart JFG (2000) Unpublished results
100. Vermeer HJ, Kamerling JP, Vliegenthart JFG (2000) Tetrahedron Asymm 11:539
101. Otamari M, Nilsson KG (1999) Int J Biol Macromol 26:263
102. Ohlson S, Jungar C, Strandh M, Mandenius C-F (2000) TIBTECH 18:49
103. Ohlson S, Strandh M, Nilshans H (1997) J Mol Recog 10:135
104. Strandh M, Persson B, Roos H, Ohlson S (1998) J Mol Recog 11:188
105. Erbel P, Haseley SR, Kamerling JP, Vliegenthart JFG (2000) Unpublished results
106. Satoh A, Matsumoto I (1999) Anal Biochem 275:268

Interaction Forces with Carbohydrates Measured by Atomic Force Microscopy

Christophe Tromas, Ricardo García

Instituto de Microelectrónica de Madrid, CSIC, Isaac Newton 8, 29760 Tres Cantos, Madrid, Spain
E-mail: rgarcia@imm.cnm.csic.es

In this contribution we review the use of an atomic force microscope for measurement of intermolecular forces. Intermolecular force measurements are achieved by studying the dependence of the force between the probe (tip) and the sample as a function of the separation, usually known as force-distance curves. The local character of the tip-sample interaction in force microscopy makes this technique unique for single molecule characterization. A precise intermolecular force measurement requires the controlled attachment of the molecules of interest to the probe and the sample. Here, we describe the functionalization of both surfaces by means of self-assembled monolayers. A precise control of the chemical environment is also required. We describe the method to measure single intermolecular interactions between carbohydrates molecules via self-assembled monolayers of neoglycoconjugates.

Keywords. AFM, Carbohydrates, Intermolecular interactions, Force measurements, Single molecule interactions

Topics in Current Chemistry, Vol. 218
© Springer-Verlag Berlin Heidelberg 2002

1
Introduction

The invention of the scanning tunneling microscope by Binnig and Rohrer in 1982 [1] has created a new type of microscope and local probe generically known as scanning probe microscopes to image and/or modify a variety of surfaces such as semiconductors, metals, polymers, and biomolecules. Scanning-probe microscopes scan a stylus in a raster fashion over the surface of the sample while monitoring an interaction between the end of the stylus (tip) and the surface. Different types of interactions can be measured in probe microscopy such as a tunneling current, an evanescent wave, or a variety of forces. According to the type of interaction used to image the surface a different microscopic technique is defined. The generic acronym is (SXM) where S stands for scanning, X stands for the specific interaction (i. e., T for tunneling, NO for near-field optics microscope, or F for a force microscope), and M for microscopy.

The atomic force microscope (AFM) also known as SFM was invented in 1986 by Binnig et al. [2]. It has emerged as the most versatile and flexible of the scanning-probe microscopes. Its versatility comes from the observation that with any two interacting surfaces there is always a force between them. These forces could have different physical origins. There could be van der Waals, ionic, chemical, capillary, magnetic, or electrostatic forces. An AFM could be operated in different environments such as air, liquid, or ultra-high vacuum. In the imaging mode, force microscopy provides a three-dimensional image of a surface with atomic and/or nanometer-scale resolution. All of those properties have made AFM a very powerful tool for characterization of surfaces.

The sensitivity, local character, and ability to control with high precision the strength of the physical interaction involved has led force microscopy to other applications beyond imaging of surfaces. For example, force microscopes are used to modify surfaces at atomic and nanometer-scales and to fabricate new types of nanolectronic devices [3]. It can also be used to map compositional variations in heterogeneous samples [4] and to measure specific interactions giving rise to a force spectroscopy [5].

In 1994 two different groups demonstrated the potential of the AFM to measure adhesion forces between individual avidin-biotin pairs [6,7]. Those experiments opened the route to develop the AFM as a tool for intermolecular force measurements.

Other techniques have also been applied to study surface interactions. Among them the surface force apparatus (SFA) stands by its sensitivity in the measurement of forces [8]. However, the SFA is limited by the geometry and nature of material used. The interaction area involves several mm^2 and surfaces should be transparent to light. The SFA has a vertical resolution of 0.1 nm and a force resolution of 10 nN. On the other hand, the nanometer size of the AFM probe allows one to measure the interaction of a small number of – or even single – molecules. It has the potential of simultaneous topography and spectroscopy characterization.

In this contribution we review the operation and applications of an atomic force microscope for intermolecular force measurements. Special emphasis is placed on the description of force vs distance curves, the key element to measuring adhesion forces. Some examples of adhesion force measurements involving lactose derivatives are also included.

2
Atomic Force Microscopy

2.1
Description of an Atomic Force Microscope

The Atomic Force Microscope (AFM) [2, 9–12], also known as Scanning Force Microscope (SFM), is a scanning probe microscope where a tip is attached to the end of a cantilever. In a first approximation the cantilever acts as a spring that is characterized by a force constant k. The changes in the cantilever deflection are optically monitored by means of a laser beam, reflected in the back side of the cantilever, and detected on a split photodiode.

Cantilevers are fabricated by microelectronic techniques and they are usually made of silicon or silicon nitride. The tips are integrated in the cantilever. Usually they have a pyramidal profile with a curvature radius ranging between 5 and 30 nm.

The sample is usually placed in a sample holder on top of a piezo-scanner. The scanner generates the lateral and vertical displacement of the sample with respect to the tip. The electrical signals coming out of the photodiode are processed in a computer and displayed as images in a monitor. All those elements are shown in Fig. 1.

Two basic AFM modes can be considered: contact and dynamic modes. Historically, the first AFM mode that has been developed is the contact mode. In this mode, a feedback loop acting on a vertical piezoelectric scanner keeps the can-

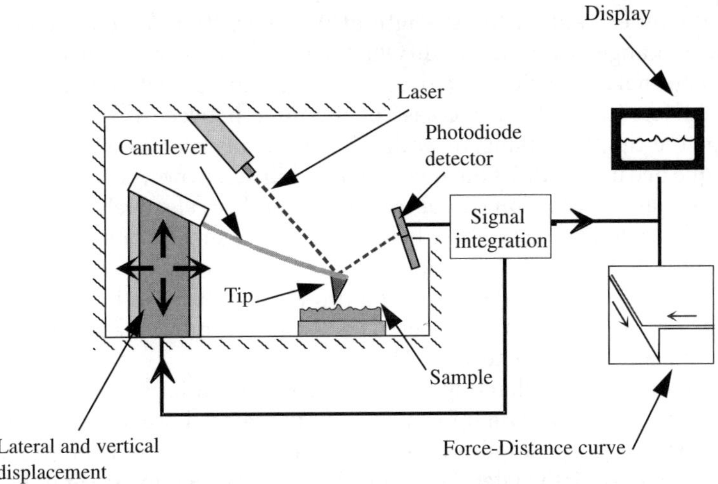

Fig. 1. Scheme of an atomic force microscope configuration

tilever deflection, and so the tip-sample force, at a user-specified set point level. In this way, the tip always remains in contact with the sample. The force F between the tip and the surface can be determined from Hook's law, $F = -kz$, where z is the cantilever deflection and k the force constant of the cantilever. The applied forces, which depend on the material and the setpoint level, range from a few piconewtons to several nanonewtons. The vertical displacement of the scanner is recorded in each point of the scanned area to form a topographical image of the sample surface. In this mode, which can be used in ambient or liquid environment, "atomic resolution" images can be obtained.

In the dynamic mode, the cantilever is oscillated at a frequency near its resonance frequency, while oscillation amplitude [13] (amplitude modulation mode) or the frequency of the cantilever [14] (frequency modulation) are monitored. The feedback loop keeps the oscillation amplitude or the frequency shift at a constant value so that the interaction between the tip and the surface remains the same along the surface. In dynamic AFM there is only an intermittent contact between the tip and the surface, or no contact at all; as a consequence the friction forces are minimized. Furthermore, the normal forces are lower than in contact mode, resulting in a higher lateral resolution and less sample damage.

2.2
Imaging, Modification, and Spectroscopy

Currently a force microscope has three main applications: imaging, modification, and compositional mapping of surfaces. In the imaging mode, the basic AFM use, the probe is scanned in a raster fashion across the sample keeping a constant force, amplitude, or frequency shift [9–12]. For example, Fig. 2 shows a representative AFM image of several antibodies deposited on a mica surface [15]. The molecules show several morphologies according to the orientation of the domains with respect to the support. Some of them show the characteristic Y shape of antibodies.

On the other hand, if the strength of the interaction force between tip and sample is changed by either modifying the tip-surface separation or by increasing an external force (electrostatic, etc.) the tip can be used to modify selectively the surface [16, 17]. An increase of the force between tip and sample may damage the tip, the sample, or both. This is usually seen as a drawback of the technique because it limits the imaging possibilities. However, the monitoring of the tip-surface interaction can be applied to modify selectively a given surface. This is illustrated in Fig. 3 where two images of the surface of a biological membrane (purple membrane) are shown before and after the tip was approached to establish mechanical contact with the proteins of the membrane [18]. In this case, the tip has acted as a nanometer-size knife to remove the proteins from the membrane and to write the letter M.

One of the more challenging applications in force microscopy is the development of a method to obtain information about material properties, i.e., to develop force spectroscopy applications. Several approaches have been proposed. In contact mode AFM, the lateral (frictional) force between the tip and the sample has been used to map compositional variations in Langmuir-Blodget films and

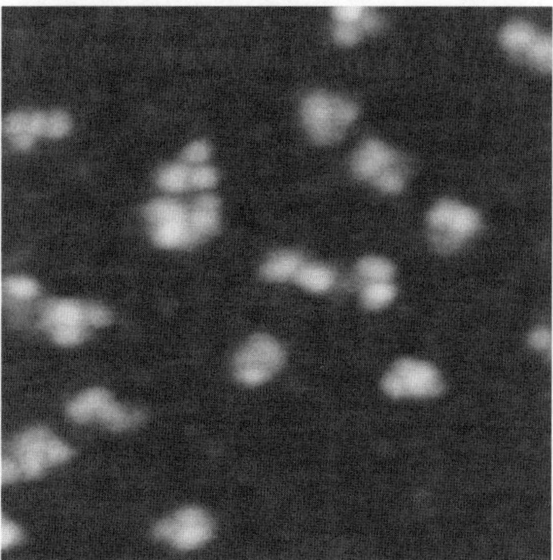

Fig. 2. Tapping-mode AFM image of anti-human serum albumin (a-HSA). The molecules show several morphologies according to their orientation with respect to the support. The image was obtained in air and at room temperature

Fig. 3. Manipulation of a protein membrane by STM. The letter is generated by scanning at 20 pA. It implies the removal of the proteins in the modified region

semiconductor structures. In dynamic AFM the phase shift between the external driving force and the tip response has also been applied to characterize material properties [19, 20]. Although those methods have been extremely efficient to provide qualitative information about the changes in the surface composition, they have been unable provide quantitative information of the properties of the material.

A different application of a force microscope, although closely related to force spectroscopy, is its use to determine single molecular interactions [5–7, 21–24]. A key element in performing those measurements is the study of the dependence of the force vs the tip-sample separation, which are usually known as force vs distance curves.

2.3
Force vs. Distance Curves

Monitoring the change of the cantilever deflection vs the tip-sample separation is the experimental procedure to quantify tip-surface interactions. A force curve is obtained by performing an approaching/retraction loop, where the tip is first moved towards the sample until mechanical contact is established and then withdrawn. Figure 4 shows an ideal force vs distance curve for a system with no specific adhesion forces.

When the tip-sample interaction is negligible, there is no cantilever deflection (position 1 in the approaching curve). Approaching long-range attractive forces may produce a slight bending of the cantilever until a sudden jump is

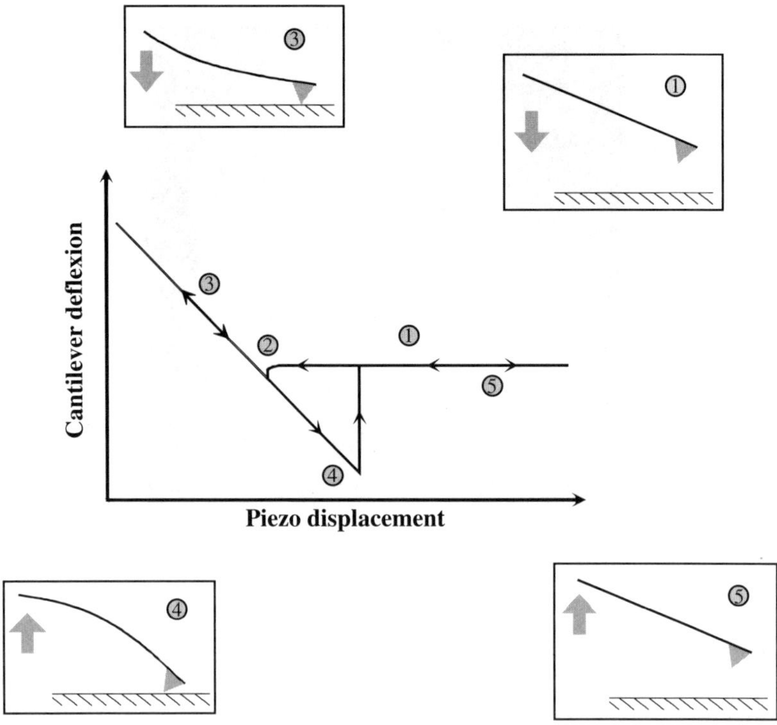

Fig. 4. AFM force distance curve: ① tip approach; ② jump to contact; ③ contact between the tip and the sample surface (repulsive forces); ④ adhesive forces and jump off contact; ⑤ tip withdrawing

observed, jump-to-contact (position 2). This jump-to-contact happens when the force gradient of the interaction equals the force constant of the cantilever. From there on a linear dependence of the cantilever deflection on the sample displacement is observed (position 3). In this scheme the deformation of the sample has not been considered. In the retraction curve the displacement where the jump out to contact is observed does not coincide with the jump-to-contact point (position 4). Several factors contribute to this effect, chief among them being the existence of adhesion forces. Finally when the tip is far from the sample the cantilever recovers the initial deflection (position 5).

In fact, two transformations are required to convert the above experimental curves into *genuine* force curves. First, the cantilever deflection (a distance) must be converted into a force. This is usually done by multiplying the cantilever deflection by its the force constant, $F \approx k \, \Delta z$. In addition, the instantaneous tip-sample separation must be obtained from the piezo displacement and cantilever deflection. This may be a time-consuming and delicate operation.

The presence of specific tip-surface interactions introduces additional features in the above scheme. Multiple intermolecular bonds between the molecules attached to the tip and those to the sample may be formed. As a consequence several step-like discontinuities may be observed in the retraction curve. Each of them is the consequence of breaking a given number of intermolecular bonds (Fig. 5).

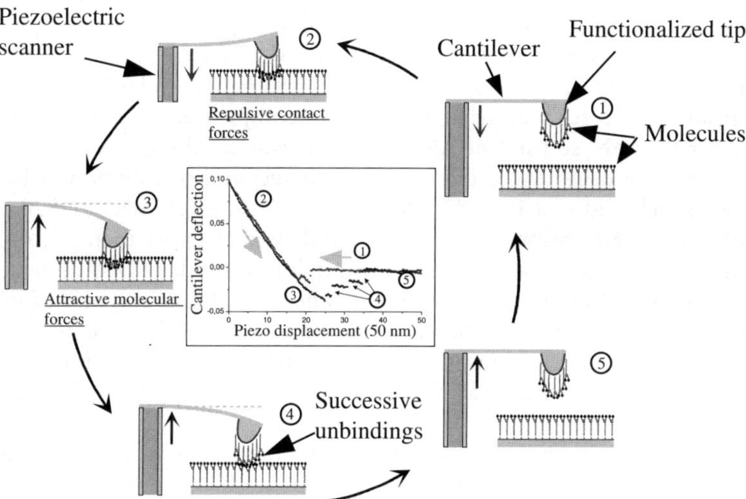

Fig. 5. Force distance curve with functionalization of tip and sample: ① the tip is approached toward the surface; ② contact repulsive forces; ③ tip withdrawal: attractive forces due to the interaction between the molecules on the tip and on the surface are observed; ④ stepwise profile corresponding to successive molecule unbindings; ⑤ after a complete withdrawal, a new loading unloading cycle is performed and the molecules bonds reform

3
Examples of Intermolecular Forces Measurement by AFM

The first experimental measurement of intermolecular forces by AFM was performed by Florin et al. [6, 25]. They used Si_3N_4 tips functionalized with avidin and the sample were agarose beads functionalized with biotin, desthiobiotin, or iminobiotin. The avidin molecules were bound to the tip through bovine serum albumin (BSA), which nonspecifically adsorbs to the silicon nitride surface of the tip. In this way, no covalent binding (for example through a linker ended by a thiol group, as presented before) was required to functionalize the tip. The interaction forces between the tip and the beads, functionalized with avidin and biotin respectively, were found to be about 20 nN. So, to determine single molecule interaction forces, the number of bonds between the tip and the surface were reduced by blocking most of the biotin on the agarose bead with free avidin. By plotting the measured forces in an histogram, regular peaks were observed, showing that these forces are quantized. These adhesion forces were found to be entire multiples of 160 ± 20 pN, which were interpreted as the binding force between a single pair of avidin and biotin molecules. These experiments were repeated with agarose beads functionalized with desthiobiotin and iminobiotin. The single molecule interaction forces where found to be 125 ± 20 and 85 ± 15 pN, respectively.

Lee et al. [26] have measured the interaction forces between single strands of DNA molecules. DNA complementary oligonucleotides $(ACTG)_5$ and $(CAGT)_5$ were covalently attached to a mica surface and to a spherical probe respectively. Force-distance curves, performed in 0.1 mol/l NaCl solution, revealed four distinct groups of forces: 1.52 ± 0.19, 1.11 ± 0.13, 0.83 ± 0.11, and 0.48 ± 0.1 nN. The average adhesive force measured between noncomplementary oligonucleotides was 0.38 ± 0.33 nN, so the 0.48 nN force value was attributed to a non-specific interaction. The other three force groups correspond to interaction between DNA oligonucleotides of 12 (0.83 ± 0.11 nN), 16 (1.11 ± 0.13 nN), and 20 (1.52 ± 0.19 nN) base pairs respectively. In a second type of experiment, the authors have investigated the intrachain forces within long strands of DNA. Tip and surface sample were functionalized with cytosine and a homopolymer of inosine, with an average length of 160 bases, was bound to one of the cytosine surfaces. The force distance curves showed a 0.46 ± 0.25 nN rupture force, which appeared only after a 240 nm withdrawal. This delayed effect was due to the length of the inosine polymer.

Misevic and coworkers [27, 28] were the first to measure association between carbohydrates by AFM. They have quantified the binding force between cell adhesion proteoglycan (AP) of the marine sponge *Microciona prolifera* under the influence of various physiological conditions. The AP molecules were linked to the tip and the surface through a covalent binding with a self-assembled monolayer (SAM) of active succinimide groups previously deposited on these surfaces. Observation of AP molecules by AFM showed that they are 200 nm diameter rings with about 20 irradiating arms 180 nm long. The curvature radius of the tip was about 50 nm. As a consequence, only one AP molecule is assumed to be attached to the end of the tip. Approach and retract cycles showed multiple steps on the retraction curve at a distance from 0 to 200 nm above the surface. This delayed interaction was interpreted as an interaction between the stringlike arms of the molecules, which were left above the surface during the experiment. The magnitude of a single step, that is the interaction between two single arms, was about 40 ± 15 pN, while the maximum and the average adhesion forces were 400 pN and 125 pN respectively. They were interpreted as interactions between ten and three pairs of AP arms respectively. The AP-AP interaction has been found to be Ca^{2+}-dependent, as the binding force and the probability of interaction were reduced in a lower calcium concentration. Furthermore, magnesium ions were not able to replace Ca^{2+}. Finally, it was demonstrated that the adhesion forces measured by AFM originate from AP-AP interaction, using a monoclonal antibody (mAb) block to inhibit the specific AP-AP interaction.

Antigen and antibody interactions have been studied by Harada et al. [29]. They used a tip functionalized with the Fab fragment of an antiferritin antibody to measure the adhesion forces with a surface coated with ferritin molecules. The force-distance showed a stepwise profile during unloading due to multiple unbinding events. The distribution forces ranged from 35 pN to 450 pN. A Fourier analysis has suggested a quantization in the intermolecular forces with a period of 63 pN. This force has been attributed to the single unbinding event between individual antigen and antibody molecules. The authors have also estimated the work involved in a single unbinding event. The work needed to break the antigen-antibody binding has been found to be approximately 19 kJ/mol ($\cong$ 4.5 kcal/mol). The value was compared with the free energy change derived from Isothermal Titration Calorimetry (ITC) experiments (9 kcal/mol). The AFM value is smaller than the free energy calculated form the ITC experiments. The authors stated that direct comparisons between those values should be done carefully. It has been suggested that AFM force measurements measure the enthalpic activation barrier to protein-protein interaction [25, 30].

Supramolecular host-guest interactions between ferrocene moieties and β-cyclodextrin receptors in aqueous environment have been studied by Schönherr et al. [31]. They functionalized a gold coated silicon nitride tip with ferrocene moieties, and a gold surface (Au(111)) with β-cyclodextrin heptasulfide receptors. Previous AFM studies have shown that those β-cyclodextrin receptors equipped with seven sulfide units form highly ordered molecular monolayer on gold surface. They also showed that the cavities were exposed to allow complexation with external guests. Force-distance showed a stepwise profile, characteristic of molecular interaction forces. A statistical analysis revealed a quantized rupture force of 56 ± 10 pN.

Fernandez and co-workers have used the AFM to study the unfolding and folding of single protein molecules [32, 33]. The force-extension curve displayed a characteristic saw-tooth pattern with the number of peaks corresponding to the number of domains stretched between the substrate and cantilever tip. The unfolding of the different protein domains could be described with the worm-like chain model. The mean force at which the domains unfold was 137 pN and the mean interval between peaks was 25 nm.

An experiment designed to measure protein-protein interactions was performed by Baumgartner et al. [34]. They have measured specific *trans*-interactions and conformational changes of recombined cadherin dimers in aqueous solutions. Unbinding forces of 35–55 pN were measured.

4
Atomic Force Microscopy Studies of Structure and Interactions of Carbohydrates

The use of AFM to image or study carbohydrate molecules is very recent and just a few results have been published so far. In Sect. 3 we have mentioned force spectroscopy experiments with proteoglycans of natural systems [27, 28]. Gaub and coworkers [35] have measured the deformation of single polysaccaharide chains. At low forces the deformation of dextran filaments was found to be dominated by entropic forces (Kuhn length of 0.6 nm). At higher forces the dextran filaments underwent a distinct conformational change. The polymer stiffened and the segment elasticity was dominated by the bending of bond angles. The conformational change was found to be reversible. In another study [36] single polysaccharide molecules were covalently anchored between a surface and an AFM tip and then stretched until they became detached. By using different surface chemistries for the attachment, the rupture of two covalent bonds were measured. They reported that the silicon-carbon bond ruptured at 2 nN, and the sulfur-gold bond ruptured at 1.4 nN.

Recognition forces between protein-membrane receptor gangliosides have been measured by Luckham and Smith [37]. The strength of the adhesion between cholera toxin B and its receptor, the ganglioside GM_1 was measured at about 90 pN.

Fritz et al. [38] have studied the rupture forces, elasticity, and kinetics of the P-selectin/P-selectin glycoprotein ligand-1 interaction. Adhesion forces of up to 165 pN were reported. Their study pointed out that rupture force and life time of the complexes depend on the applied force per unit of time. This effect was related to the intrinsic molecular elasticity and the external pulling velocity.

AFM images of several polysaccharides such as scleroglucan have been described by Brant [39]. In this contribution there is also a short analysis of the potential force spectroscopy studies. AFM images of bundles of polysialic acid units (a component of the neural cell adhesion molecule N-CAM) were presented by Toikka et al. [40].

A rather different application for a carbohydrate molecule was proposed by Shigekawa et al. [41]. By performing molecular manipulations of cyclodextrin necklace with a scanning tunneling microscope tip they have devised a molecular abacus.

5
Protocol for Intermolecular Force Measurements

In this section we will provide a brief description of the experimental protocol we have used to study interaction between carbohydrate molecules. A proper quantitative intermolecular force measurement requires a well-defined and stable chemical environment. In the framework of a project on the application of AFM to study carbohydrate interactions we have used self-assembled monolayers (SAM) on gold of structurally well-defined carbohydrates molecules. SAM on gold provides a strategy to have 2D funtionalized surfaces with a known attachment of the molecules to the tip and the flat surface. Here we describe the steps followed for structural well-defined functionalized surfaces and to achieve accurate force measurements.

5.1
Fluid Cell

One of the main advantages of force microscopy is its ability to provide nanometer resolution images of surfaces in air environments. However, air environments are not suitable for performing reliable quantitative force measurements. The spontaneous formation of a water meniscus between tip and sample gives rise to a capillary force of several nN that makes hard or impossible the measurement of intermolecular forces of tens of pN. A fluid cell provides a stable environment to perform force vs distance curves and to eliminate meniscus formation. In a fluid cell compatible with force microscopy [42], the sample and the tip-cantilever ensemble are immersed in fluid (see Fig. 6). The fluid cell also allows one to study the interaction in physiological conditions.

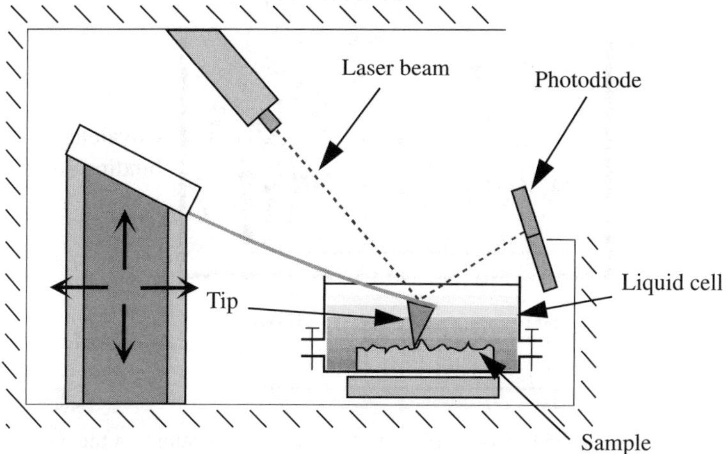

Fig. 6. Schematic of an AFM liquid cell. The design of the cell allows to change easily the liquid

5.2
Deposition of Carbohydrates by Self-Assembled Monolayer Processes

The carbohydrates relevant to this work are rather molecules of about 1.5 nm in size. Reliable force measurement of small molecules requires a procedure to deposit them on the tip and on a flat surface, the sample hereafter. By chemical synthesis the carbohydrates have been attached to one of the ends of an aliphatic chain (neoglyconjugates). At the other end of the chain there is an SH group. The thiol group is required to establish a covalent bond with a gold atom. This requires that both tip and sample are covered by a thin film of gold. Then tip and sample are immersed for several hours in a solution of the neoglyconjugates. A self-assembled process will form a single monolayer of the molecules on the surfaces [43]. The covalent bond between Au and S has a binding energy several times higher than the binding energies of the adhesion processes relevant to this study; as a consequence the forces involved will not destroy the monolayer (Fig. 7).

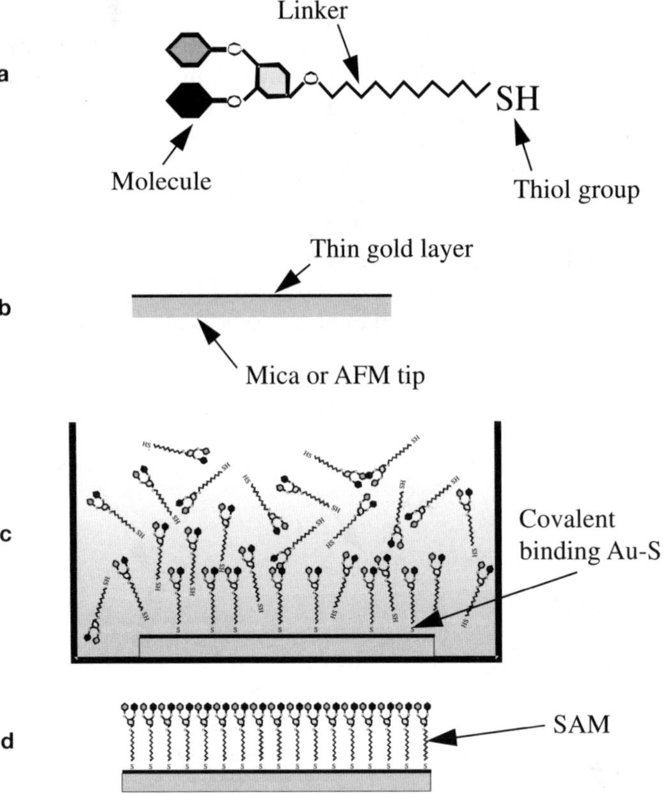

Fig. 7 a–d. Protocol for the functionalization of the tip and the sample: **a** the molecule of interest are synthesized with a linker ended in a thiol group; **b** a thin gold layer is deposited on the tip and the sample; **c** the gold surfaces are immersed in a molecule solution – covalent bonds form between the sulfur atoms and the gold surfaces; **d** after rinsing, a self-assembled monolayer (SAM) is obtained

5.3
Calibration of the Force Microscope

The curves obtained during a tip loading and unloading cycle display the photodiode voltage vs the vertical displacement of the piezoelectric scanner. In order to convert the photodiode voltage variation into a force, two calibration stages must be performed. First, voltage variations have to be converted into a cantilever deflection. Second, to convert the cantilever deflection into a force, the cantilever spring constant has to be determined.

In the region of a force vs distance curve where tip and sample are in contact, the vertical displacement of the piezo equals the cantilever deflection. We are assuming that sample deformation is negligible in comparison with the cantilever deflection. Then, the slope of the force curve in that region is one and proportionality between photodiode voltage and deflection can be established.

Once the photodiode sensitivity has been calibrated, the cantilever deflection can be precisely measured. The interaction force can be then easily determined using Hooke's law:

$$F = k_c \Delta Z \tag{1}$$

with k_c the cantilever spring constant and ΔZ the cantilever deflection. However, the nominal value of the spring constant as given by the manufacturer can differ substantially from the real value. Each individual cantilever spring constant should be calibrated. There are three different methods to measure the cantilever constant. The theoretical method uses the Young's modulus and the Poisson's ratio, and the shape and dimensions of the cantilever to calculate its spring constant [44]. In the static method, a known force is applied to the cantilever while monitoring its deflection with the photodiode [45, 46]. This force can be a gravity force (a mass is attached to the tip) or an elastic force (the tip is pressed on a calibrated cantilever). Dynamic methods are based on the dynamic behavior of the cantilever. The preferred dynamic method records the thermal response of the cantilever [47, 48]. Then using the equipartition theorem provides the value of the spring constant.

6
Measurement of Lactose-Lactose Interaction Forces

We report here, as an application of the above protocol, the measurement of interaction forces between lactose molecules by AFM. These experiments take place in a more ambitious project the purpose of which is the determination of intermolecular forces between glycosphingolipids, and in particular between Lewis[X] antigens. Those molecules, organized in patches at the outer cell membrane, are involved in many biological processes [49, 50] and are supposed to be responsible for cell recognition via carbohydrate-carbohydrate association [50, 51]. As the lactose molecule is the common part of most of glycosphingolipid molecules, the study of single lactose molecule interaction is the preliminary stage of this project.

Two lactose derivative have been synthesized: 11-thioundecyl-β-lactoside **1** and 11-thio-3,6,9-trioxa-undecyl-β-lactoside **2**. Gold deposited AFM tips and mica surfaces have been functionalized with those molecules, according to the protocol described in Sect. 5.2. The experiments have been performed in a liquid cell, in water, and in calcium solution at physiological concentration (10 mmol/l). Commercial Si_3N_4 AFM cantilevers (Olympus) with a nominal spring constant of 0.05 N m^{-1} have been used. The real spring constant has been calibrated for each cantilever according to the thermal response method [47,48]. The unloading rate during force distance curve experiments was 100 nm s^{-1}, and a constant contact force of 200 pN was maintained for 2 s before unloading to ensure a chemical equilibrium between the molecules on the tip and on the surface. For each tip-sample pair, more than 400 force vs distance curves were measured over 20 different zones on the sample surface. Furthermore, each experiment has been repeated with different tips and different samples.

Figure 8 summarizes force distance curves obtained in lactose-lactose experiments. In the case of molecule **1** (Fig. 8a, b), no specific interactions were observed. It clearly demonstrates that there are no specific lactose-lactose adhe-

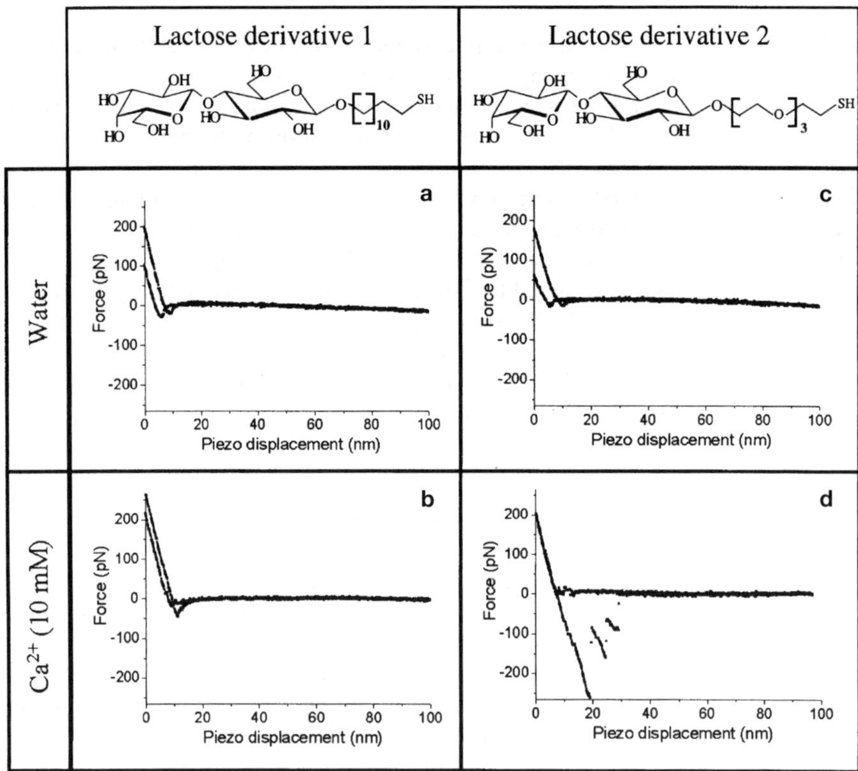

Fig. 8 a–d. Typical profiles of force distance curves: **a, b** tip and surface functionalized with the lactose derivative **1** in water and 10 mmol/l Ca^{2+} solution respectively; **c, d** tip and surface functionalized with the lactose derivative **2** in water and 10 mmol/l Ca^{2+} solution respectively

sion forces, either in water or in calcium solution. When repeating the same experiments in water with the lactose derivative **2**, no interaction is observed (Fig. 8c). However, in calcium solution, the unloading part of the force-distance curves shows the typical stepwise profile, characteristic of intermolecular interaction (Fig. 8d). These steps correspond to quick cantilever movement due to the rupture of molecules binding. The height of the steps has to be proportional to the unbinding force. However, as the unbinding events may involve several pairs of molecules, a statistical analysis has been performed to determine the adhesion force between two single molecules. The distribution of unbinding forces, measured over 300 curves, is represented in the histogram at Fig. 9a.

The force distribution histogram shows the existence of preferred values of the force. This underlines the presence of individual interaction events. To reinforce and to measure the periodicity of the force distribution we have used the method proposed by Florin et al. [6]. The autocorrelation function of the data is calculated after subtracting the envelope function of the histogram (Fig. 9b). This envelope has been determined through a six-order binomial filter.

The autocorrelation function (Fig. 9b) presents a strong periodicity of 8 ± 2 pN. This periodicity is attributed to the interaction force between two single molecules of lactose derivative **2**. The experiments have been repeated with different tips and samples. In all cases similar results were obtained.

The experiments with the first lactose derivative **1** did not show the existence of interaction forces between lactose molecules. The adhesion forces observed in the case of the second molecules cannot be attributed to specific carbohydrate-carbohydrate interactions. The only difference between molecule **1** and **2** is the linker; thus, the measured adhesion forces should be due to a linker-linker interaction or to a lactose-linker interaction. To confirm this hypothesis, two control experiments have been performed in water and in calcium solution. The tip was functionalized exclusively with the tetraethylene glycol molecule (TEG), while the sample was functionalized with the tetraethylene glycol molecule (first control) and the lactose derivative **2** (second control). Adhesion forces were

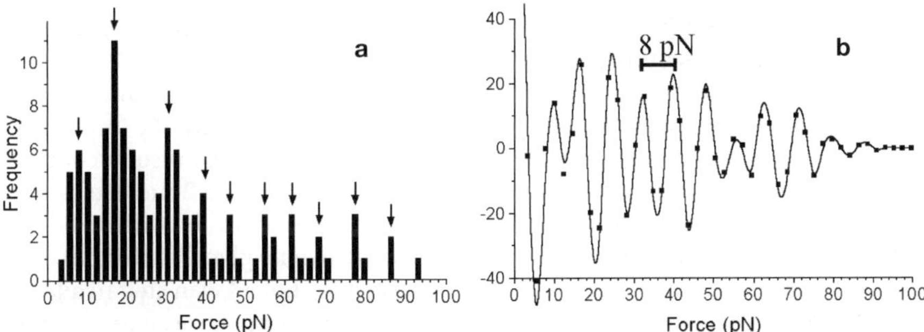

Fig. 9a,b. Analysis of the distribution of the unbinding forces measured between lactose derivative **2** molecules in calcium solution: **a** histogram of the force distribution; **b** autocorrelation function (after subtracting the histogram envelop). A periodicity of 8 pN is observed on the autocorrelation function. This corresponds to the unbinding force between two molecules

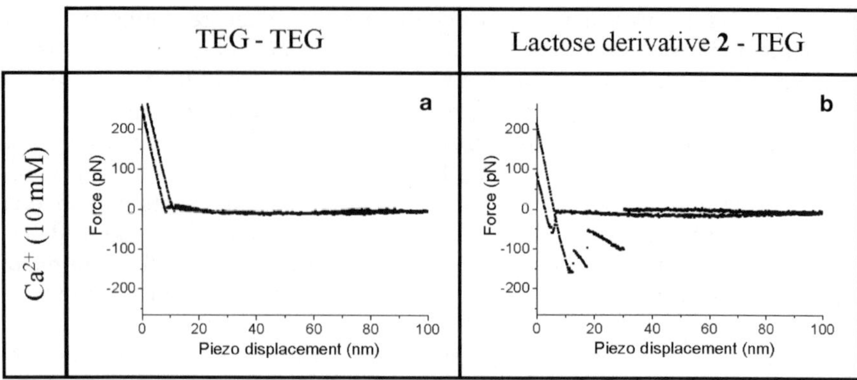

Fig. 10a, b. Typical profiles of the force distance curves of two experiments: a tip and sample functionalized with the tetra ethylene glycol chain in 10 mmol/l Ca^{2+} solution; b tip functionalized with the tetra ethylene glycol chain and sample functionalized with the lactose derivative 2 in 10 mmol/l Ca^{2+} solution

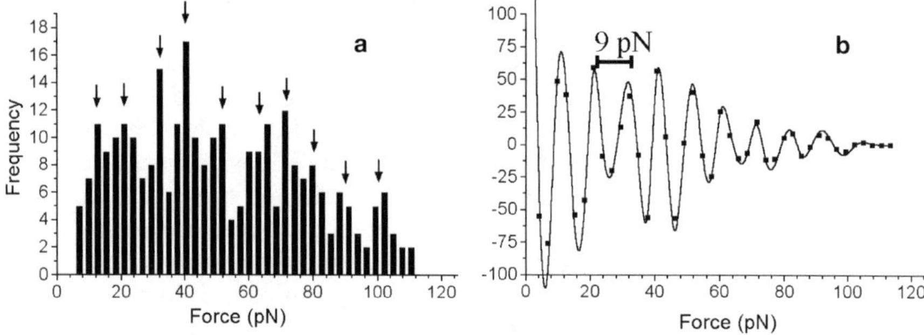

Fig. 11a, b. Analysis of the distribution of the unbinding forces measured between lactose derivative 2 and tetra ethylene glycol molecules in calcium solution: a histogram of the force distribution; b autocorrelation function (after subtracting the histogram envelop). A periodicity of 9 pN is observed on the autocorrelation function

observed only in the control experiment between lactose derivative 2 and tetra ethylene glycol in calcium solution (Fig. 10).

The force distribution and the corresponding autocorrelation function are presented in Fig. 11. The observed value, 9 ± 2 pN, coincides with the force measured between single molecules of lactose derivative 2. These results confirm that the adhesion forces measured between lactose derivative 2 molecules was not due to a specific carbohydrate-carbohydrate interaction but to a calcium dependent interaction between lactose and tetra ethylene glycol molecules.

The above experiment illustrates that the ascription of a given step in a force vs distance curve to an intermolecular force requires very precise control experiments to ensure the specificity of the observed interaction.

A similar procedure was applied to study specific interaction forces between individual Le^X antigen molecules. This study has demonstrated the ability and selectivity of Le^X antigen for self-recognition [52].

7
Summary

We have described the use of the atomic force microscope for intermolecular force measurements. Precise measurements require a stable chemical environment that is usually provided by using fluid cells. Force vs tip displacement curves are essential to measure adhesion forces. In most cases the steps observed in force curves are the consequence of multiple interactions. To extract the value corresponding to a single interaction requires a statistical analysis that involves the use of autocorrelation functions.

We have also described measurements involving lactose derivatives. These experiments showed an interaction force of 8 ± 2 pN between the lactose and the tetraethylene glycol molecule. The experiments also illustrate some of the difficulties that appear when the steps observed in force curves are attributed to intermolecular forces.

Acknowledgement. This work was supported by the Ministerio de Educación y Cultura of Spain (PB98–0471) and TMR project of the European Union (ERBFMRXCT98.0231).

8
References

1. Binnig G, Rohrer H (1982) IBM J Res Dev 30:353
2. Binnig G, Quate CF, Gerber C (1986) Phys Rev Lett 56:930
3. García R, Calleja M, Rohrer H (1999) J Appl Phys 86:1898
4. Overney RM, Meyer E, Frommer J, Güntherodt HJ, Fujihira M, Takano H, Goth Y (1994) Langmuir 10:1281
5. Janshoff A, Neitzert M, Oberdörfer Y, Fuchs H (2000) Angew Chem Int Ed 39:212–3237
6. Florin EL, Moy VT, Gaub HE (1994) Science 264:415
7. Lee GU, Chrisey LA, Colton RJ (1994) Science 266:771
8. Israelachvili JN (1992) Intermolecular and surface forces. Academic Press, London
9. Rugar D, Hansma P (1990) Phys Today 43:23
10. Bustamante C, Keller D (1994) Phys Today 48:32
11. Hansma H, Pietrasanta L (1998) Curr Opin Chem Biol 2:579–584
12. Procedures in scanning probe microscopies (1998) Wiley, Chichester
13. Martin Y, Williams CC, Wickramasinghe HK (1987) J Appl Phys 61:4723
14. Albrecht TR, Grütter P, Horne D, Rugar D (1991) J Appl Phys 69:668
15. San Paulo A, García R (2000) Biophys J 78:1599
16. Binnig G, Rohrer H (1999) Rev Mod Phys 71:S324–S330
17. Nyffenegger RM, Penner RM (1997) Chem Rev 97:1195–1230
18. García R, Tamayo J, Soler JM, Bustamante C (1995) Langmuir 11:2109
19. Tamayo, J, García R (1996) Langmuir 12:4430
20. Bar G, Thomann Y, Whangbo MH (1998) Langmuir 14:1219
21. Samori B (2000) Chem Eur J 4249–4255
22. Bustamante C, Macosko JC, Wuite GJL (2000) Nature 1:130
23. Viani M et al. (2000) Nature Struct Biol 7:644–646
24. Fisher TE, Marszalek PE, Fernandez JM (2000) Nat Struct Biol 7:719–724
25 . Moy VT, Florin EL, Gaub HE (1994) Science 266:257–259
26. Lee GU, Chrisey LA, Colton RJ (1994) Science 266:771–773
27. Dammer U, Popescu O, Wagner P, Anselmetti D, Güntherodt HJ, Misevic GN (1995) Science 267:1173

28. Misevic GN (1999) Microsc Res Technol 44:304
29. Harada Y, Kuroda M, Ishida A (2000) Langmuir 16:708
30. Chilkoti A, Boland T, Ratner BD, Stayton PS (1995) Biophys J 69:2125
31. Schönherr H, Beulen MWJ, Bügler J, Huskens J, Van Veggel FCJM, Reinhoudt DN, Vancso GJ (2000) J Am Chem Soc 122:4963
32. Oberhauser AF, Marszalek PE, Erickson HP, Fernandez JM (1998) Nature 393:181
33. Carrion-Vazquez M, Oberhauser AF, Fowler SB, Marazalek PE, Broedel SE, Clarke J, Fernández JM (1999) Proc Natl Acad Sci USA 96:3694
34. Baumgartner W, Hinterdorfer P, Ness W, Raab A, Vestweber D, Schindler H, Drenckhahn (2000) Proc Natl Acad Sci 97:4005–4010
35. Rief M, Oesterhelt F, Heymann B, Gaub HE (1997) Science 275:1295–1297
36. Grandbois M, Beyer M, Rief M, Clausen-Schauman H, Gaub HE (1999) Science 283:1727–1730
37. Luckman PF, Smith K (1998) Faraday Discuss Chem Soc 111:307–320
38. Fritz J, Katopodis A, Kolbinger F, Anselmetti D (1998) Proc Natl Acad Sci USA 95:12,283–12,288
39. Brant DA (1999) Curr Op Struct Biol 9:556–562
40. Toikka J, Aalto J, Hÿrinen J, Pelliniemi LJ, Finne J (1998) J Biol Chem 273:28,557–28,559
41. Shigekawa H, Miyake K, Sumaoka J, Harada A, Komiyama M (2000) J Am Chem Soc 122:5411–5412
42. Molecular Imaging (1999) PicoSPM, Molecular Imaging, Tempe, USA
43. Ulman A (1996) Chem Rev 96:1553 and references cited therein
44. Sader JE, White E (1994) J Appl Phys 74:1
45. Senden TJ, Ducker WA (1994) Langmuir 10:1003
46. Gibson CT, Watson GS, Myhra S (1996) Nanotechnology 7:259
47. Hutter JL, Bechhoefer J (1993) Rev Sci Instrum 64:1868
48. Walters DA, Cleveland JP, Thomson NH, Hansma PK, Wendman MA, Gurley G, Elings V (1996) Rev Sci Instrum 67:3583
49. Hakomori S, Nudelman E, Levery S, Solter D, Knowles BB (1981) Biochem Biophys Res Commun 100:1578
50. Eggens I, Fenderson B, Toyokuni T, Dean B, Stroud M, Hakomori S (1989) J Biol Chem 264:9476–9484
51. Hakomori S (1991) Pure Appl Chem 63:473
52. Tromas C, Rojo J, De la Fuente JM, Barrientos AG, García R, Penadés S (2001) Angew Chem Int Ed Eng (in press)

Recognition Processes with Amphiphilic Carbohydrates in Water

Guangtao Li, Marie-Françoise Gouzy, Jürgen-Hinrich Fuhrhop

Freie Universität Berlin, FB Biologie, Chemie, Pharmazie,
Institut für Chemie/Organische Chemie, Takustrasse 3, 14195 Berlin, Germany
E-mail: fuhrhop@chemie.fu-berlin.de

The article deals with carbohydrate interactions in aqueous media. Carbohydrate head groups of glycolipids not only form the hydrated surface of micelles and vesicles. They also stick together in hydrogen-bonded cycles and attach to each other via stereoselective hydrophobic interactions. Even in water supramolecular structures are thus formed, e. g., quadruple helices or chiral boron ester fibers. Such structures collapse on addition of the enantiomers to form planar bilayer or monolayer crystals ("chiral bilayer effect"). Furthermore, the length of a carbohydrate head groups often determines the curvature of molecular assemblies in water and may drastically change the relative surface areas. In water-amphiphile interfaces hydrogen bonds are found experimentally to be more stable by several orders of magnitude as compared to bulk solution. A theoretical model, which implies a change of "electrostatic potential", gives only about a factor of four. Boronate groups on the surface of hydrophobic liquid crystals or porphyrin macrocycles also interact strongly with monosaccharides. Most surprisingly and possibly important is the recent finding, that pyranoses and model compounds stick to hydrophobic nanometer gaps and do not equilibrate with bulk water over periods of days. Some new results on water-carbohydrate interactions in glycoprotein and cyclodextrin cavities are also briefly discussed for a comparison.

Keywords. Stereoselective recognition, Glycolipid assemblies, Saccharide entrapment, Water-filled nanometer pores

Topics in Current Chemistry, Vol. 218
© Springer-Verlag Berlin Heidelberg 2002

1
Introduction

An article on "Amphiphilic Carbohydrates as a Tool for Molecular Recognition in Organized Systems" by P. Boullanger appeared 1997 in the same series [1]. It described the properties of amphiphilic carbohydrate derivatives with respect to spherical vesicles, micelles, and liquid crystals as well as recognition processes of oligosaccharides that are bound to glycoproteins or fixated in monolayer surfaces. Nothing was written there about micellar fibers made of open-chain saccharides, chirality effects, or solid state NMR studies which characterize conformational changes in crystals and supramolecular assemblies with respect to solutions. We shall discuss these subjects and give an overview of recent results and the techniques applied to characterize the molecular complexes and assemblies containing carbohydrate amphiphiles. This applies particularly to water, which is bound in hydrophobic membrane gaps and interacts with dissolved saccharides. The extraordinary stability of hydrogen bonds at interfaces is also discussed. A few comparisons between recognition processes in cyclodextrins or lectins with those in micellar assemblies will be given.

2
Carbohydrates in Water-Filled Hydrophobic Membrane Gaps

Molecular dynamic calculations on glucose in aqueous solution indicate, that water molecules flow around the *equatorial* OH groups in a disorderly manner, whereas the *axial* OH group of in the α-anomer tends to bind water more tightly [2–4]. Solvation effects, in particular hydration, effect the rotamer distributions of the CH_2OH group, but the calculated hydration sphere does not support the idea, that the environment of β-glucose should resemble a fairly ordered structure, e.g., the tridimite ice lattice based on a tetrahedral binding of four hydrogen atoms to oxygen or pentagonal water with one side of the oxygen atoms interacting with hydrophobic sites [5,6]. As long as the carbohydrates are in bulk aqueous solutions, no particular effects of the carbohydrates become apparent. No particular gelation effects of mono- or disaccharides in water are known. Gelation by helical starch assemblies just follows the same rules as observed for helical proteins such as collagen or nucleic acids: the fibers form a coherent net by sticking together at crossing points and disentangle at temperatures around 70 °C.

Nevertheless, β-glucose and β-cellobiose fit perfectly into the hexagonal ice structure. This became evident experimentally in hydrophobic volumes of a few cubic nanometers, when 0.1 mol/l glucose solutions were trapped in 1–2 nm wide hydrophobic membrane gaps with a steroid [7] or porphyrin [8–10] bottom. Such gaps on gold electrodes are perfectly permeable to ferricyanide ions, when filled with water, as was shown by cyclic voltammetry. If, however, a 0.1 mol/l molar solution of cellobiose or ascorbic acid was filling such membrane gaps, no ferricyanide ion could pass. Furthermore, and most important, both water-soluble compounds were not washed out of the nanometer-cube of water entrapped in membranes and in full contact with bulk water volumes without carbohydrates. No equilibration took place within several hours. If, however, 1 mol% of maltose was added to the cellobiose, no blocking effect was observable. It is thought that pyranoses with only *equatorial* substituents adsorb to the hydrophobic membrane surface, forming a stable, well organized surface monolayer there, a process which takes hours before it is complete; they cannot be removed by water and also immobilize surface water (Fig. 1).

Ascorbic acid has the same effect. Since it does not fit into hexagonal (= diamond-like) ice structures, it was proposed that water on hydrophobic surfaces may also appear in pentagonal clusters [5, 6], such as found in crystalline clathrates of sulfonium salts with bulky hydrocarbon surfaces [5]. The effect of carbohydrate adsorption on alkane surfaces in presence of water has to the best of our knowledge not been studied so far.

A related problem concerns the solidification of water on the surface of antifreeze proteins (AFPs). Such proteins bind to specific ice surfaces and thereby inhibit the growth of ice particles. In a similar manner these proteins also bind to carbohydrates [11–13]. In many cases Ca^{2+}-ions are involved in both ice binding and carbohydrate binding. Neither the binding site for water in AFPs nor the water structure within the hydrophobic nanometer-gaps is, however, known yet.

There are no specially stable or rigid hydration spheres in bulk water solutions of carbohydrates, but close to hydrophobic surfaces both the carbohydrates and the water molecules become exceedingly immobile (see also Sect. 8).

Since several natural products provide large hydrophobic surfaces, e.g., steroids or porphyrins, one may also expect strong binding interactions between these molecules and saccharides in an aqueous environment. To the best of our knowledge, this has not been verified so far.

ms-Tetranaphthol porphyrins adsorb monosaccharides selectively. However, this occurs only in chloroform solution and circular dichroism (CD) effects are small [14]. A recent review on carbohydrate recognition cites many receptor systems, which work in chloroform or methanol, but no receptor takes non-amphiphilic carbohydrates in water [15]. Charged boronate groups are needed to bind monosaccharides in water to such molecules (next section; see Fig. 7 and Table 1).

3
Micellar and Vesicular Tubules with Carbohydrate Surfaces

If glyconolactones are opened with long-chain amines, one obtains glyconamides with all asymmetric centers located in a flexible chain. Figure 2 gives the three most important examples, namely *N*-alkyl glucon-, mannon-, and galactonamides. In DMSO their conformation is dominated by the *gauche* effect and 1,3-repulsion. The *gauche* effect means that neighboring (= 1,2)-OH groups favor a *gauche* orientation. The repulsive 1,3-effect, on the other hand, causes deformation of the *all-anti* conformers by steric repulsion [16]. Recent echo double resonance experiments on the solid fibers using ^{15}N- and ^{13}C-labeled gluconamide have established a $_2G$-sickle [16c].

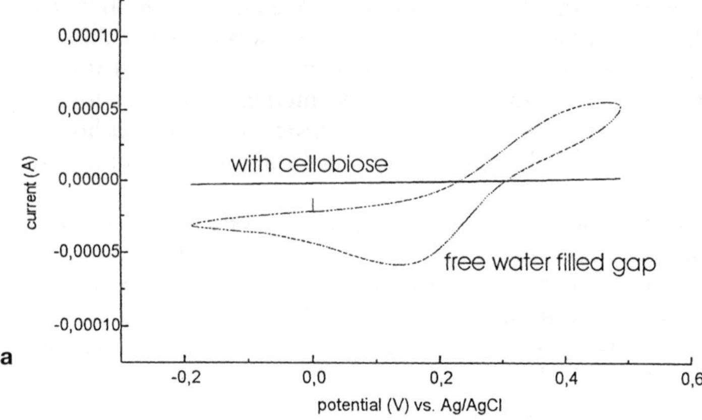

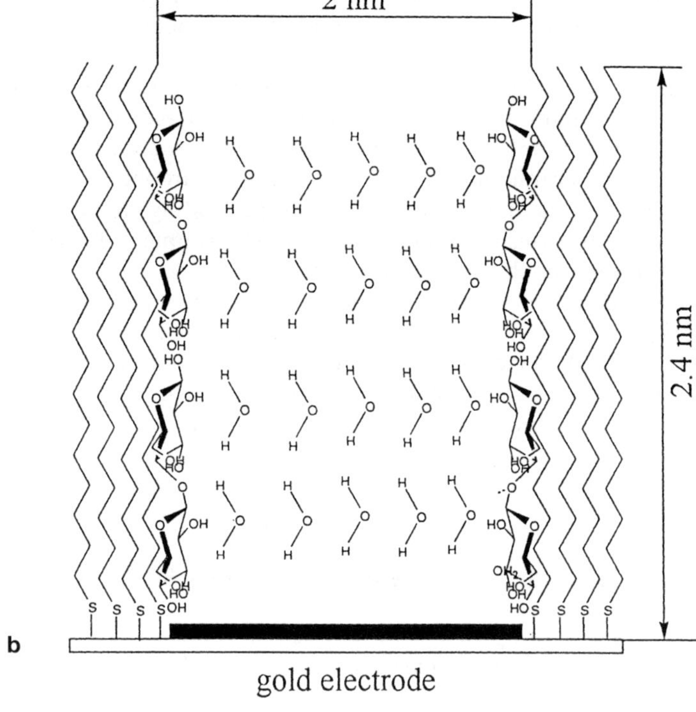

Fig. 1 a, b. a Typical cyclic voltammogram of ferricyanide using a membrane coated gold electrode with 2-nm gaps before and after addition of cellobiose. **b** Model of the cellobiose coating. About 6–10 molecules of cellobiose are present within one gap [10]

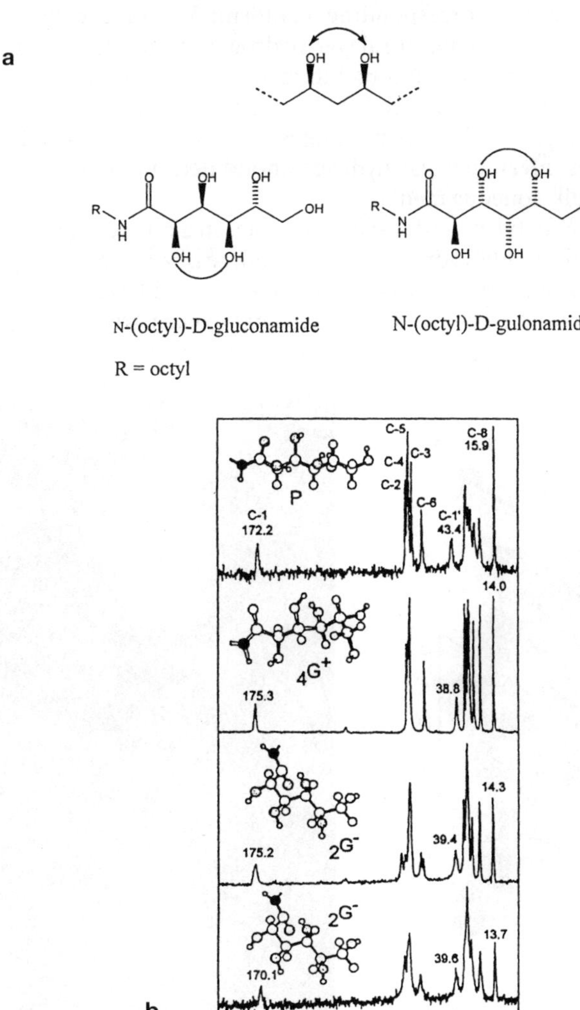

Fig. 2a,b. **a** Model of the gauche and 1,3-repulsion effects. **b** Typical conformers of open-chain *N*-octyl-glyconamides in different environments and the corresponding CP-MAS ^{13}C-NMR spectra [16a–c] from top to bottom: gluconamide and D-gulonamide crystals, D,L-gluconamide crystallites and D-gluconamide fibers. The extraordinary chemical shift of the terminal methyl group (= (8)) in the top spectrum indicates a head-to-tail arrangement of the crystal. All other assemblies are tail-to-tail. None of the solid materials contains water

In crystals, homodromic hydrogen bond cycles between the OH-groups have been found to enforce tight packing as well as high-energy *all-anti* conformers [17–19]. In aqueous solution, however, fast formation of linear amide hydrogen bond chains leads to the formation of fibers and a large variety of molecular conformations as obtained by solid state CP-MAS ^{13}C-NMR has been correlated with the structure of supramolecular assemblies as characterized by transmission electron microscopy (TEM) coupled with computerized image analysis. Figure 3 displays the most spectacular case of a quadruple helix of N-D-octyl-gluconamide and the corresponding conformation of the monomer. Most important is a homodromic intralayer hydrogen bond cycle, which connects molecules and supports the binding interactions between the secondary amide groups. In bulk water, the helices stick together at crossings and form a gel [18–27]. Other glyconamides form tubules, twisted ribbons, or rolled-up sheets [26]. In all cases intermolecular hydrogen bond determine the supramolecular structures in bulk aqueous media.

If one mixes N-octyl-D- and -L-gluconamides in a ratio of 1:1, then no helices or gels are found. Highly hydrated semicrystalline bilayers precipitate. In case of a *pseudo-racemic* mixture of two gluconamides with different chain lengths, one first obtains left- and right-handed helices next to each other, but again within a

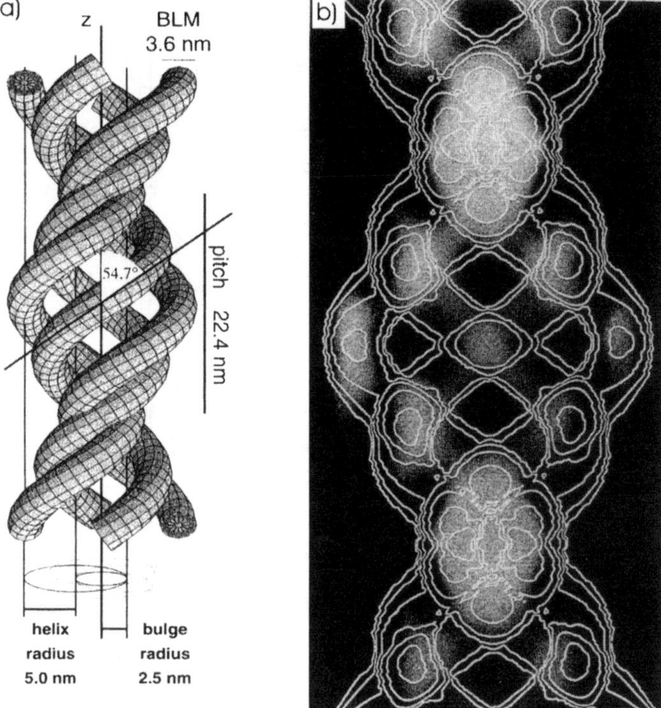

Fig. 3a,b. a Molecular model of the helix. **b** Solid state electron micrograph and image analysis of an N-octyl gluconamide quadruple helix as obtained by self-organization in water [17a,b]

few minutes *pseudo-racemic* platelets are found [25, 26]. The pure enantiomers tend to form head-to-tail crystals with all chiral centers in identical orientation [20–22]. Since the rearrangement of the tail-to-tail orientation in the fibers to head-to-tail in crystals is highly unlikely in water, the fibers with very large surface energies are stable enough to survive for several months under appropriate conditions ("chiral bilayer effect") (Fig. 4). This effect in chiral membranes is fundamental to the stability of aqueous gels and thereby to life, since living systems are predominantly made of chiral fibers (proteins, cellulose, DNA) in aqueous environments.

Similar effects are also found in organic solvents ("organogels") if one applies less polar derivatives, but the molecular assemblies are always multilayered and ill defined. Their formation is purely entropy-driven [28]. Dibenzylidene-D-sorbitol in glycol-water mixtures is a commercially used antiperspirant gel [29] (Fig. 5).

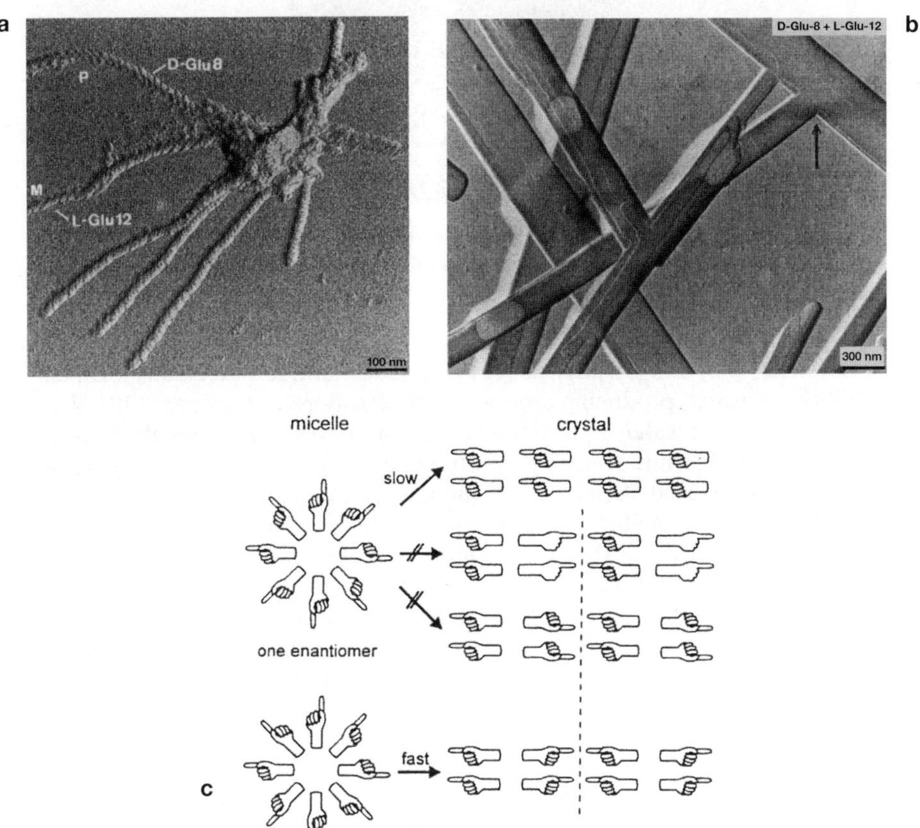

Fig. 4a–c. The chiral bilayer effect: **a** left and right-handed helices are formed at first from *N*-octyl-D and *N*-dodecyl-L-gluconamides; **b** after an hour or so pseudoracemic crystallites are formed from the helices – the original gel has decomposed; the suspended helices are replaced by precipitating crystallites; **c** model of the chiral bilayer effect – chiral amphiphiles crystallize only slowly to head-to-tail crystals – fibers are therefore long-lived [17, 25]

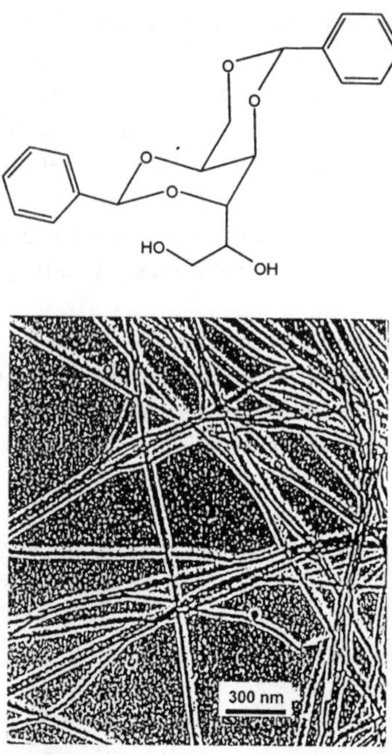

Fig. 5. Electron micrograph of dibenzylidene-D-sorbitol fiber from glycol-water mixtures as a typical example for organogels. The fibers have a thickness of about 400 nm and no fine structure [29]

Carotenes [30] and porphyrins [31] with gluconamide side-chains form chiral fibers in water, producing strong CD-effects. Boronate anions – $B(OH)_3^-$ – react with diols in water to form molecular complexes. pH-Values above 10 are usually required. If, however, amphiphilic copolymers with neighboring ammonium groups are used, then a surface monolayer efficiently adsorbs mono- and disaccharides from the bulk solution even at pH 6–8. The lower limit of detection by measurements of the limiting molecular area of the copolymer is in the order of 10^{-5} mol/l [32] (Fig. 6).

The enlargement of the area by monosaccharides occurs in the order none < Glc < Fru < Gal < Man, which suggests that the position of OH-groups is responsible for the binding to the boronic acid. The known order for phenylboronic acid groups is *α,γ-trans* ≫ *α,β-trans*. Glucose is usually least efficient. In case of lipid AD, a copolymer of a methacryl amide with an acryl amide bearing phenyl-boronic acid substituents, the monosaccharides Glc, Gal, and Fru decreased its surface area at alkaline pH. This points presumably to the formation of a 1:2 complex or bridging of two phenylboronic acid groups by one of given monosaccharide molecules [33].

An amphiphilic bis-pyridinium porphyrin with two phenylboronate substituents adsorbs carbohydrates in water containing a small amount of methanol at pH 10. Strong CD-effects are observed in the Soret band region (Fig. 7). Depending on the direction of the 2-OH group (up or down) of the

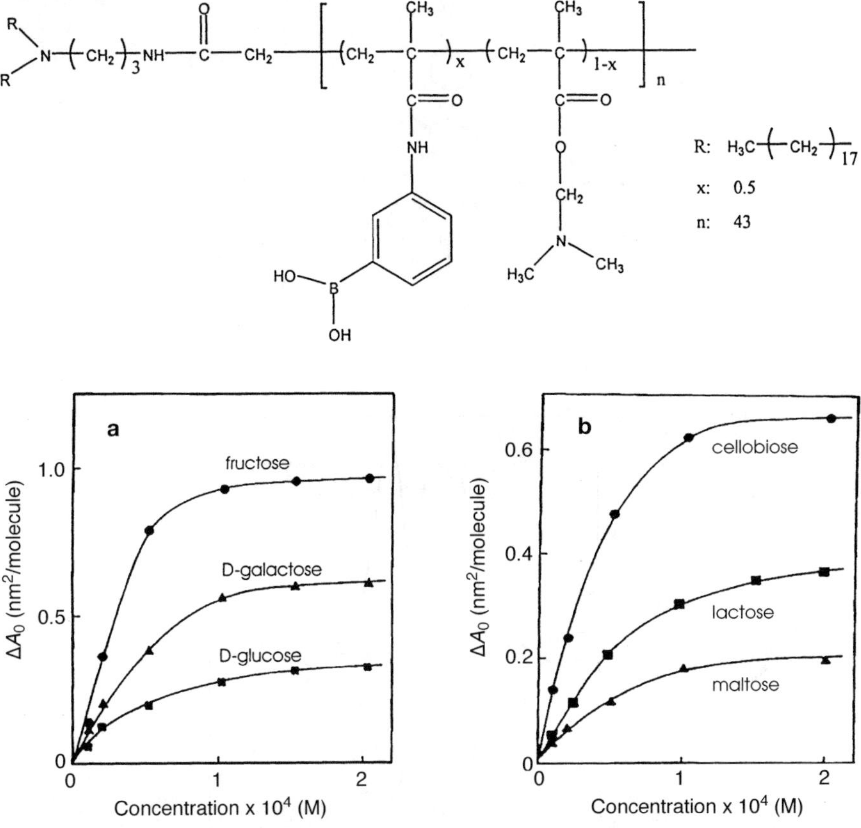

Fig. 6 a, b. Surface pressure-area isotherms of boronate polymers in presence of carbohydrates in the subphase. The monolayers expand upon addition of: **a** monosaccharides; **b** disaccharides [32]

monosaccharide, positive or negative exciton coupling bands were observed; similar effects are known with peptidic porphyrins [34–36] (Table 1).

Aldosylamines, e.g., the lactose derivative given below, coordinate to copper(II)-ions in bulk water and giant vesicles ($\varnothing$ ~2 μm) are formed (Fig. 8a), whereas the metal free amphiphiles give much smaller microvesicles and lamellar aggregates [37]. Pentasaccharides, as head groups on the other hand, produce large circular platelets in bulk water (Fig. 8b). The extension of the head groups does not allow concave curvature of the vesicle interior. Only micelles, which contain exclusively concave surfaces or planar sheets, are allowed [38].

The structural effects of carbohydrate head groups in molecular assemblies are thus manifold:

1. Secondary amide groups lead to hydrogen-bond chains producing extended non-covalent fibers.
2. Bends in open-chain carbohydrate head groups enforce helices.

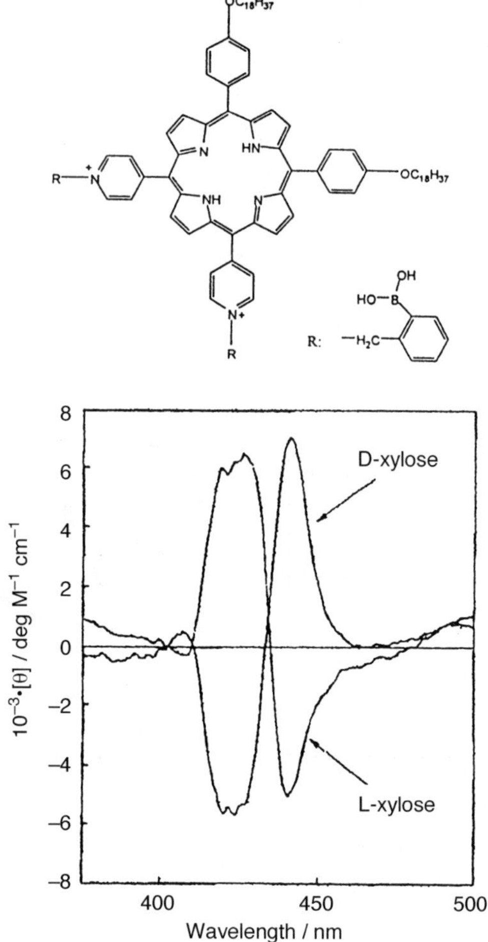

Fig. 7. CD-spectra of water-soluble boronate porphyrins in the presence of D- or L-xylose [34]

Table 1. Changes of CD-bands of porphyrin liquid crystals upon addition of aqueous monosaccharide solutions

Monosaccharide	Direction of 2-OH	λ_{min} ([θ]) of Cotton effects [a]
D-Fructose	Up	426 ($+1.0 \times 10^4$)
L-Fructose	Down	425 (-5.7×10^3)
D-Xylose	Down	427 (-5.7×10^3)
L-Xylose	Up	427 (-6.4×10^3)
D-Glucose	Down	410 (-4.0×10^3)
L-Glucose	Up	410 ($+3.4 \times 10^3$)
D-Ribose	Down	432 (-2.1×10^4)

[a] 0.3 vol.% methanol, pH 10.0, with 0.10 mol/l carbonate buffer, [monosaccharide]
= 1.00×10^{-2} mol/l

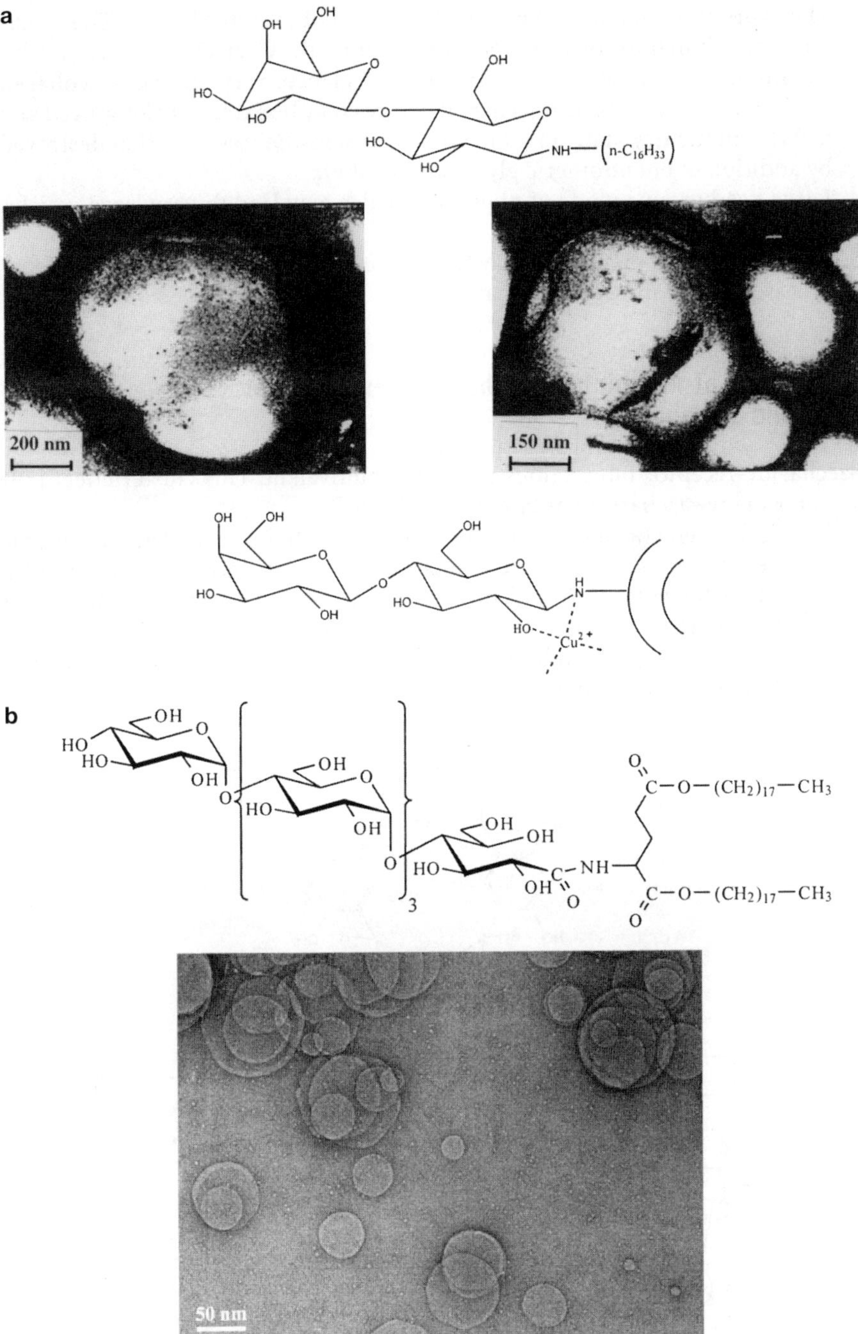

Fig. 8 a, b. **a** *N*-Hexadecyl-lactosylamine gives giant vesicles in water upon addition of copper(II) ions [37]. **b** A glucose pentasaccharide as a head group does not allow a concave vesicle surface because of steric overcrowding. Planar disks are therefore formed in water [38]

3. The hydrophobic effect induces the formation of multiple helices. These may look very similar to covalent polyamides under the electron microscope. Gluconamide quadruple helices look, at first glance, very similar to collagen triple helices. Such chiral micellar structures may be extremely long-lived and isolable in the dry state. In aqueous suspensions helices are often destroyed by addition of enantiomeric glyconamide fibers.
4. Large head groups made of oligosaccharides may lead to very large vesicles or even planar disks.
5. Complexation of metal ions may then enforce more curvature in the molecular assembly by introducing charges.

4
Monolayers of Amphiphilic Carbohydrates on Water and Solid Subphases

Saccharide-receptor interactions are often multivalent. This cluster effect [39] requires many saccharides as lateral neighbors.

Rigid calixarene LB-monolayers with long alkyl chains fixated at the indium-tin oxide (ITO) electrode/water interface bind various monosaccharides from bulk water with high and differentiated binding constants. Changes of electrode potentials are detected at carbohydrate concentrations in water as low as 10^{-5} mol/l (Fig. 9). In bulk organic solvents the same binding process is dominated by hydrogen bonding [40]; on electrode surfaces in aqueous envi-

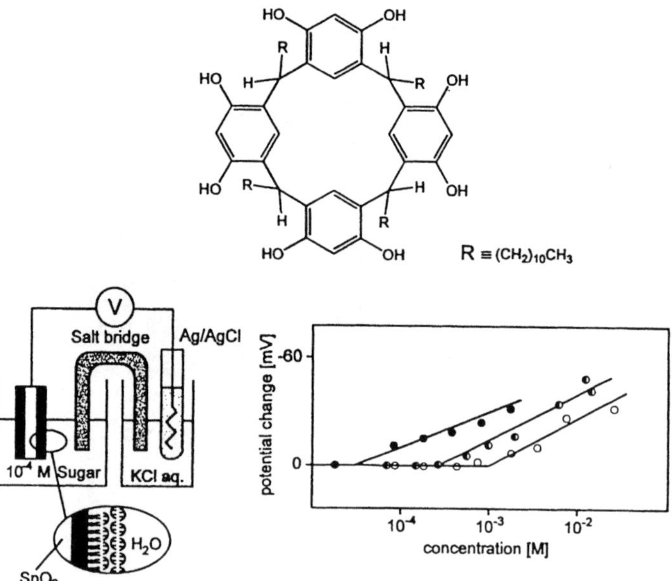

Fig. 9. Amphiphilic calixarenes form monolayers on tin oxide electrodes, which then differentiate between water dissolved monosaccharides (*from left to right:* ribose, galactose, glucose) [40]

ronments the special polarity of the electrode/water interface is presumably responsible for the favorable adsorption of the carbohydrates [41].

The hydrophobic calixarene was substituted with ethylamine on the phenol units and condensed with lactose lactone to give a polyp-like structure. This compound is tightly adsorbed by quartz surfaces. Water cannot remove it, but aqueous amine solutions lead to desorption. The experimental occupation area is 3.5 nm^2/molecule and saturation binding is achieved within 10 s. The cluster compound obviously forms a closely packed monolayer on quartz. The quartz-adsorbed polyp binds 8-anilinonaphtalene-1-sulfonate (ANS) from water with a binding constant of 2×10^5 M^{-1}, which was shown by fluorometric titration at 479 nm (Fig. 10) [42, 43]. It has also been shown that water-soluble monosaccharides are tightly bound to hydrophobic gaps of steroid size in LB-monolayers [7, 10] (see Fig. 1).

$(1 \rightarrow 3)$-β-D-Glucans which are substituted on every third unit by β-$(1 \rightarrow 6)$-D-glucose ("scleroglucan") produce triple helices with a pitch of 1.8 nm and six backbone glucose residues per helical turn in each strand. The triple helix is very stiff and the glucans are useful as aqueous viscosity control agents. The triple helix decomposes in DMSO to form a single strand coil. Renaturation in water is incomplete and leads to circular polysaccharide structures with diameters of many nanometers [44], which are reminiscent of DNA toroids [45]. The proposed mechanisms of formation, namely formation of linear clusters and subsequent surface diminution by rolling-up, are also similar, the glucose substituents playing the role of the main-chain connecting link (Fig. 11).

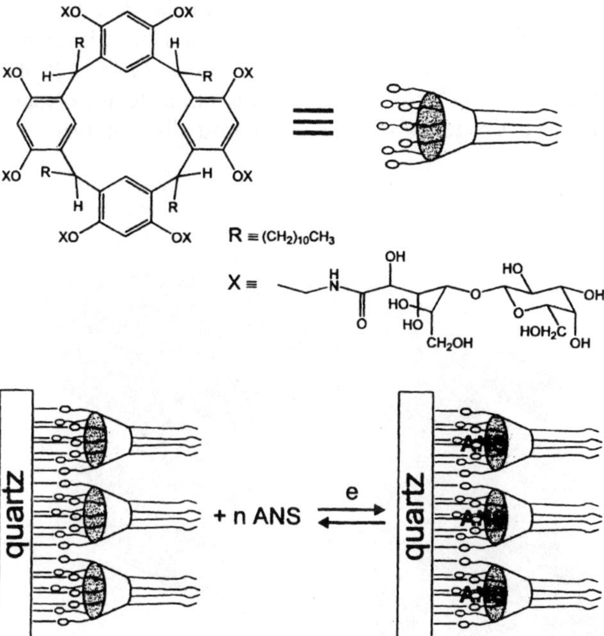

Fig. 10. The calixarene polyp with eight galactose arms binds tightly to quartz surfaces and binds the water-soluble 8-anilinonaphthalene-sulfonate (ANS) fluorescence dye [42]

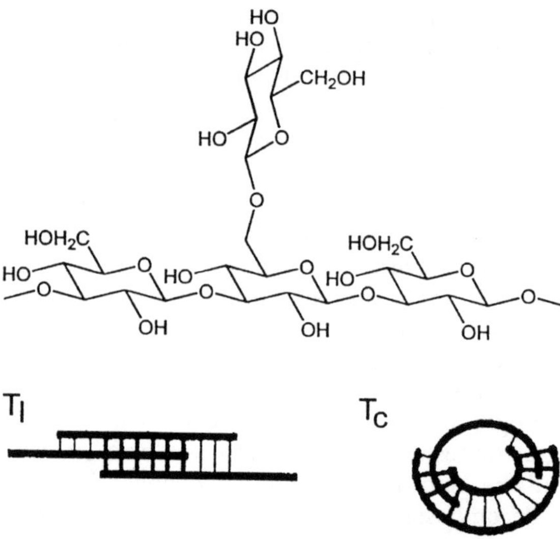

Fig. 11. Repeating unit of scleroglucan and a model of a triple helix in water and of a toroid formed after denaturation in DMSO and renaturation in water [44]

Cellulose fibers, which are partially hydrophobized by trimethyl silyl substituents as well as cellulose itself, align on water surfaces to form two-dimensional crystals, which are quite regular in the direction of the polymer chain. In case of alkylated cellulose ("hairy rods") multilayer grids of molecular size can be synkinesized [17b], if the direction of the LB-plate is changed by 90° in each application [46] (Fig. 12). Similar cholesteric polysaccharides have also been used as hydrogel nanoparticles to bind albumin protein in water [47].

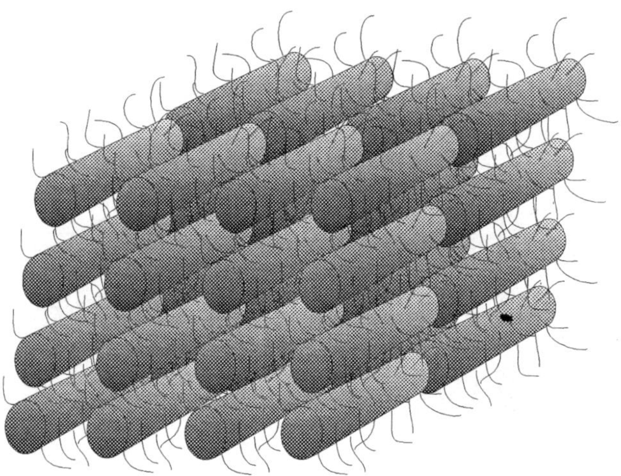

Fig. 12. Molecular grid made of fatty acid esters of cellulose (hairy rods) [46]

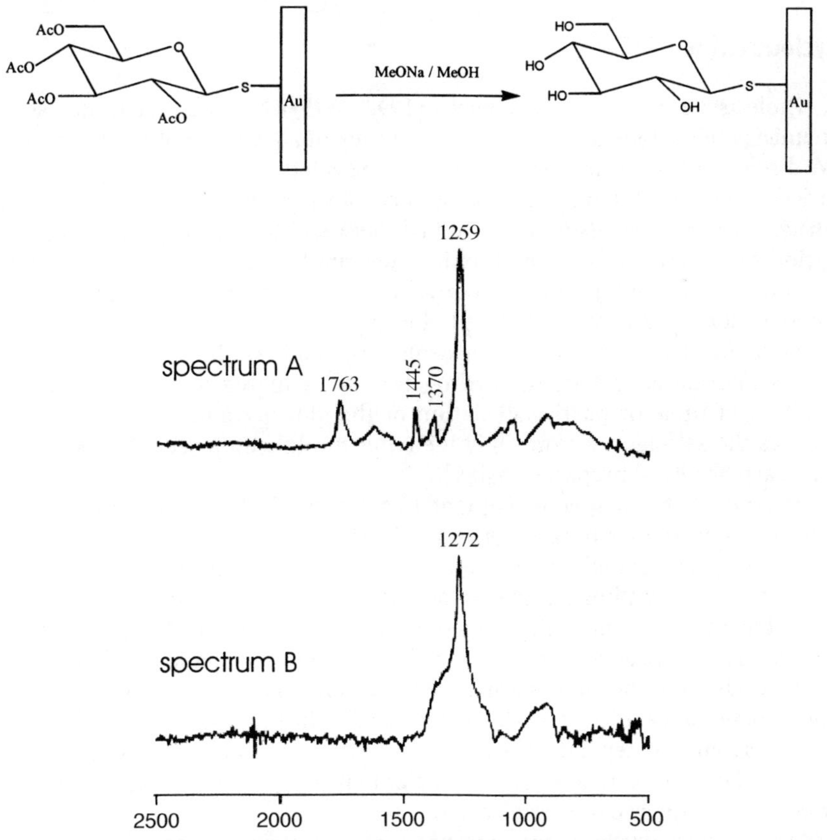

Fig. 13. Self-assembled monolayers of glucose pentaacetate on gold have been saponified without detachment as shown by the intensity of the 1260–1270 cm^{-1} IR-absorption [48]; spectrum A – acetate 1763 cm^{-1}, spectrum B – practically change of C-OAc (1259 cm^{-1}) to C-OH (1272 cm^{-1}), but no change of intensity

Thio-Glc, thio-Glc(ac)$_4$, and thio-manno have been bound covalently by self-assembly monolayers (SAM) to smooth gold surfaces. The tetraacetate was saponified in situ without significant loss of the monolayers (Fig. 13). Exposure of the mannose monolayer to Con A led to a specific binding interaction as shown by surface plasmon resonance [48].

These observations on the recognition processes and reactions on planar surfaces show that dipolar interactions are strongly enhanced with respect to reactions of small molecules with Con A in bulk volumes. Nevertheless, the acid-base reactions with reagents in the bulk solution are hardly slowed down by repulsive charge interactions or hydration forces. Even interactions between polymers occur quickly and there is no special protection against the attack of reactive ions. This situation is very different to the one observed in hydrophobic membrane gaps filled with water, which are disconnected from the bulk phase (see Sect. 2).

5
Cyclodextrins

A whole issue of Chemical Reviews (1998, 98(5)) has been dedicated to the chemistry of cyclodextrins and gives a survey of all aspects of their chemistry. We discuss only a few more recent results on cyclodextrin complexes in water in order to compare them to micellar systems. Recognition processes in bulk and surface water have also been studied between the hydrophobic cavity of cyclodextrins and various amphiphilic guest molecules. Only a few new results in aqueous solutions are given in order to discuss this important hydrophobic cavity in context with other hydrophobic gaps.

γ-Cyclodextrin is the most water-soluble of all cyclodextrins (23 g/100 ml), but its solution is hazy due to hydrogen-bonded aggregates. Rise of pH to 10, addition of urea, or partial alkylation of the glucopyranose hydroxyl groups reduces the self-aggregation. Only the latter modification seems to be feasible for pharmaceutical preparations[49].

^{1}H- and ^{13}C-NMR spectroscopy of 1S,4S- and 1R,4R-camphor in D$_2$O solutions of α-cyclodextrin have, for example, shown that two cyclodextrin molecules entrap one camphor molecule with a free energy difference of 0.3 Kcal/mol. The 1S,4S-camphor complex is more stable ($\Delta\Delta G = -8$ kcal/mol) [50]. D- and L-leucine give an enantioselectivity of 33 or a $\Delta\Delta G$ of about 2 kcal/mol [51]. For a large variety of substituted acids [52] and various aromatic and heterocyclic compounds [53], thermodynamic measurements have been performed and a phenomenological theory has been developed. The binding energy to cyclodextrins was found to depend strongly on ΔA of the guest, where ΔA is the decrease in nonpolar surface area (in Å^2/molecule) that is exposed to solvent before and after entering the cyclodextrin cavity.

Another new development concerns surface recognition processes by cyclodextrin sulfides on smooth gold surfaces. Plasmon surface polariton spectroscopy and cyclic voltammetry (Fig. 14) allow at first directly to measure the self-assembly kinetics of the thiolated cyclodextrins, which occurs in a three-step process: physisorption, binding/orientation, and finally adlayer formation [54].

Replacing the primary alcohol groups on the cyclodextrin rim by 2-hydroxy ethylamino groups allows anionic head groups, e.g., phosphate, to bind efficiently to the surface and to bury a hydrophobic core in the cyclodextrin center [55]. If a water-insoluble polymer, e.g., an acrylate, carries methyl or butyl side chains, α-cyclodextrin may entrap these alkyl groups and solubilize the polymer efficiently [56].

Comparable SAMs have also been prepared on silver colloids. The thiolated cyclodextrins were then applied to bind an azo-dye, which changed its color upon complexation. It was found that the equilibrium constant on the colloid surface was larger by a factor of about 5×10^3 than that of the free system in aqueous solution [57] (Fig. 15). The cyclodextrin gap again reacts differently from the hydrophobic membrane gaps discussed in Sect. 2. Exchange reactions are fast; there is very little hydration water involved in complexation. Cyclodextrin monolayers also act, however, as efficient blockers for ion transport to electrode surfaces.

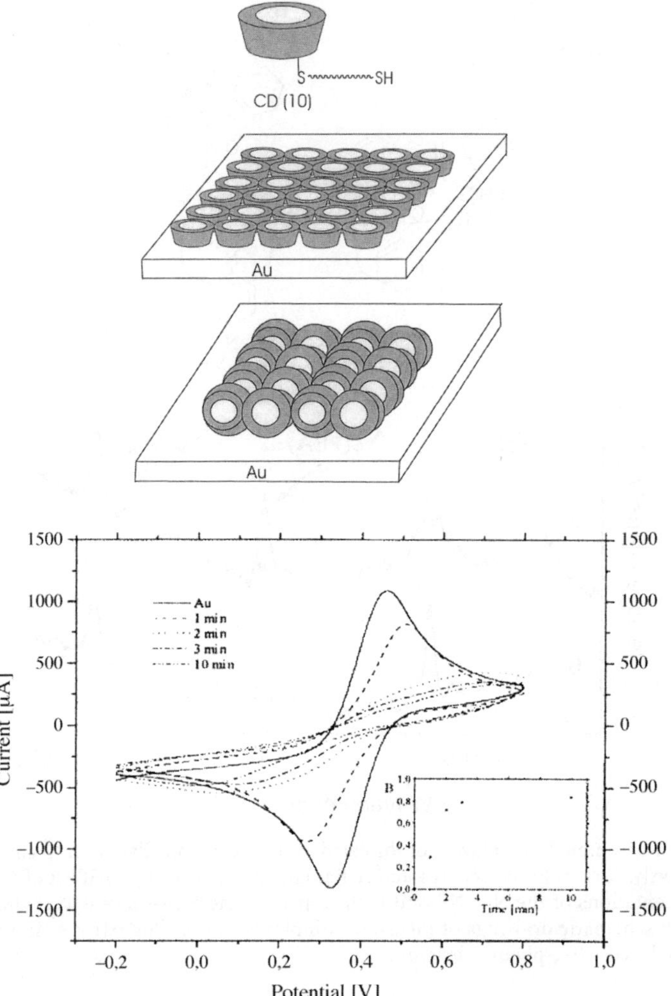

Fig. 14. Cyclic voltammogram of ferricyanide ions in bulk water during the self-assembly of cyclodextrin-hydrosulfide on gold [54]. *Insert:* calculated hindrance B vs self assembly time

A major effect of cyclodextrins consists in separating an aqueous solute from its hydration sphere by "fitting". A related effect has also been observed in simple micelles with an oversized head group region. Dodecylmalono-bis-*N*-methylgluconamide is an amphiphile with a wide head group and a narrow hydrophobic core. Upon addition to SDS micelles, it expels 12–24 water molecules/mol from the polar shell to a degree where the mixed micelles become "dry", i.e., the number of water molecules per surfactant molecule falls below that needed to hydrate the sodium counterions. As a result a massive decrease of polarity was observed as characterized by changes of hyperfine splitting constants of a nitroxide label [58–60] (Fig. 16).

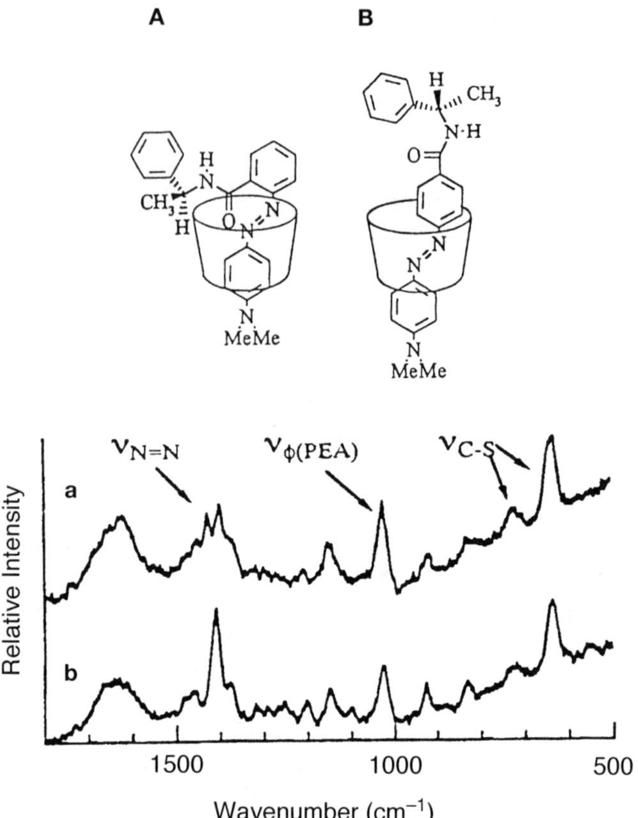

Fig. 15. Self-assembled monolayers of thiolated cyclodextrin on silver entrap chiral dyes stereoselectively. Surface Enhanced Resonance Raman Spectra (SERR). a) SERR of A, b) SERR of B. The orientations of the N = N bond with respect to the Ag-surface is very different [57]. A) and B): schematic drawings of inclusion complexes in α-cyclodextrin – A: o-methyl red, B: p-methylyzed-(S)-1-phenylethylamine

6
Lectins and Glycolipids

"Fitting by dehydration" is also found in protein-carbohydrate complexes. The best-characterized protein-carbohydrate interactions occur in plant lectins. Other carbohydrate-binding proteins include immunoglobulins and enzymes.

Lectins recognize terminal monosaccharide units in covalent polymers and in non-covalent fibers only connected by hydrogen bond chains or in vesicles stabilized by the hydrophobic effect. Extremely large assemblies, such as proteins or vesicles, are accommodated. 4-Deoxy or 3-deoxy derivatives of the recognized sugar moieties have little binding activity to lectins. It is this chiral center, the edge opposite to the glycosidic site, which points to the bulk water phase and is responsible for molecular recognition. Crystallography shows that in a

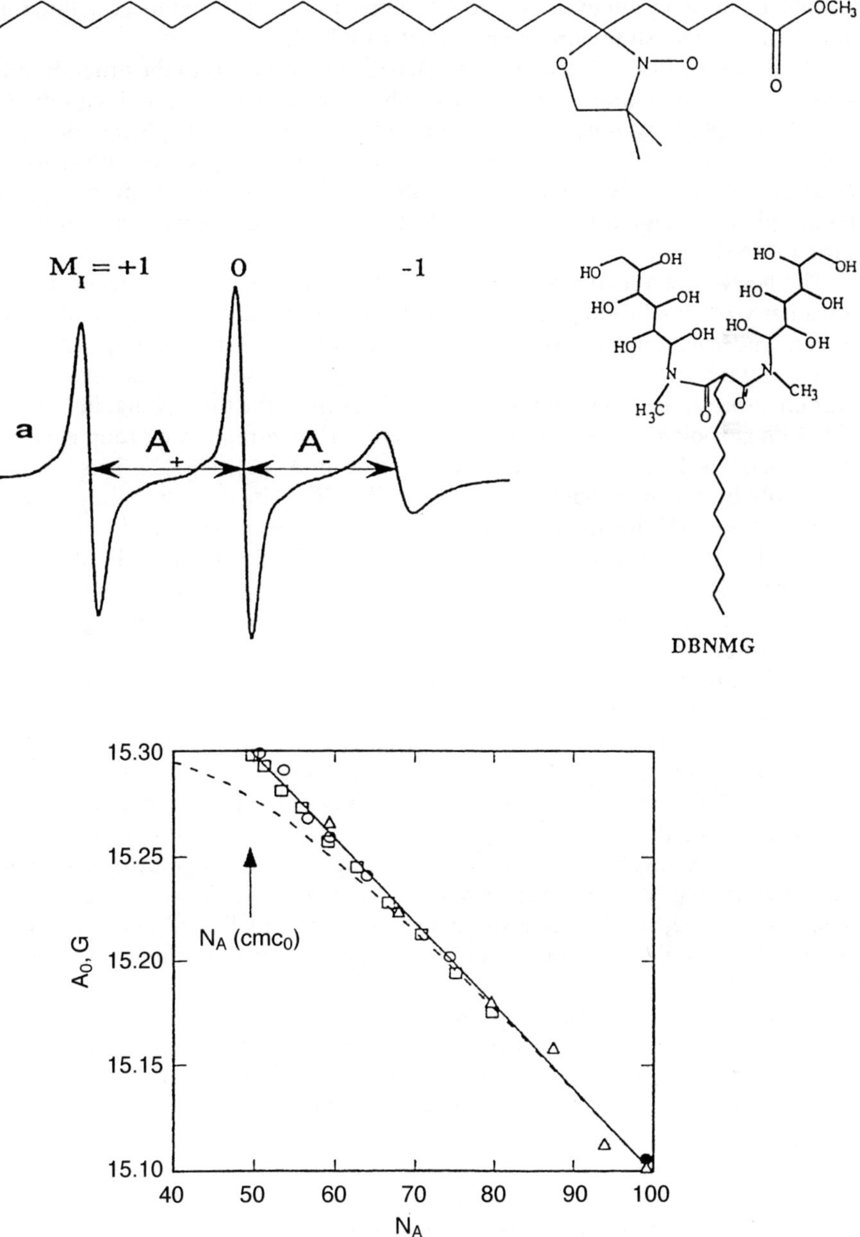

Fig. 16. ESR spectrum of a nitroxide labeled amphiphile upon "drying" of a carbohydrate-derivative (DBNMG) micelle by SDS addition. The coupling constant becomes smaller by 0.2 Gauss [58–60]

mannose binding protein the 3- and 4-OH groups directly bind to a calcium ion and form an extensive hydrogen bond network [61].

The binding affinity of lectins to saccharide monomers is in the order of millimolar. It rises to nanomolar when carbohydrate clusters are provided [39, 40, 61]. The simplest example is the smooth surface of vesicles with glucose or mannose head groups [62] or with glutamate based trivalent glycosides with spacers. Most of the lectins also agglutinate erythrocytes, which gave them their early name "phytohemagglutinines". Some lectins are glycoproteins, some are pure proteins [63].

The best-known lectin is concanavalin A and comes from soybeans. It is a tetramer with one binding site each for two carbohydrate molecules and manganese and calcium ions. More than one half of Con A is made of β-pleated sheets and it has been proposed that these sheets can also stereoselectively bind carbohydrate assemblies. This argument comes from the finding that lipid vesicles with carbohydrate and with tetraacetate carbohydrate head group agglutinate equally well and equally selectively with Con A [62] (Fig. 17). This finding can easily be accommodated with a planar binding site corresponding to a β-pleated sheet [62] not with a narrow cleft, where four protons and four acetyl groups should be clearly distinguishable. Con A also binds to glycolipid monolayers on gold surfaces with two binding sites [64]. X-ray studies of Con A-mannose complexes [65] and calorimetric studies [66] also indicate participation of several water molecules in lectin-carbohydrate binding.

S-Type animal lectins (galectins) have been crystallized in free form, where five water molecules are bound to the binding site, and with galactose, galactosamine and lactose, where the carbohydrate replaces water. Lactose co-crystals were obtained by soaking native protein crystals with 5 mmol/l lactose, cocrystals with galactose were grown with the addition of 50 mmol/l galactose to the crystallization buffer. In hGal-7 five amino acids (His 49, Asn S1 and G2, Arg 53, and Glu 72) are involved in carbohydrate recognition through hydrogen bond interactions. Trp 69 may be involved in stacking interactions. In the case of galactose O1, O2, and O3 are also involved in water-medicated interactions, in the lactose complex the glucose unit replaces the water on the surface [65c] (Fig. 18).

X-ray analysis of crystalline proteins cannot detect disordered water localized in hydrophobic gaps. ^{1}H-NMR experiments showed, however, a high occupancy with water and residency times from nanoseconds to 200 μs [67].

From the few recent examples given for carbohydrate-protein complexes, one may conclude that the surface of the binding site for monosaccharides is also highly hydrated, even in crystals containing the carbohydrate guest. If, however, a disaccharide binds, then the second carbohydrate unit sticks out of the enzyme gap and does not retain any water molecules in the solid state. Enzyme gaps for monosaccharides are just fitting one molecule and they produce a strong dipolar environment, which attracts water molecules. Cellobiose and related compounds fitting into ice-like water assemblies totally immobilize the entrapped water volume for days [7 – 10].

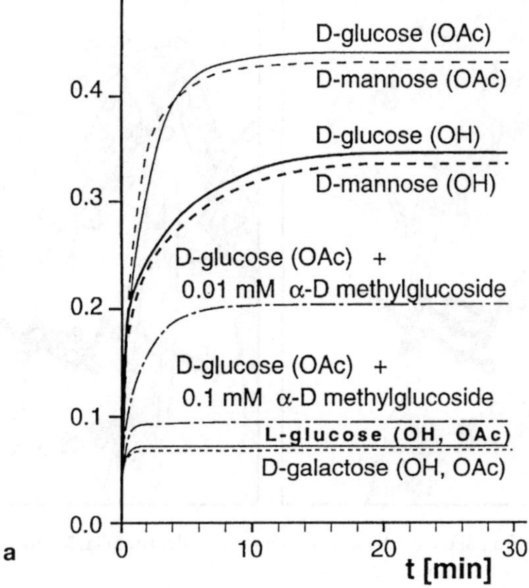

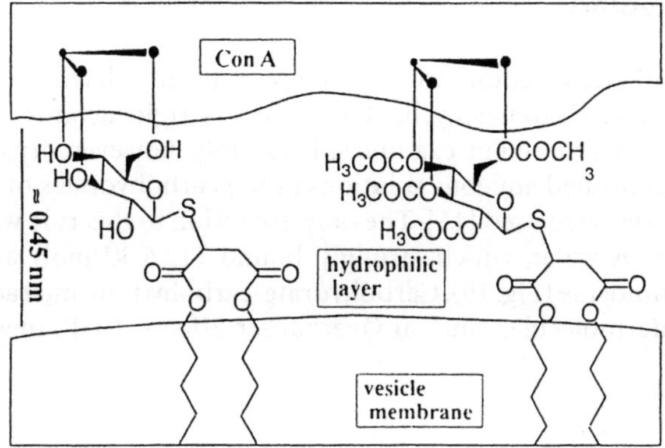

Fig. 17a, b. a Vesicles made of double chain glycosides are precipitated stereoselectively by Con A, the turbidity of the solution rises within a few seconds. L-Glucose and D-galactose show no response, but the tetraacetates of D-glucose and D-mannose are recognized like the parent alcohols. **b** Model of a planar binding site on Con A [62]

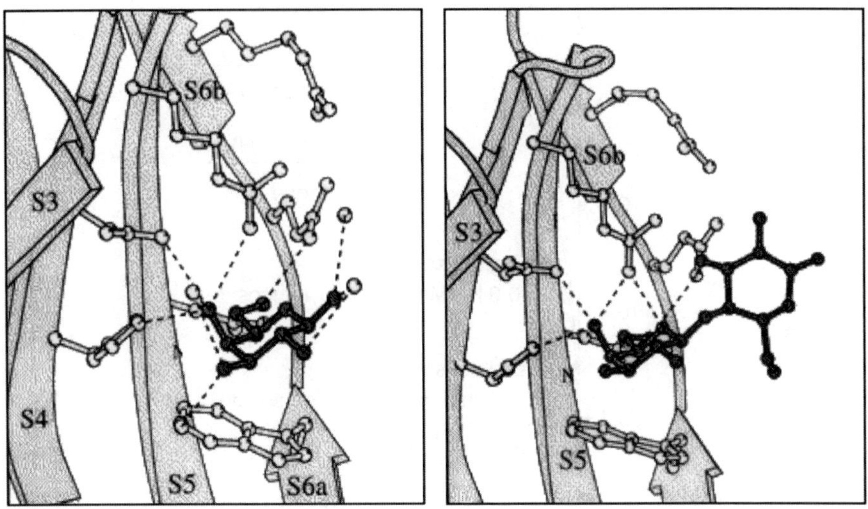

Fig. 18 a, b. Binding of: **a** galactose; **b** lactose to human galectin hGal 7. The outer galactose ring in b replaces several water molecules in a [66]

7
Model Receptors

Several synthetic receptors for the monosaccharides have been described in a recent review. Calixarenes, cyclic steroid dimers, rigid lactams and cyclodextrins were the most prominent examples. Invariably, however, organic solutions or highly concentrated aqueous solutions of the carbohydrates or their hydrophobic derivatives were used [15]. The only exception to this rule was a macrocyclic glycophane in water, which strongly bound (12.5 kJ/mol) an axial α-nitrophenyl-mannoside (Fig. 19). Carbohydrate-carbohydrate interactions were indicated by intermolecular Nuclear Overhauser Effects (NOE) in water [68, 69].

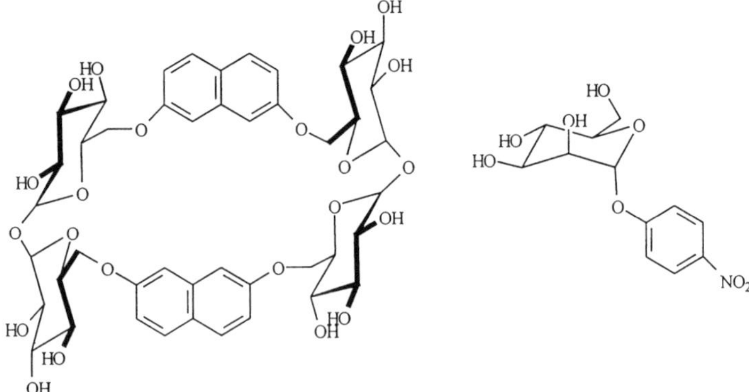

Fig. 19. Components of a molecular complex that forms in water [67, 68]

Complexation in water has obviously been achieved here by strong lipophilic interactions. Stereoselectivity may then be obtained by fitting or non-fitting hydrogen bonding to eventual OH-groups of the substrate. Such interactions may then stabilize or remove the "ice-bergs" around the solute in hydrophobic gaps (Sect. 2) and control dipolar interactions with the walls of the receptor molecule.

8
Conclusion

Carbohydrate recognition in water usually implies hydrogen bonding to strongly basic sites, in particular the oxygen atoms of boronates or glutamic acid, and the nitrogen atoms of histidine or arginine. A hydrophobic stacking interaction with aromatic systems, e.g., tryptophane, may help. Binding constants are in the order of millimolar. Another important factor is "fitting", often accompanied by dehydration in a hydrophobic environment.

Much tighter binding is obtained in interfacial regions of water and lipids. Again hydrogen bonding is responsible for stereoselectivity [70], but the strength of the binding and the longevity of the assemblies is dominated by the immobility of surface water, in particular in hydrophobic environments. This effect has been demonstrated repeatedly by kinetic and thermodynamic data [7–10, 41]; it has been semiquantitatively calculated using a hydrocarbon wall model and water/hydrocarbon interaction potentials as a basis. Ice-like water structures were then used to interpret the wall/water interface [70–72]. Nothing is, however, known about the enormous stabilizing effects of carbohydrates on such thin surface layers of water. A free energy calculation using a reaction field theory has also been used for guanidinium phosphate in bulk water and at the water/air interface. In the monolayer a binding energy of 34 kJ/mol was estimated, in bulk water 0.9 kJ/mol [70–72]. Hydrogen bond formation is favored on the water side of the interface as compared with that in bulk water. The lipid phase modulates an electrostatic potential in the neighborhood of the interface and thus stabilizes hydrogen bonds. This quantum mechanical model does, however, neither explain the molecular origin of the strong binding forces observed nor does it relate to the observed immobility of aqueous solutes [7–9]. Immobilization of 3–4 layers of water on hydrophobic planar surfaces has, however, only been concluded from computer models [73, 74].

Such surface water is realized in Nature by the β-pleated sheet of proteins. Binding sites for membranes on protein surfaces have, however, only been demonstrated by stereoselective precipitation reactions [30], not by crystal structures. In model systems such planar binding sites are easily accessible by holey LB-layers on solid surfaces. Again, they have only been characterized by qualitative binding experiments with monolayers on water or on smooth gold electrodes [7–10]. Their characterization by solid state NMR-spectroscopy is, however, within experimental reach if the surface monolayers containing rigid gaps are transferred to relatively large colloidal particles with a diameter of about 30 nm. Crystallization of complexes between membrane structures of glycolipids (not the monosaccharides!) and lectins or of glycoproteins and lectins should also reveal strong interactions between sheet-like regions. Hydrogen bonding within gaps of small, water-soluble hosts are not promising, because they provide no volume for the immobilization of water.

The "chiral bilayer effect" finally gives an example for strong binding of enantiomers leading from highly curved fibers or spheres to sheets in water. Recognition between two high energy membrane structures leads directly to low energy crystal sheets [17, 18, 23].

To summarize: there are many exciting observations concerning water-carbohydrate interaction in the environment of non-polar protein or membrane surfaces and gaps in such surfaces. The given models, however, are macroscopic only. The change of "hydration forces" by adding "structure makers" has been postulated [75], and frozen water arrays over non-polar surfaces have been discussed [76]. No consequences with respect to exchange rates with bulk water were, however, drawn. Time scales above milliseconds have, to the best of our knowledge, not been considered in any model so far. The molecular structures of the water clusters, the reason for their longevity, and the arrangement of the carbohydrates within these cluster are not known; only detailed NMR experiments can help to understand the phenomena, which means that uniform samples must be prepared first in milligram quantities. Nevertheless it became experimentally evident that some saccharides stuck to hydrophobic surfaces like glue only in aqueous media.

Acknowledgement. Financial support by the Deutsche Forschungsgemeinschaft, the SFB 448 "Mesoscopic Systems", the TMR network "Towards An Understanding Of Carbohydrate Recognition in Aqueous Media", the Fonds der Deutschen Chemischen Industrie, and by the FNK of the Free University is gratefully acknowledged.

9
References

1. Boullanger P (1997) Top Curr Chem 187:275
2. Qiang L, Brady JW (1996) J Am Chem Soc 118:12,276
3. Woods RJ (1996) In: Lipkowitz KB, Boyd DB (eds) Application of molecular modeling techniques to the determination of oligosaccharide solution conformations. VCH Publishers New York, vol 9, chap 3
4. Ravindranathan S, Feng X, Karlsson T, Widmalm G, Levitt MH (2000) J Am Chem Soc 122:1102
5. Jeffrey GA, McMullan RK (1962) J Chem Phys 37:2231
6. Dickerson RE, Geis I (1969) The structure and action of proteins. Harper and Row, p 18
7. Fuhrhop J-H, Bedurke T, Gnade M, Schneider J, Doblhofer K (1997) Langmuir 13:455
8. Fudickar W, Zimmermann J, Ruhlmann L, Roeder B, Siggel U, Fuhrhop J-H (1999) J Am Chem Soc 121:9539
9. Fuhrhop J-H, Ruhlmann L, Messerschmidt C, Fudickar W, Zimmermann J, Röder B (1998) Pure Appl Chem 70:2385
10. Li G, Gouzy MF, Fuhrhop J-H (in preparation)
11. Yeh Y, Feeney RE (1996) Chem Rev 96:601
12. Ewart KV, Li Z, Yang DSC, Fletcher GL, Hew CL (1998) Biochemistry 37:4080
13. Loewen MC, Gronwald W, Sönnichsen FD, Sykes BD, Davies PL (1998) Biochemistry 37:17745
14. Mizutani T, Kurahashi K, Murakami T, Matsumi N, Ogoshi H (1997) J Am Chem Soc 119:8991

15. Davis AP, Wareham RS (2000) Angew Chem Int Ed 38:2978
16. (a) Svenson S, Schäfer A, Fuhrhop J-H (1994) J Chem Soc Perkin Trans 2:1023; (b) Svenson S, Kirste B, Fuhrhop J-H (1994) J Am Chem Soc 116:11,969; (c) Sack I, Macholl S, Fuhrhop J-H, Buntkowsky G (2000) Phys Chem Chem Phys 2:1781
17. (a) Köning J, Böttcher C, Winkler H, Zeitler E, Talmon Y, Fuhrhop J-H (1993) Acta Cryst B49:375; (b) Fuhrhop J-H (1994) Molecular assemblies and membranes. In: Stoddart JF (ed) The synkinetic approach. Monographs in Supramolecular Chemistry, Royal Soc Chem, London
18. Fuhrhop J-H, Endisch C (2000) Molecular and supramolecular chemistry of natural products and model compounds. Marcel Dekker, New York
19. Fuhrhop J-H, Rosengarten B (1997) Synkinetic natural compound chemistry. Synlett 1015
20. Müller-Fahrnow A, Hilgenfeld R, Hesse H, Saenger W (1988) Carbohydr Res 176:165
21. André C, Luger P, Fuhrhop J-H, Rosengarten B (1993) Acta Cryst B49:375
22. André C, Luger P, Svenson S, Fuhrhop J-H (1993) Carbohydr Res 241:47
23. Fuhrhop J-H, Schnieder P, Rosenberg J, Boekema E (1987) J Am Chem Soc 109:3387
24. Fuhrhop J-H, Svenson S, Böttcher C (1990) J Am Chem Soc 112:4307
25. Fuhrhop J-H, Schnieder P, Boekema E (1988) J Am Chem Soc 110:2861
26. Fuhrhop J-H, Boettcher C (1990) J Am Chem Soc 112:1768
27. Böttcher C, Boekema EJ, Fuhrhop J-H (1990) J Microscopy 160:173
28. Terech P, Weiss RG (1997) Chem Rev 97:3133
29. Yamasaki S, Tsutsuimi H (1997) Bull Chem Soc Jpn 67:906
30. Fuhrhop J-H, Krull M, Schulz A, Möbius D (1990) Langmuir 6:497
31. Fuhrhop J-H, Demoulin C, Rosenberg J, Böttcher C (1990) J Am Chem Soc 112:2827
32. Niwa M, Sawada T, Higashi N (1998) Langmuir 14:3916
33. Kitano H, Kuwayama M, Kanayama N, Ohno K (1998) Langmuir 14:165
34. Shinkai S, Takeuchi M (1998) In: Ungaro R, Dalcanale E (eds) Supramolecular science: where it is and where it is going. Kluwer Academic Publishers, Dordrecht
35. Arimori S, Takeuchi M, Shinkai S (1996) J Am Chem Soc 118:245
36. Ogoshi H, Mizutani T (1998) Acc Chem Res 31:81
37. Bhattacharya S, Acharya SNG (2000) Langmuir 16:87
38. Takeoka S, Sou K, Böttcher C, Fuhrhop J-H, Tsuchida E (1998) J Chem Soc, Faraday Trans 94:2151
39. (a) Connolly DT, Townsend RR, Kawaguchi K, Bell WR, Lee YC (1982) J Biol Chem 257:939; (b) Lee RT, Lee YC (1988) Biochem Biophys Res Commun 155:1444
40. Kobayashi K, Asakawa, K, Kato Y, Aoyama Y (1992) J Am Chem Soc 114:10,307
41. Kurihara K, Ohto K, Tanaka Y, Aoyama Y, Kunitake T (1991) J Am Chem Soc 113:444
42. Fujimoto T, Shimizu C, Hayashida O, Aoyama Y (1997) J Am Chem Soc 119:6676
43. Fujimoto T, Shimizu C, Hayashida O, Aoyama Y (1998) J Am Chem Soc 120:601
44. McIntire TM, Brant DA (1998) J Am Chem Soc 120:6909
45. Böttcher C, Endisch C, Fuhrhop J-H, Caterall C, Eaton M (1998) J Am Chem Soc 120:12
46. Schaub M, Wenz G, Wegner G, Stein A, Klemm D (1993) Adv Mater 5:919
47. Nishikawa T, Akiyoshi K, Sunamoto J (1996) J Am Chem Soc 118:6110
48. Revell DJ, Knight JR, Blyth DJ, Haines AH, Russell DA (1998) Langmuir 14:4517
49. Szente L, Szejtli J, Kis GL (1998) J Pharm Sci 87:778
50. Dodziuk H, Ejchart A, Lukin O, Vysotsky MO (1999) J Org Chem 64:1503
51. Liu Y, Zhang Y-M, Qi A-D, Chen R-T, Yamamoto K, Wada T, Inoue Y (1997) J Org Chem 62:1826
52. Gadre A, Connors KA (1997) J Pharm Sci 86:1210
53. Connors KA (1996) J Pharm Sci 85:796
54. Weisser M, Nelles G, Wohlfart P, Wenz G, Mittler-Neher S (1996) J Phys Chem 100:17,893
55. Ghosh M, Zhang R, Lawler RG, Seto CT (2000) J Org Chem 65:735
56. Harada A, Adachi H, Kawaguchi Y, Kamachi M (1997) Macromolecules 30:5181
57. Yamamoto H, Maeda Y, Kitano H (1997) J Phys Chem B 101:6855
58. Bales BL, Messina L, Vidal A, Peric M, Nascimento OR (1998) J Phys Chem B 102:10,347
59. Bales BL, Shahin A, Lindblad C, Almgren M (2000) J Phys Chem B 104:256

60. Bales BL, Howe AM, Pitt AR, Roe JA, Griffiths PC (2000) J Phys Chem B 104:264
61. Lee YC, Lee RT (1995) Acc Chem Res 28:321
62. Fuhrhop J-H, Arlt M (1990) Angew Chem 102:699, Angew Chem Int Ed Engl 29:672
63. Lis H, Sharon N (1998) Chem Rev 98:637
64. Mann DA, Kanai M, Maly DJ, Kiessling LL (1998) J Am Chem Soc 120:10,575
65. (a) Naismith JH, Field RA (1996) J Biol Chem 271:972; (b) Loris R, Maes D, Poortmans F, Wyns L, Bouckaert J (1996) J Biol Chem 271:30614; (c) Leonidas DD, Vatzaki EH, Vorum H, Celis JE, Madsen P, Acharya KR (1998) Biochemistry 37:13,930
66. Swaminathan CP, Surolia N, Surolia A (1998) J Am Chem Soc 120:5153
67. Ernst JA, Clubb RT, Zhou H-X, Groneborn AM, Clore GM (1995) Science 267:1813
68. Coterón JM, Vicent C, Bosso C, Penadés S (1993) J Am Chem Soc 115:10,066
69. Jiménez-Barbero J, Junquera E, Martinpastor M, Sharma S, Vicent C, Penadés S (1995) J Am Chem Soc 117:11,198
70. Ariga K, Kunitake T (1998) Acc Chem Res 31:371
71. Sakurai M, Tamagawa H, Inoue Y, Ariga K, Kunitake T (1997) J Phys Chem B 101:4810
72. Tamagawa H, Sakurai M, Inoue Y, Ariga K, Kunitake T (1997) J Phys Chem B 101:4817
73. Vossen M, Forstmann F (1994) J Chem Phys 101:2379
74. Gompper G, Hauser M, Kornysheva (1994) J Chem Phys 101:3378
75. Israelachvili J, Wennerström H (1996) Nature 379:219
76. Lemieux RU (1996) Acc Chem Res 29:373

Artificial Receptors as Chemosensors for Carbohydrates

Tony D. James[1], Seiji Shinkai[2]

[1] Department of Chemistry, University of Bath, BA2 7AY, UK
E-mail: T.D.James@Bath.ac.uk
[2] Department of Chemistry and Biochemistry, Graduate School of Engineering,
Kyushu University, Fukuoka 812-8581, Japan
E-mail: seijitcm@mbox.nc.kyushu-u.ac.jp

As the chemistry of saccharides and related molecular species plays a significant role in the metabolic pathways of living organisms, detecting the presence and concentration of biologically important sugars in aqueous solution is necessary in a variety of medicinal and industrial contexts. The recognition of D-glucose is of particular interest, for example in the monitoring of diabetics. Recent research provides clear evidence that tight control of blood sugar levels in diabetics sharply reduces the risk of long term complications, which include blindness, kidney failure, heart attacks and even gangrene and amputation of the limbs. Current enzymatic detection methods of sugars offer specificity for only a few saccharides; additionally, enzyme based sensors are unstable in harsh conditions. Phenylboronic acid has been known for 120 years. However, it took until 1959 for the first quantitative evaluation of saccharide boronic acid interactions. Boronic acids react with 1,2 or 1,3 diols of saccharides to form five- or six-membered cyclic esters in non-aqueous or basic aqueous media. The stable boronic acid-based saccharide receptors offer the possibility of creating saccharide sensors 'chemosensors' which are selective and sensitive for any chosen saccharide.

Keywords. Boronic acid, Molecular recognition, Supramolecular chemistry, Chemosensors, Saccharide, Carbohydrate

Topics in Current Chemistry, Vol. 218
© Springer-Verlag Berlin Heidelberg 2002

Nature uses only the longest threads to weave her patterns, so each small piece of her fabric reveals the organisation of the entire tapestry.

Richard P. Feynman

1
Introduction

This chapter will discuss one particular class of artificial receptor for carbohydrates – the 'boronic acids'.

Some excellent reviews exist in the literature covering the use of boronic acids in supramolecular chemistry up till 1996 [1–5]. This chapter will cover some of the earlier work also covered in these earlier reviews, but will in general present only the newer developments in the field.

For many years boron-containing compounds have been important in organic synthesis as protective groups and coupling agents [6, 7]. However, it is only recently that aromatic boronic acid derivatives have been employed in the construction of receptors for molecules such as saccharides.

Phenylboronic acid **1** was first synthesised by Michaelis and Becker in 1880 [8, 9] but 74 years passed before the first binding studies of diols (or saccharides) with boronic acid appeared in the literature, when in 1954 Kuivila et al. noticed that boronic acids solubilised saccharides and polyols, and postulated the formation of a cyclic ester [10]. This result agreed with the known ability of borates

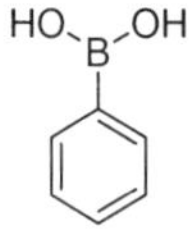

1

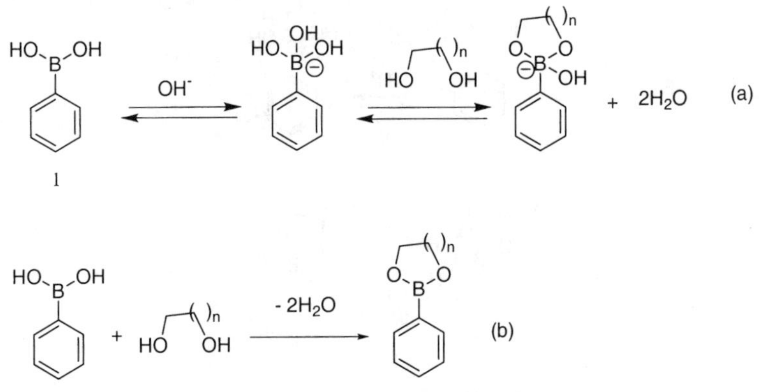

Scheme 1a, b. Ester formation with boronic acids: **a** reaction in basic aqueous media; **b** reaction in aprotic media in the presence of a dehydrating agent

to complex with polyhydroxyl compounds [11]. Subsequently it was found that arsenite and tellurate also formed complexes with polyols [12, 13]. Lorand and Edwards published the first quantitative evaluation of saccharide boronic acid interactions in 1959 [14].

Despite this, the structure of the boronic acid and saccharide complex in aqueous solution is still disputed [15, 16]. Boronic acid covalently reacts with 1,2 or 1,3 diols to form five- or six-membered cyclic esters in non-aqueous or basic aqueous media (Scheme 1). The adjacent rigid *cis* diols of saccharides form stronger cyclic esters than simple acyclic diols such as ethylene glycol. With saccharides the choice of diol used in the formation of a cyclic ester is complicated by the possibility of pyranose to furanose isomerisation of the saccharide moiety. Lorand and Edwards determined the selectivity of phenylboronic acid towards saccharides [14] (Table 1). This selectivity order seems to be retained by all monoboronic acids, and will be referred to throughout this chapter. At this point it is worth clarifying the meaning of pK_a when referring to boronic acids. The easiest way to think about the pK_a is to consider a solvent water molecule bound to the sp^2 boronic acid; at high pH the water is deprotonated and forms a tetrahedral sp^3 boronate. On saccharide binding and formation of a cyclic boronate ester the pK_a of the boronic acid is enhanced, or in other words the 'ester' is more acidic than the 'acid'. The enhanced acidity is due to a bond angle

Table 1. Stability constants (log K) of polyol complex with phenylboronic acid **1**

Polyol	Boronic acid **1** (log K)
D-Fructose	3.64
D-Galactose	2.44
D-Mannose	2.23
D-Glucose	2.04
Ethylene glycol	0.44

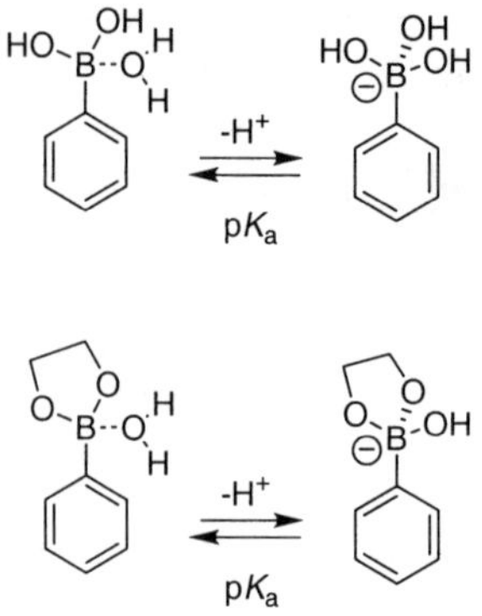

Scheme 2. The pK_a of boronic acids. The 'ester' is more acidic than the 'acid'

compression on formation of a cyclic boronate ester. Boronic acids have a 120° (sp^2) bond angle but on the formation of a cyclic ester the bond angle is reduced to 108°. Obviously, the compression of bond angle from 120° to 108° makes the change in hybridisation from sp^2 to sp^3 on deprotonation occur more readily (Scheme 2).

As the chemistry of saccharides and related molecular species plays a significant role in the metabolic pathways of living organisms, detecting the presence and concentration of biologically important sugars (glucose, fructose, galactose, etc.) in aqueous solution is necessary in a variety of medicinal and industrial contexts. The recognition of D-glucose is of particular interest, since the breakdown of glucose transport in humans has been correlated with certain diseases: renal glycosuria [17, 18], cystic fibrosis [19], diabetes [20, 21] and also human cancer [22]. Industrial applications range from the monitoring of fermenting processes to establishing the enantiomeric purity of synthetic drugs. Current enzymatic detection methods of sugars offer specificity for only a few saccharides; additionally, enzyme-based sensors are unstable in harsh conditions. Stable boronic acid-based saccharide receptors offer the possibility of creating saccharide sensors which are selective and sensitive for any chosen saccharide.

The development of boronic acid based-molecular receptors has been hampered by synthetic difficulties. Boronic acids can exist as a mixture of acid and anhydride creating difficulties in analysis. Incombustible residues complicate elemental analyses of boronic acid-based molecules [23]. The cleavage of boronic acids by copper salts [24], silver salts [23] and even by hot water has been reported [23]. This chapter will not discuss these synthetic aspects, and will instead provide a survey of the development of boronic acid related compounds as saccharide sensors.

Boronic acids rapidly and reversibly form cyclic esters with diols albeit in basic aqueous media as shown in Scheme 1. Many synthetic systems that attempt to mimic natural systems have employed hydrogen bonding as the main binding force. Such systems have met with great success in non-hydrogen bonding solvents, which are non-competitive with the guest for the binding pocket. Synthetic hydrogen bonding systems may yet evolve and make a successful transition into water [25]. However, if the transition cannot be made, all is not lost; synthetic systems need not be structural mimics of natural systems.

2
CD Receptors

Optical activity stemming from chirality is manifested by practically all-natural products such as nucleic acids, sugars, proteins, etc. Chiroptical properties have been known since the mid-nineteenth century and are among the most important physical properties for both the food and drug related industries. Increasingly strict regulations have applied pressure on industry to produce enatiomerically pure drugs because of the possible devastating deleterious effects of the wrong enantiomer. The induced CD (circular dichroism) properties of molecular complexes with chiral guest species such as saccharides upon chiral induction are important in providing a method to determine the chirality of the guest. This is significant for non-chromophoric host molecules.

The presence of hydroxy groups with slight reactivity differences creates problems in designing saccharide receptors. However, the high number of hydroxyl groups is also an advantage if these hydroxy groups are used in the recognition/binding process.

2.1
Homogeneous Systems

When two boronic acid units are arranged in a specific orientation, a saccharide may be bound in a 1:1 fashion by its head and tail. The co-operative binding of two boronic acids creates a rigid cyclic complex. The asymmetric immobilisation of two chromophoric benzene rings by ring closure with saccharides can be read out by CD spectroscopy. Diboronic acid **2** forms small cyclic complexes with monosaccharides and some disaccharides in basic aqueous media as monitored by CD spectroscopy [26–28] (Scheme 3). D- and L-saccharides give positive and negative exciton coupling respectively with the highest association constant being 19,000 $dm^3\,mol^{-1}$ for glucose. Among the D-saccharides tested, D-galactose was the only saccharide to show a negative exciton coupling in the CD spectrum. Table 2 summarises the CD spectral data and association constants of different saccharides with **2**. This work has shown that the absolute configuration of saccharides can be conveniently predicted from the sign of the CD spectra of **2**. The 1,2-diol is widely accepted as the primary binding site of glucose (α-form) with boronic acid [14]. The pyranose structure of the D-glucose

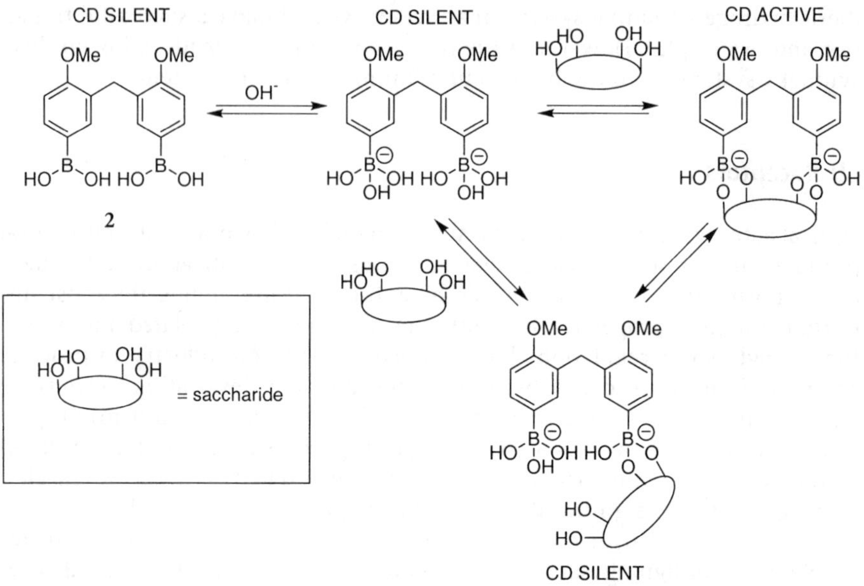

Scheme 3. Formation of a 1:1 saccharide:diboronic acid (2) CD active complex

Table 2. Absorption and CD maxima of **2** monosaccharide complexes[a]

Saccharide	UV λmax (nm)	CD		Stoichiometry	K (dm^3 mol^{-1})
		λmax (nm)	$[\theta]$max[b] (deg cm^2 dmmol^{-1})		
D-Glucose	274	275	−5300	1:1	19,000
	200	205	+231,000		
		190	−214,000		
D-Mannose	272	274	−400	1:1	60
	200	205	+69,000		
		191	−23,000		
D-Galactose	273	276	+410	1:1	2,200
	200	205	−22,000		
		191	+19,000		
D-Talose	272	275	−3700	1:1	4,600
	200	205	+247,000		
		190	−196,000		
D-Fructose	274	−[c]	−[c]	−	−
	200	−[c]	−[c]		

[a] 25 °C, [**2**] = 1.00 or 2.00 × 10^{-3} mol dm^{-3}, pH 11.3 with 0.10 mol dm^{-3} carbonate buffer.
[b] $[\theta]$max Values are calculated for 100% complexation values from the 1:1 complex region.
[c] No CD observed.

diboronic acid **2** complex was suggested using ^{1}H NMR. However, a recent investigation into the structure of this complex by ^{11}B, ^{13}C and ^{1}H NMR analysis suggests that the complexes contain the furanose form of D-glucose [15].

Diboronic acid **3** has a greater distance between the two boronic acid moieties and was expected to give a strong association with disaccharides [27, 29]. However, association constants for disaccharides were much less than those for the complex of monosaccharides and **2**. CPK models revealed that the distance between the two boronic acids in **3** is shorter than the distance between the head and tail hydroxy groups of the disaccharides. Accordingly, the CD activity was much weaker and was inadequate for the estimation of association constants. The 3,3′-stilbenediboronic acid species **4** similarly gave a CD active species upon disaccharide complexation [30]. Diboronic acid **5**, which has a larger spacer between the two boronic acid moieties, gave better complexes with disaccharides [31]. The extended chromophore facilitates the detection of CD bands and the allosteric switching of selectivity by *cis-trans* isomerisation of **5** has been investigated.

Nature relies on allosteric (or co-operative) interactions in many biological functions. The transport of sugars across a cell membrane into the cell is controlled by such interactions. In an attempt to mimic these functions, a diboronic acid saccharide-binding unit was combined with simple crown ether. The strong co-operative binding of **2** prompted the design of such an allosteric system **6** [32]. The control of the angle between two phenyl rings of **2** by some secondary effect could change the sugar binding ability of **2**. Having a very similar binding site to **2**, and with greater flexibility in rotation of the phenyl rings, **6** exhibits an even larger association constant for glucose (31,000 dm^3 mol^{-1}) in basic methanolic aqueous solutions (CD measurements). The binding of metal ions to the crown ether was monitored by ^{1}H NMR spectroscopy. Metal ion binding to the crown ether induces a twist in the two boronic acid groups resulting in the reduced affinity with glucose (negative allosterism).

2.2
Heterogeneous Systems

Some chiral molecules are known to form chiral helical aggregates [33]. Likewise, chirality can be induced in aggregates of non-chiral molecules by interactions with chiral species [1, 34]. Helical chiral aggregates are found when aggregates of a non-chiral tetraboronic acid porphyrin **7** are treated with monosaccharides [35]. Induced chirality of the aggregate can be monitored by CD spectroscopy. Furthermore, the sign of the exciton coupling of the sugar complexed aggregate can be predicted by the structural orientation of the complex. The chromophoric boronic acid derivative **8** was also found to aggregate in mixed solvents (water:dimethyl formamide) [36, 37]. The aggregate was CD active when glucose was added to the aggregate. The aggregative properties of other di- and tetra-boronic acid porphyrins systems have also been investigated [38, 39].

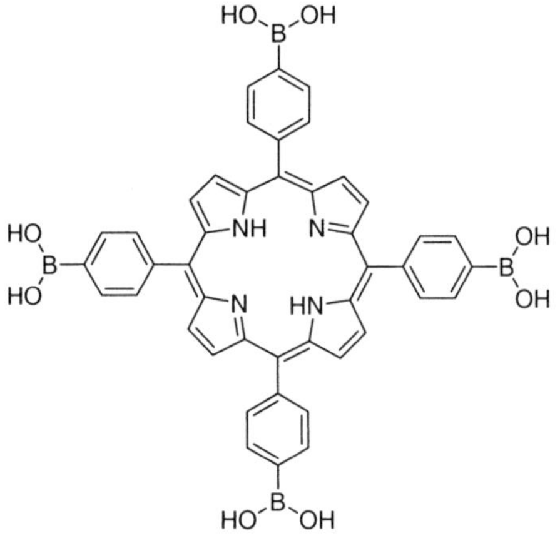

7

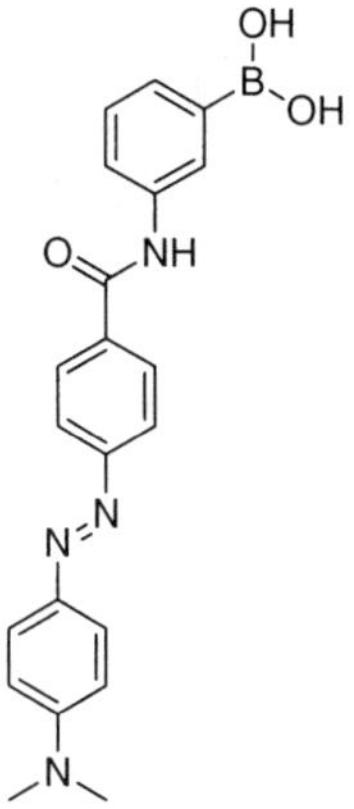

8

CD spectroscopic measurements are useful for the detection and identification of optically pure saccharides and for determining the optical purity of a sample. However, this method is only appropriate to the detection of optically pure saccharides. The expensive equipment required for CD measurements also hampers the practical applications of this method.

3
Fluorescent Receptors

Fluorescence sensors for saccharides are of particular interest in a practical sense. This is in part due to the inherent sensitivity of the fluorescence technique. Only small amounts of a sensor are required (typically 10^{-6} mol/l) offsetting the synthetic costs of such sensors. Also, fluorescence spectrometers are widely available and inexpensive. Fluorescence sensors have also found applications in continuous monitoring using an optical fibre and intracellular mapping using confocal microscopy.

3.1
Homogeneous Systems

3.1.1
PET Receptors

Photoinduced electron transfer (PET) has been widely used as the preferred tool in fluorescent sensor design for atomic and molecular species [40–45]. PET sensors generally consist of a fluorophore and a receptor linked by a short spacer. The changes in oxidation/reduction potential of the receptor upon guest binding can alter the PET process creating changes in fluorescence.

The first fluorescence PET sensors for saccharides were based on fluorophore boronic acids. Yoon and Czarnik showed that 2- and 9-anthryboronic acid [46] **9** and **10** could be used to detect saccharides. However, the fluorescence change was small (I (in the presence of saccharide)/I_0 (in the absence of saccharide) = ca. 0.6). The pK_a of the fluorophore boronic acids are shifted by saccharide present in the medium. The extent of the effect is in line with the inherent selectivity of phenylboronic acid [14]. The PET from the boronate anion is believed to be the source of the fluorescence quenching. Although, 2-anthrylboronic acid displays only a small fluorescence change eight aromatic boronic acids have been screened [47, 48] and it was determined that **11** and **12** are more suitable candidates for saccharide detection.

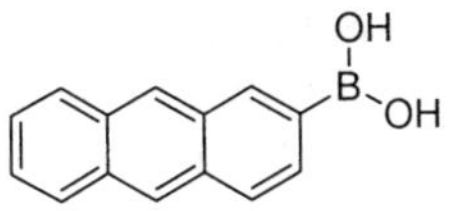

9

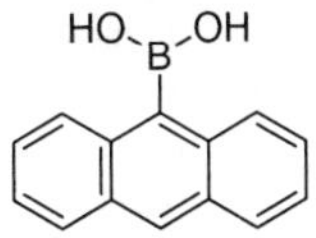

10

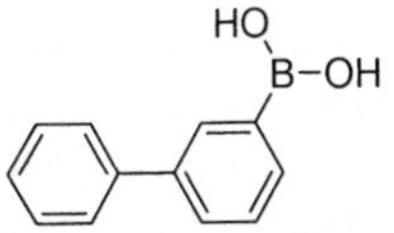

11

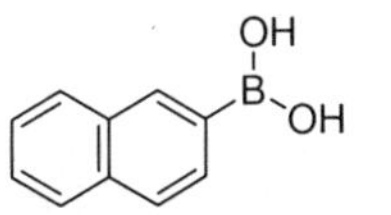

12

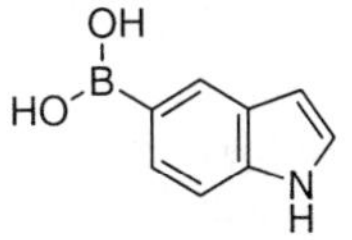

13

Aoyama and coworkers has also shown that 5-indolylboronic acid **13** undergoes fluorescence quenching upon complexation with oligosaccharides [49]. In the determination of the pK_a of the boronic acid, ^{11}B NMR shifts were consistent with the fluorescence studies. The stability constants of monosaccharides were similar to the inherent selectivity of phenylboronic acid [14]. Oligosaccharides, on the other hand, gave relatively lower stability constants, although higher oligomers showed an increased stabilisation relative to lower oligomers due to a secondary interaction with the indole N-H.

With the systems outlined above, facile boronic acid saccharide complexation only occurs at the high pH required to create a boronate anion. To overcome these disadvantages molecular fluorescence sensors that contain a boronic acid group and an amine group were developed. The boronic acid group is required to bind with and capture sugar molecules in water. The amine group plays two roles in the system. First, for biological systems the physiological pH is neutral. Boronic acids with a neighbouring amine can bind strongly with sugars at neutral pH, but, simple boronic acids can only bind strongly with sugars at a high pH. Second, the fluorescence intensity is controlled by the amine. With no sugar the 'free' amine reduces the intensity of the fluorescence (quenching by photoinduced electron transfer). This is the 'off' state of the fluorescent sensor. When sugar is added, the amine becomes 'bound' to the boron centre. The boron bound amine cannot quench the fluorescence and hence a strong fluorescence is observed. This is the 'on' state of the fluorescent sensor. The system described above illustrates the basic concept of an 'off-on' fluorescent sensor for sugars.

The first of these fluorescent PET 'off-on' saccharide sensors was prepared in 1994 [50, 51]. The very large pK_a shift found upon saccharide binding provides a wide pH range for saccharide sensing (Fig. 1). The large shift of the pK_a is due to the interaction found between the boronic acid moiety and the amine group. The boronic acid-amine interaction inhibits the photoinduced electron transfer quenching process in the complex **14b** (Scheme 4). Complete separation of the amine and the acid moiety at very high pH, as in **14c**, quenched the anthracene fluorescence further. However, the fluorescence decrease is not sufficient for the calculation of the pK_a. The introduction of D-glucose remarkably enhances the fluorescence of **14** over large pH range (Fig. 1). The enhanced interaction between boronic acid and amine, upon saccharide binding, inhibits the electron transfer process giving higher fluorescence (as **14d** in Scheme 4). This increased interaction would be expected since the saccharide binding to boronic acid increases its acidity creating a more electron-deficient boron atomic centre. This simple monoboronic acid system **14** shows a selectivity order, which is inherent to all monoboronic acids [14] (Fig. 2).

The simple 'off-on' PET system was improved with the introduction of a second boronic acid group **15** [51]. With compound **15** similar equilibria to those observed with **14** exist. However, for clarity, only species arising from neutral **15** and saccharide are shown (Scheme 5). For compound **15** two possible saccharide binding modes can inhibit the electron transfer process, so giving higher fluorescence: the 2:1 complex **15a** and 1:1 complex **15b**. Due to fortuitous spacing of the boronic acid groups the diboronic acid was selective for D-glucose over

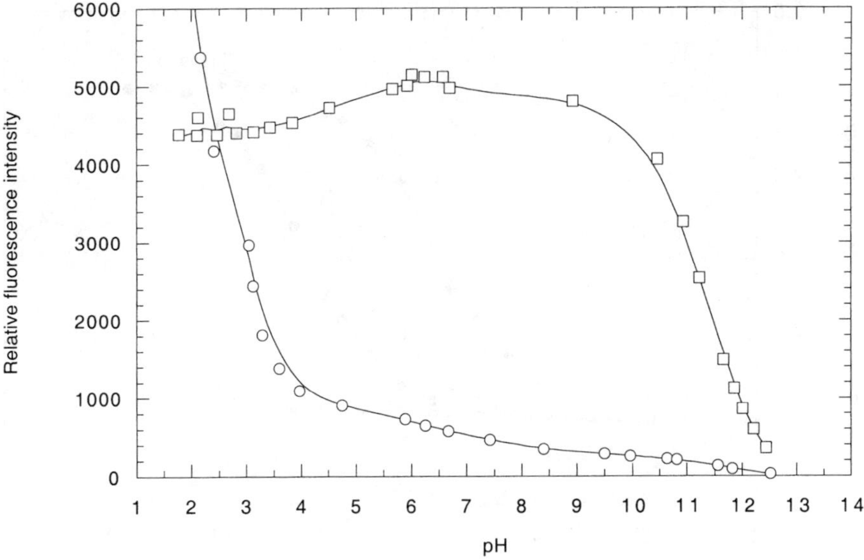

Fig. 1. Fluorescence intensity vs pH profile of 1.2×10^{-5} mol dm^{-3} of **14** in 0.05 mol dm^{-3} NaCl at 25 °C, λ_{ex} 370 nm, λ_{em} 423 nm: (□) 0.05 mol/l [D-glucose], (○) blank

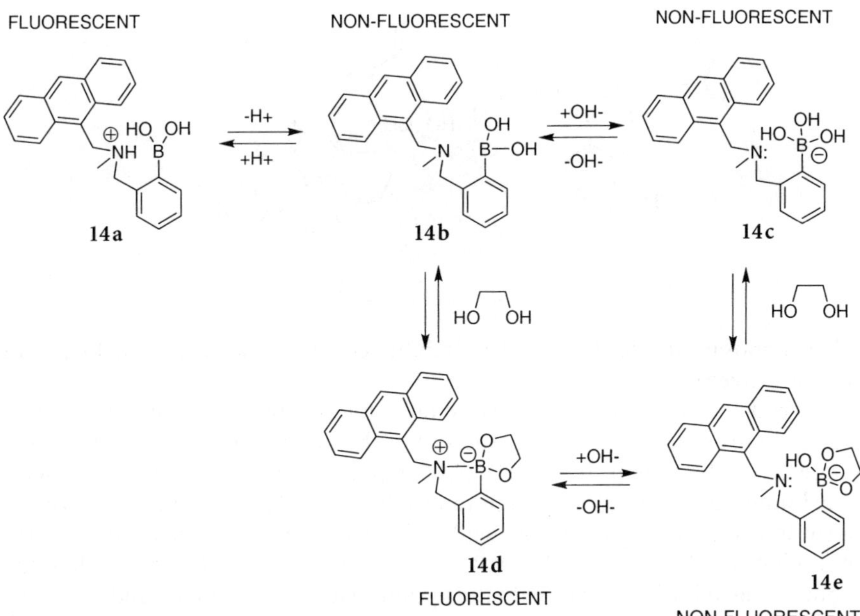

Scheme 4. The effect of saccharide complexation and pH changes on the fluorescence of monoboronic acid **14**

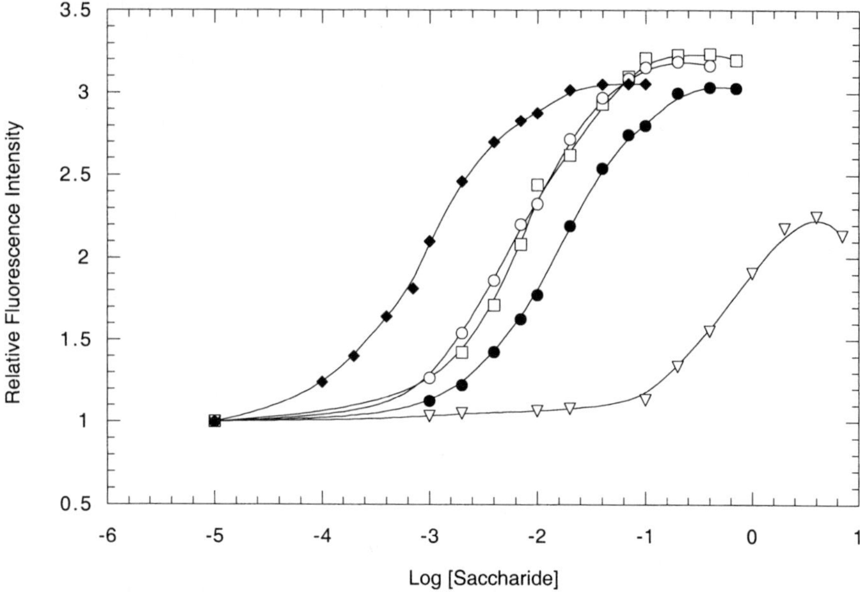

Fig. 2. Fluorescence intensity vs log [saccharide or ethylene glycol] profile of 1.0×10^{-5} mol dm^{-3} of **14** in 33.3% MeOH/H$_2$O pH 7.77 buffer at 25 °C, λ_{ex} 370 nm, λ_{em} 423 nm: (◆) D-fructose, (○) D-galactose, (□) D-allose, (●) D-glucose, (▽) ethylene glycol

other monosaccharides (Fig. 3). The stability constants calculated for **14** and **15** from are given in Table 3.

Norrild and coworkers have carried out a more detailed investigation of this system in order to confirm the structure of the bound glucose [52]. They were interested in the system since the ^{1}H NMR reported [51, 53] indicated that D-glucose bound to the receptor in its pyranose form. Norrild and Eggert had previously shown that simple boronic acids selectively bind with the furanose form of D-glucose [15]. From ^{1}H NMR observations it was concluded that the diboronic acid initially binds with the pyranose form of D-glucose and over time the bound glucose converts to the furanose form. The Norrild experiments were carried our using 1:1 D-glucose and diboronic acid, whereas the fluorescence titrations reported [51, 53] range from 10- to 100-fold excess of D-glucose.

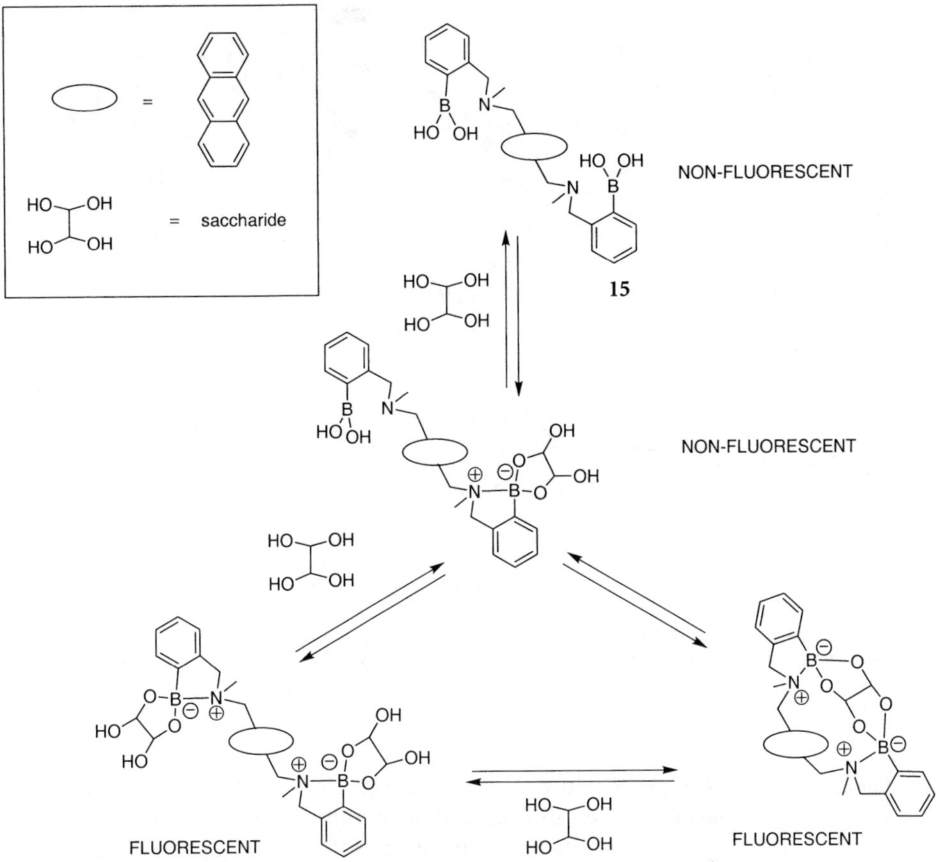

Scheme 5. The effect of saccharide complexation on the fluorescence of diboronic acid 15

Table 3. Stability constants (log K) for monosaccharide and ethylene glycol complex with boronic acid **14** and **15** in pH 7.77 (0.05 mol dm^{-3} phosphate buffer, 33% methanolic aqueous solution)

Saccharide	Boronic acid **14** log K[a]	Boronic acid **15** log K[a]
D-Glucose	1.8	3.6
D-Fructose	3.0	2.5
D-Allose	2.5	2.8
D-Galactose	2.2	2.2
Ethylene glycol	0.4	0.2

[a] Measurements were done in 33% methanolic aqueous solutions.

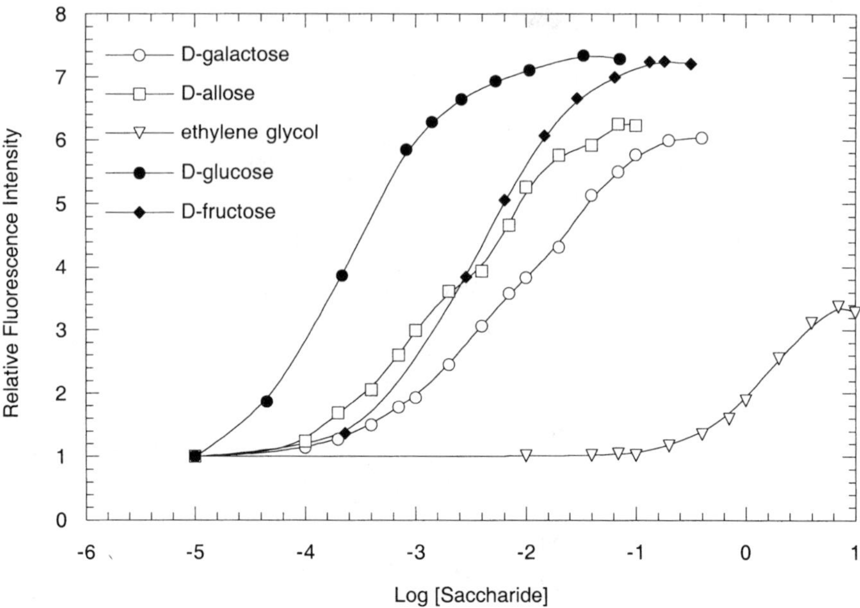

Fig. 3. Fluorescence intensity vs log [saccharide or ethylene glycol] profile of 1.0×10^{-5} mol dm^{-3} of **15** in 33.3 % MeOH/H$_2$O pH 7.77 buffer at 25 °C, λ_{ex} 370 nm, λ_{em} 423 nm: (◆) D-fructose, (○) D-galactose, (□) D-allose, (●) D-glucose, (▽) ethylene glycol

Therefore, in the D-glucose fluorescence titrations it is possible that the furanose form will be replaced by a new molecule of the pyranose form as it forms (water at 31 °C contains 0.14 % of D-glucose as the furanose form). On a somewhat related theme it has been shown that the simple monoboronic acid **16** can selectively signal the furanose form of saccharides [54].

Work by Irie et al. on the control of intermolecular chiral 1,1′-binapthyl fluorescence quenching by chiral amines [55] and the use of 1,1′-binaphthyl in the recognition of chiral amines by Cram [56] were the inspiration behind the design of **17** (*R* or *S*). Chiral recognition of saccharides by **17** (*R* or *S*) utilises both steric and electronic factors [57]. The asymmetric immobilisation of the amine groups relative to the binaphthyl moiety upon 1:1 complexation of saccharides by D- or L-isomers creates a difference in PET. This difference is manifested in the maximum fluorescence intensity of the complex. Steric factors arising from the chiral binaphthyl building block are chiefly represented by the stability constant of the complex. However, the interdependency of electronic and steric factors upon each other is not excluded. This new molecular cleft, with a longer spacer unit compared to the anthracene based diboronic acid **15** gave the best recognition for fructose. Table 4 shows the binding constants for some D- and L-monosaccharides. The chiral recognition of saccharides employing hydrogen bonding by polyhydroxy molecules has been reported [25]. However, discriminative detection of isomers in aqueous media by fluorescence, as far as we are aware, had not been achieved before. In this system steric factors and electronic factors bimodally discriminate the chirality of the saccharide. Competitive studies with D- and L-monosaccharides show the possibility of selective detection of saccharide isomers. The availability of both R and S isomers of this particular molecular sensor is an important advantage, since concomitant detection by two probes is possible.

Diboronic acid **18** with a small bite angle has been synthesised and has been shown to be selective for small saccharides such as D-sorbitol [58]. Conversely, diboronic acid **19** with larger spacing of the boronic acid groups looses selectivity and sensitivity [59]. However the system may be useful for the detection of saccharides in concentrated solutions, such as those encountered in the brewery and confectionery industries. Dendritic boronic acid **20**, shows enhanced binding affinities but the selectivity amongst the monosaccharides is reduced [60].

An allosteric diboronic acid **21** has been prepared where formation of a metal crown sandwich causes the release of bound saccharide [61] (Scheme 6). The allosteric concept was also applied in the two-dimensional PET sensor **22**. With

18

19

Table 4. Stability constants and fluorescence enhancements for D- and L-saccharides with 17R (or S) in pH 7.77 (0.05 mol dm^{-3}) phosphate buffer, 33 % methanolic aqueous solution)

Saccharide	D log$K$ (± 0.05)	L log$K$ (± 0.05)	D/L Fluorescence intensity ratio
Fructose	4.0 (3.7)	3.5 (4.0)	1.47 (0.69)
Glucose	3.3 (3.4)	3.1 (3.5)	1.93 (0.53)
Galactose	3.1	3.3	0.82
Mannose	<2.4	–	–

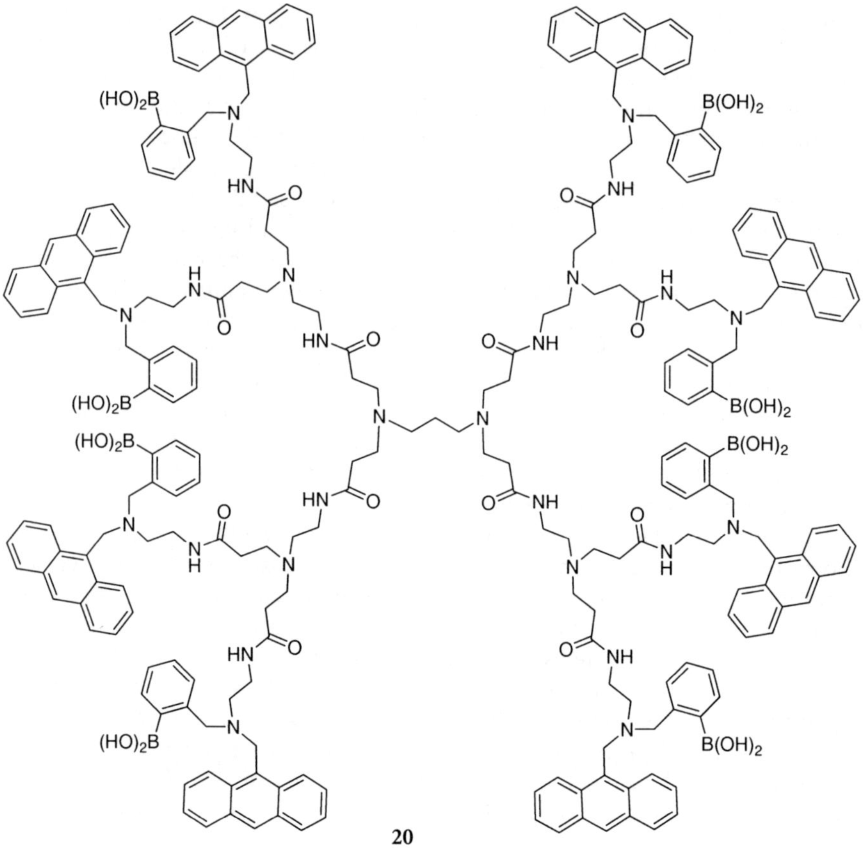

20

21

22

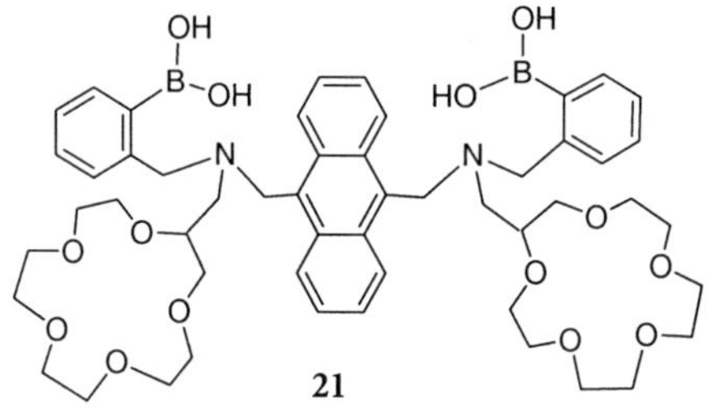

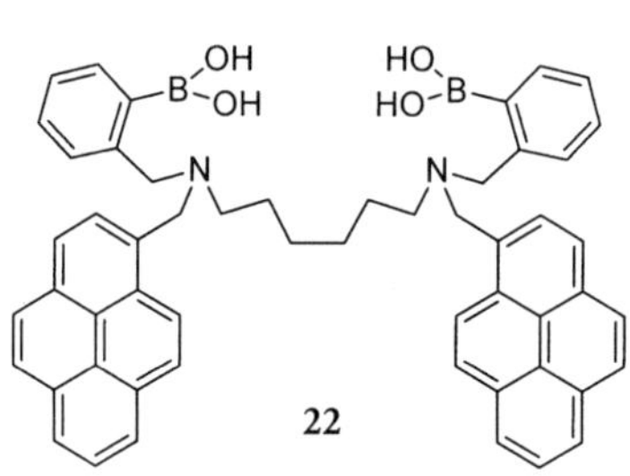

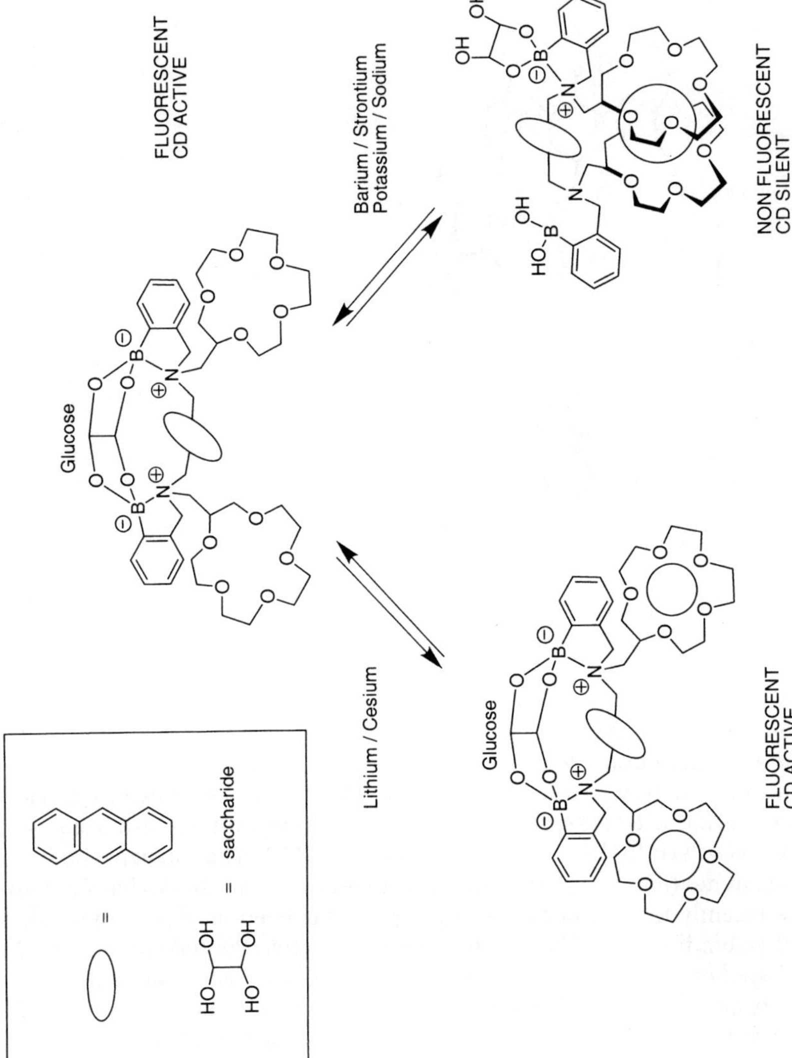

Scheme 6. The effect of 2:1 metal complex and 1:1 metal 'sandwich' complex on the CD activity and fluorescence intensity of diazacrown diboronic acid **21**

23

24 (n=0) and **25** (n=1)

the two-dimensional system, the amount of excimer can be directly correlated with the amount of non-cyclic saccharide complex formed [62].

The calixarene framework **23** has been used as a core on which to develop novel saccharide selective systems [63]. The calixarene unit has also been used to develop allosteric [64] and luminescent systems [65]. Wu has prepared borono-alkyl calixarenes (Fig. 4) and has shown that they also bind with saccharides [66].

More recently the boronic acid PET system has been used in combination with other binding sites. The D-glucosamine selective fluorescent systems **24** and **25** based on a boronic acid and aza crown ether has been explored [67, 68]. Sensors **24** and **25** consist of monoaza-18-crown-6 ether or monoaza-15-crown-5 as a binding site for the ammonium terminal of D-glucosamine hydrochloride,

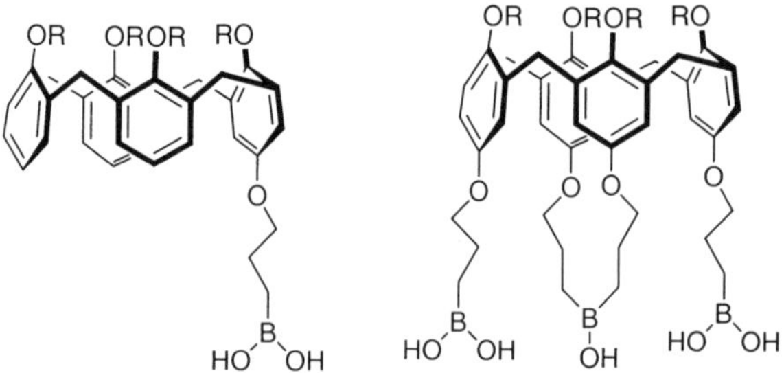

Fig. 4. Boronoalkyl calixarenes

while a boronic acid serves as a binding site for the diol (carbohydrate) part of D-glucosamine hydrochloride. The nitrogen of the azacrown ether unit can participate in PET with the anthracene fluorophore; ammonium ion binding can then cause fluorescence recovery. This recovery is due to hydrogen bonding from the ammonium ion to the nitrogen of the azacrown ether. The strength of this hydrogen-bonding interaction modulates the PET from the amine to anthracene. As explained above, the boronic acid unit can also participate in PET with the anthracene fluorophore, and diol binding can also cause fluorescence recovery. The anthracene unit serves as a rigid spacer between the two-receptor units, with the appropriate spacing for the glucose moiety. This system behaves like an *AND* logic gate, [69, 70], in that fluorescence recovery is only observed when two chemical inputs are supplied; for this system the two chemical inputs are an ammonium cation and a diol group.

Compounds **26** and **27** do not display any fluorescence enhancement with D-glucose because they have no saccharide-binding site. D-glucosamine hydrochloride binding with compounds **26** and **27** is also not observed under the experimental conditions, since, the buffer solution saturates with D-glucosamine hydrochloride before the binding event can be observed. As expected compound **14** shows fluorescence enhancement with D-glucose ($K = 67 \pm 3$ dm^3 mol^{-1}) and D-glucosamine hydrochloride ($K = 18 \pm 1$ dm^3 mol^{-1}). With D-glucose the boronic acid has a choice of binding either the 1,2- or 4,6-diols, but with D-glucosamine hydrochloride binding with just the 4,6-diol is possible. The stability constant of **14** with D-glucose is higher than that observed with D-glucosamine hydrochloride reflecting the known selectivity of boronic acids for the 1,2-diol of D-glucose. Compounds **24** and **25** show fluorescence increase with D-glucosamine hydrochloride ($K = 18 \pm 2$ dm^3 mol^{-1} and 17 ± 2 dm^3 mol^{-1} respectively), but no increase with D-glucose. This result clearly demonstrates that for a fluorescent output *both* a diol *and* ammonium group must be present in the guest. The stability of the D-glucosamine hydrochloride complex with compounds **24**, **25** and **14** are the same within experimental error ($K = 18 \pm 2$ dm^3 mol^{-1}). These results show that the azacrown ether imparts no additional stability to the D-glucosamine complex. However, D-glucosa-mine must be involved in a hydrogen bonding interaction with the secondary benzylic nitrogen of the azacrown ether in order to suppress PET. If such an interaction were not present, then a fluorescence increase would not have been observed for compounds **24** and **25** with D-glucosamine hydrochloride (Fig. 5).

26 (n=0) and **27** (n=1)

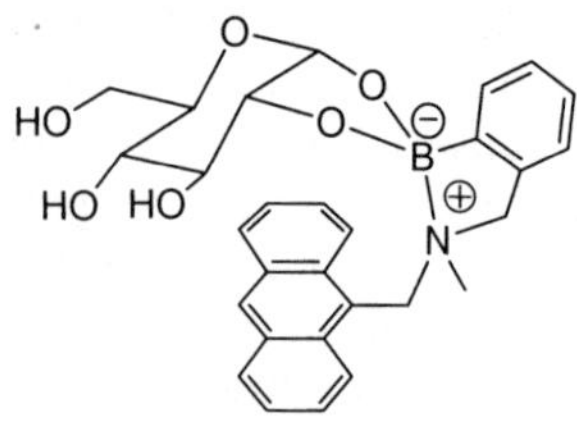

Fig. 5. Fluorescent and non-fluorescent complexes of **14** and **25** with D-glucose and D-gluco-samine hydrochloride

A D-glucuronic acid-selective fluorescent system **28** based on a boronic acid and metal chelate has also been developed [71, 72].

A new twist has recently been introduced with the development of a novel 'on-off' PET system **29** which behaves contrastingly to the conventional 'off-on' PET system [73]. In this system steric crowding on saccharide binding breaks the B-N bond found in the 'free' receptor.

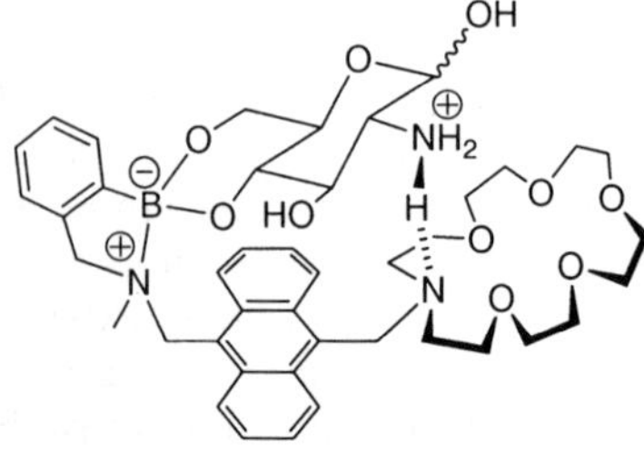

28

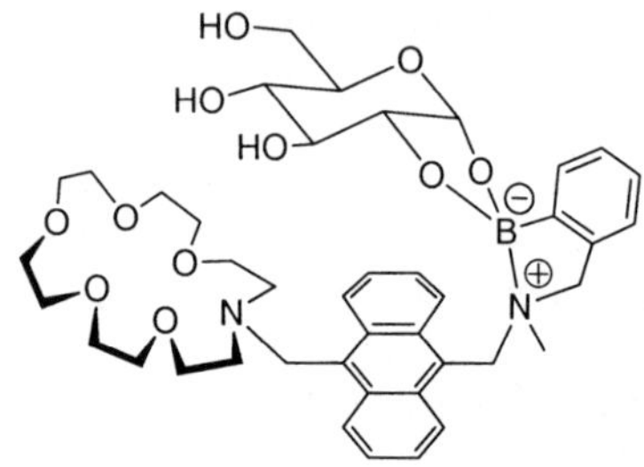

29

3.1.2
Non-PET Receptors

As shown above, diboronic acids can bind monosaccharides selectively, where the 1:1 binding creates a rigid molecular complex [26–32]. This rigidification effect can also be utilised in designing fluorescence sensors for disaccharides. Diboronic acid species **4** selectively complexes with disaccharides in basic aqueous media to create cyclic complexes which alter the fluorescence properties of the molecule [30]. It is known that excited stilbene is quenched by radiationless decay via rotation of the ethylene double bond. Obstruction of this rotation leads to increased fluorescence [74]. The rigidification of **4** by disaccharide binding increases the stilbene fluorescence. In particular, the disaccharide melibiose shows higher selectivity for **4** than other common disaccharides.

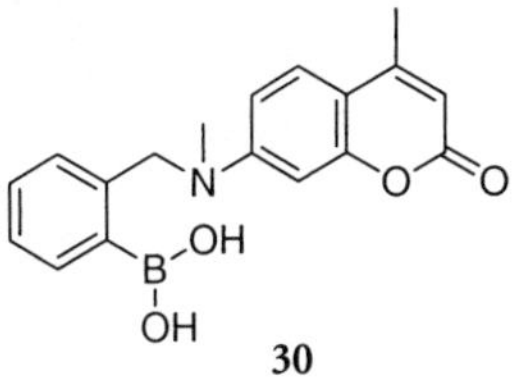

30

The amino coumarine boronic acid **30** prepared is an example of an internal charge transfer (ICT) chromophore [75]. Here both fluorescence intensity and wavelength is affected since the nitrogen is directly connected with the chromophore. Sadly, this system shows only a small shift in intensity and wavelength.

Molecular rigidification has been used to generate a fluorescence increase with cyanine diboronic acid **31** on saccharide binding [76]. Rigidification has also been used with a diboronic acid appended binaphthyl **32** to develop a chiral discriminating system [77].

Norrild and coworkers have developed an interesting diboronic acid system **33** [78]. The system works by reducing the quenching ability of the pyridine groups of **33** on saccharide binding.

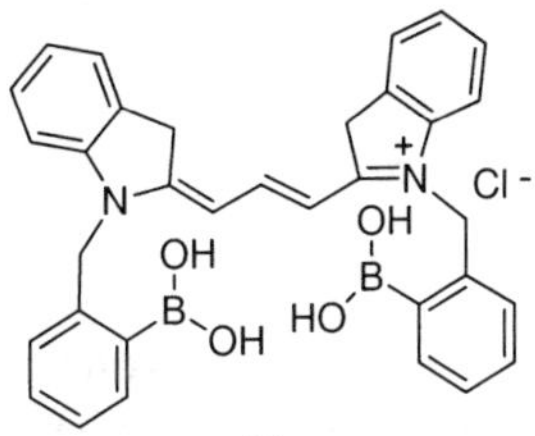

31

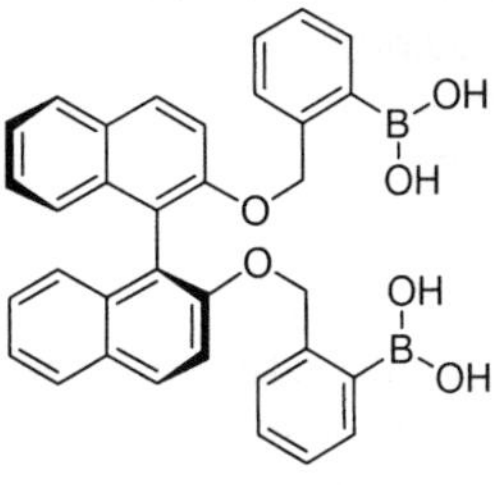

32 (R)

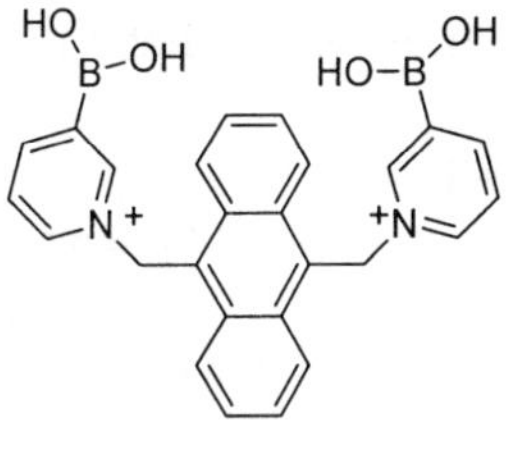

33

3.2
Heterogeneous Systems

Aggregates of fluorescent species do not exhibit fluorescence due to concentration quenching. Molecular forces, which deaggregate such systems, can cause an increase in the fluorescence intensity. The aggregates of porphyrin **34** formed in DMF:water could be dispersed by the introduction of saccharides into the system [79, 80]. When saccharides form complexes with the boronic acids they solubilise the porphyrin into the bulk phase leading to a very large fluorescence increase. Some selectivity in this process has been observed.

34

4
Coloured Receptors

Coloured sensors for saccharides are of particular interest in a practical sense. If a system with a large colour change can be developed it could be incorporated into a diagnostic test paper for D-glucose, similar to universal indicator paper for pH. Such a system would make it possible to measure D-glucose concentrations without the need for specialist instrumentation. This would be of particular benefit to diabetics in developing countries.

4.1
Homogeneous Systems

A fairly recent development has been the study of the effect of saccharides on the colour of dyes containing boronic acid functionality. Boronic acid azo dyes have been known for over 40 years where they were used for investigations in the treatment of cancer by a technique called boron neutron capture therapy (BNCT) [81, 82]. However, it has only been in the 1990s that related dyes and their interaction with saccharides have been studied. Russell [83] has synthesised a boronic acid azo dye from *m*-aminophenylboronic acid, which was found to be sensitive to a selection of saccharides.

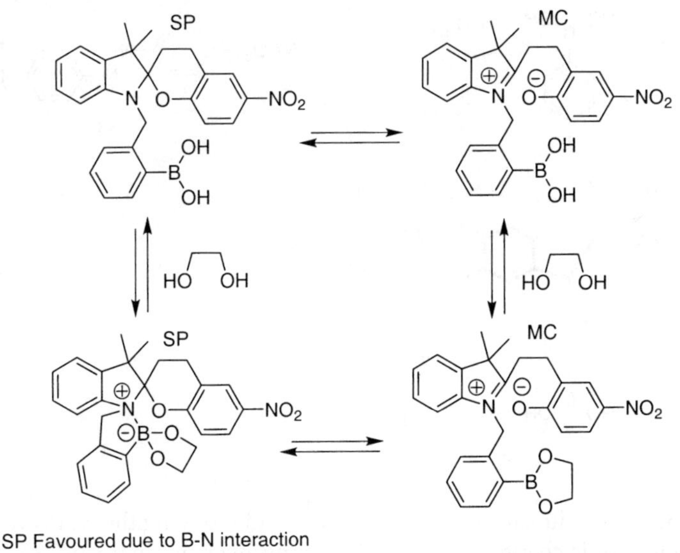

Scheme 7. The effect of saccharides on the Spiropyran vs merocyanine equilibrium

Boronic acid appended spirobenzopyrans **35** undergo changes in the absorption spectra on the addition of saccharides [84]. The added saccharides change the position of the merocyanine (MC) to spyropyran (SP) equilibrium and hence colour of the system. With added saccharide the SP structure is favoured due to a stronger B-N interaction in the saccharide complex (Scheme 7).

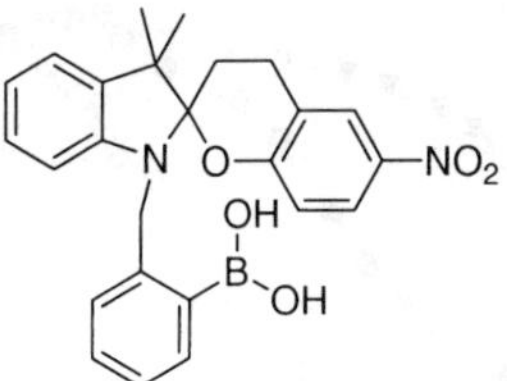

Boronic acid dye **36** has been synthesised which undergoes an absorption spectral change on addition of nucleosides [85]. The boronic acid binds with the ribose and the dimethylaminophenylazo moiety can stack with the adenine of the nucleoside. Strongin has prepared a tetraboronic resorcinarene for the visual sensing of saccharides **37**. However, for a colour change to be observed the saccharide must be heated (90 °C) in DMSO [86].

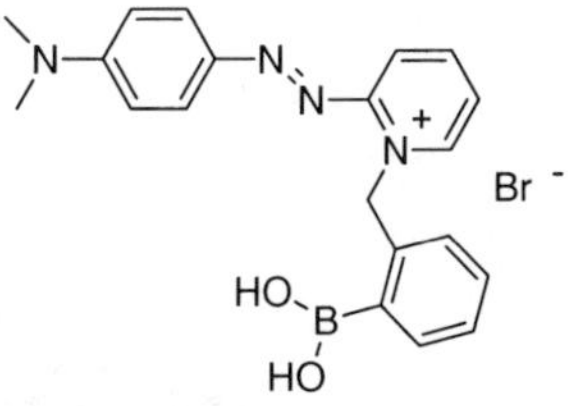

In 1994 a synthetic molecular colour sensor for saccharides was reported [87]. The designed molecular internal charge transfer (ICT) sensor, dye molecule **38**, was based on the intramolecular interaction between the tertiary amine and the boronic acid group. The electron-rich amine creates a basic environment around the electron-deficient boron centre, which has the effect of induc-

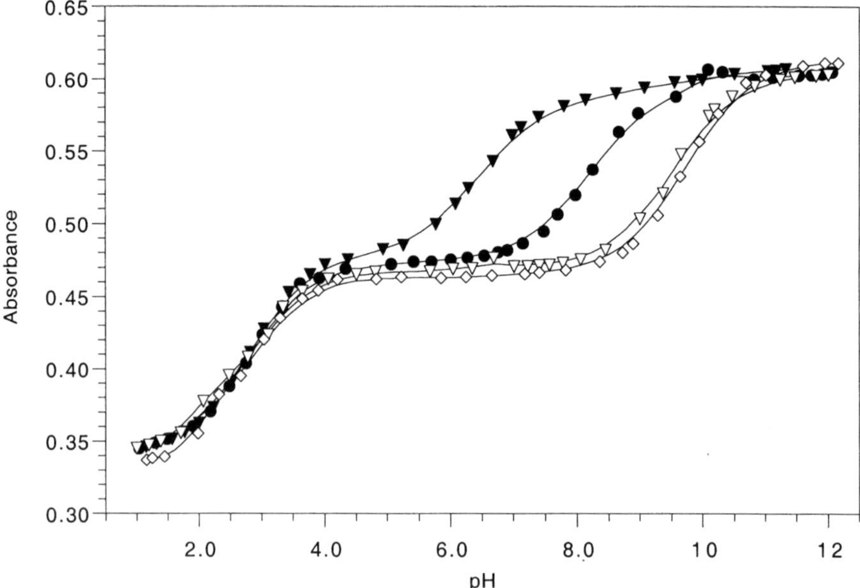

ing the boronic acid-saccharide interaction and reducing the working pH of the sensor. Electronic changes associated with this decrease in the pK_a of the boronic acid moiety on saccharide complexation were shown to be transmitted to the neighbouring amine. This creates a spectral change in the connected ICT chromophore, which can be detected spectrophotometrically. The absorption pH profile for the molecular sensor **38** in aqueous media with different saccharides is given in Fig. 6. The equilibrium processes at different pH regions and the most

Fig. 6. Absorption spectral changes vs pH of **38** with 0.05 mol dm^{-3} NaCl, λ_{max} 468 nm: with 0.05 mol dm^{-3} (◆) D-fructose, (●) D-glucose, (▽) ethylene glycol, (◇) blank

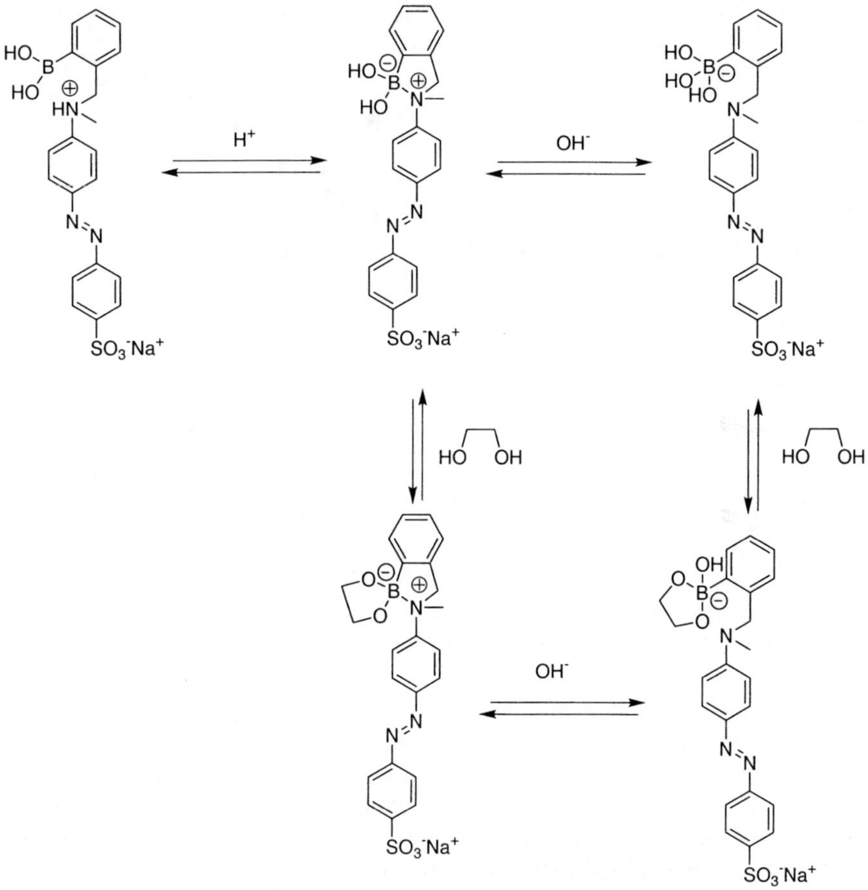

Scheme 8. The effect of saccharide complexation and pH changes on monoboronic acid **38**

important species involved are given in Scheme 8. The pK_a related to the boron-nitrogen interaction of **38** shifts on the addition of saccharides. The largest pK_a shift was found for D-fructose ($pK_a = 3.31$) and the smallest for simple diols such as ethylene glycol ($pK_a \approx 0$). The main drawback of this system is the relatively small shifts in the absorption bands of the chromophore upon saccharide binding.

Recently a diazo dye system, which shows a large visible colour change from purple to red on saccharide binding, was synthesised [88]. With azo dye **39** the wavelength maximum shifts by approximately 55 nm to a shorter wavelength

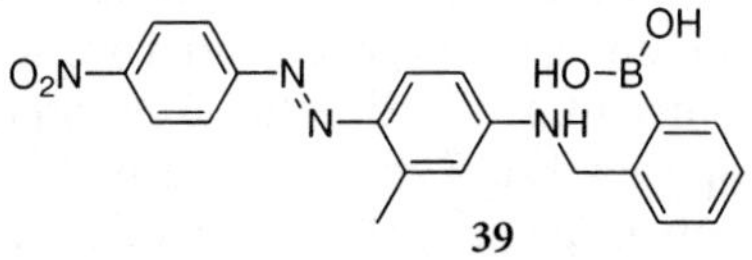

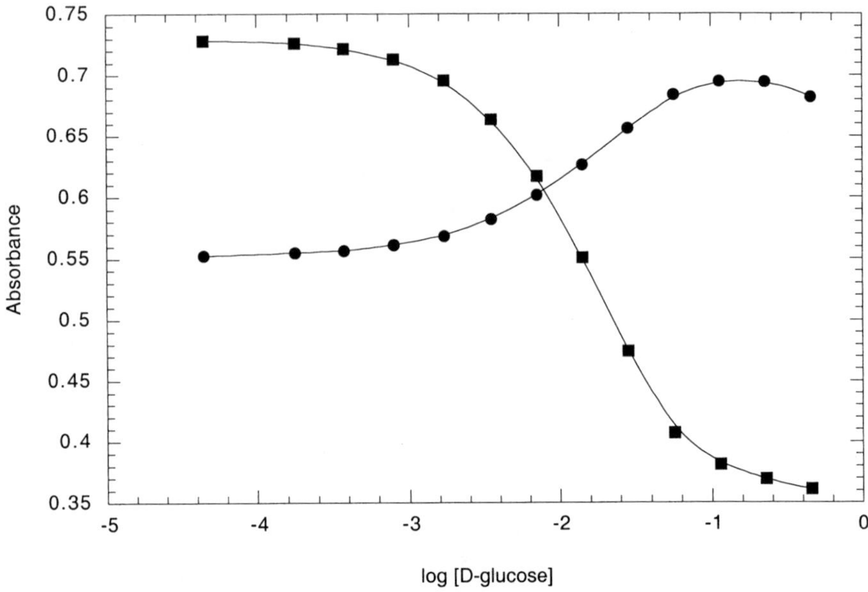

Fig. 7. Absorption at (●) 509 nm and (■) 564 nm vs log [D-glucose] profile of 5.66×10^{-5} mol dm^{-3} of **39** in 52.1 % MeOH/H$_2$O pH 11.32 buffer at 25 °C

upon saccharide complexation (Fig. 7). The concentration of the guest required to produce the change is different in each case, which is due to the different stability constants of the saccharides. The wavelength shift obtained with **39** on addition of diols is the largest observed to date. The stability constants (log$K$) of the boronic acid dye-saccharide complexes are D-fructose (3.75), D-glucose (1.85) and ethylene glycol (0.66) respectively.

Scheme 9 shows the species in equilibria responsible for the observed colour change, and consistent with the experimental results. With dye molecule **38** it is proposed that at intermediate pH a boron-nitrogen interaction is prevalent, whereas at high and low pH this bond is broken. What makes the equilibria of dye molecule **39** more interesting is the presence of the *anilinic hydrogen*, which can give rise to different species at high pH. This apparently simple modification in the molecular structure is responsible for the enhanced response of these dyes relative to those previously reported.

In the absence of saccharide, at pH 11.32, the observed colour is purple and in the presence of saccharide the colour is red. From previous work it is known that when saccharides form cyclic boronate esters with boronic acids, the Lewis acidity of the boronic acid is enhanced and therefore the Lewis acid-base interaction between the boronic acid and the amine is strengthened. This stronger B-N interaction will favour the red species over the equivalent saccharide bound purple species. The reason for this can be understood by considering species **39a** and **39b** from Scheme 9. In the presence of saccharide the B-N interaction in species **39b** is stronger than that in species **39a**. The increased B-N interaction of species **39b** will make the N-H proton of species **39b** more acidic than the cor-

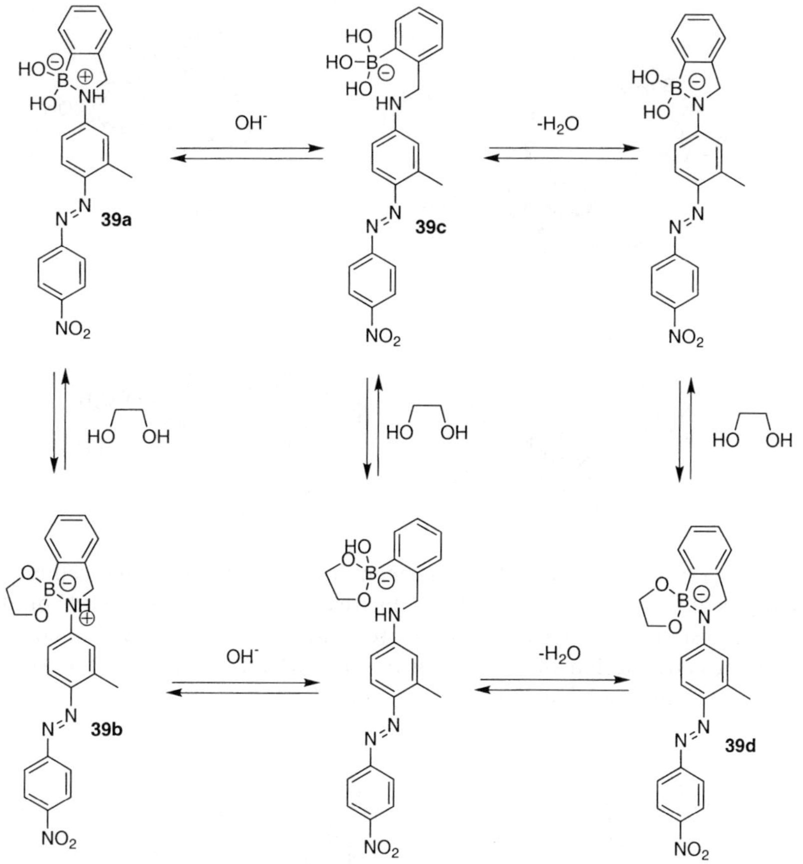

Scheme 9. The effect of saccharide complexation and pH changes on monoboronic acid **39**

responding bond in species **39a**. Therefore at higher pH, species **39b** will deprotonate to form the red species **39d**, whereas the weaker B-N bond in species **39a** is broken by hydroxide ion to form the purple coloured species **39c**.

The proposal of these equilibrium species may also explain why dye molecule **38** did not give a visible spectral shift on saccharide binding. Because the anilinic nitrogen is tertiary in nature rather than secondary, there is no possibility of deprotonation, so the high pH boron-nitrogen bond cannot be formed. Hence there is no differentiation between the equilibrium species at high pH and consequently no spectral shift is observed.

Anslyn and coworkers have recently reported two very interesting systems based on boronic acid receptors. The Anslyn systems involve a competitive spectrophotometric assay. The first system is a receptor for glucose-6-phosphate **40** [89]. The binding of glucose-6-phosphate is measured through the competitive displacement of 5-carboxyfluorescein **41**. The second system is a receptor for tartrate or malate **42** [90]. The binding of tartrate or malate is measured through the competitive displacement of alizarin complexone **43**.

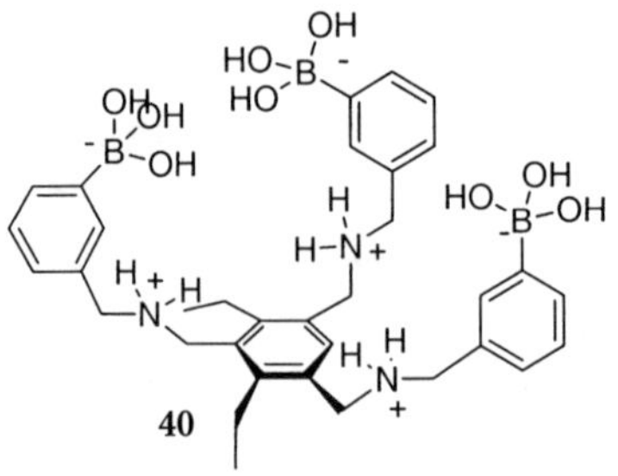

An interesting catalytic system has been developed, where enhanced rates of hydrolysis of the imine bond of **44** and **45** are observed with saccharide binding [91]. The system is not strictly a colour receptor; however the system could be used to measure saccharide concentrations by monitoring the rate of disappearance of the imine signal at 320 nm.

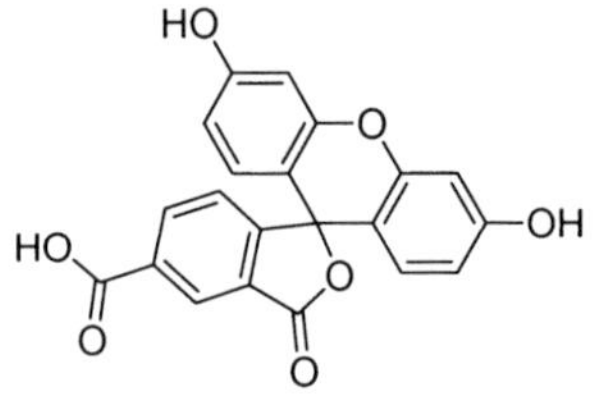

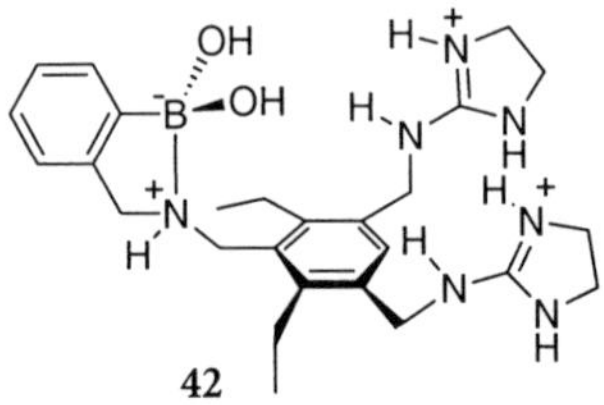

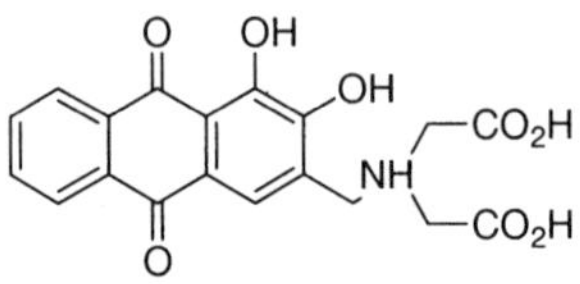

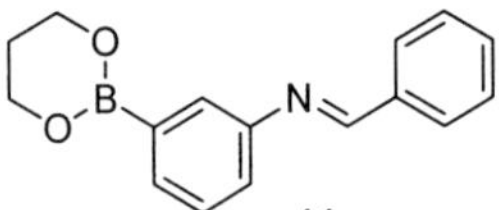

With the aid of better chromophores and better receptors it will be possible to develop selective and sensitive colour sensors. Such sensory systems will have widespread applications in industry. For example, it should be possible to provide cheap and stable 'test papers' for the detection of blood glucose.

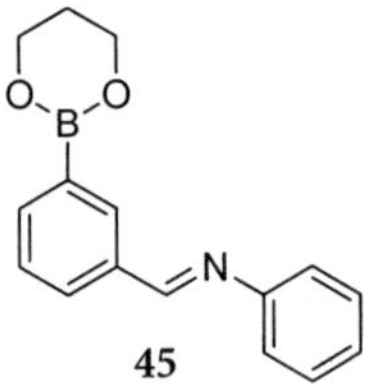

4.2
Liquid Crystalline Systems

Liquid crystals composed of chiral molecules may possess chiral helical structures in which the length of the pitch determines the colour of the liquid crystal. Influence on the pitch by inclusion species can result in visible colour changes and the spectral shifts can reflect the absolute structure of such species. A liquid crystalline composite including cholesterylboronic acid **46** changes colour upon introduction of saccharides [92, 93]. The pitch of the liquid crystal directly represents the chirality of the bound sugar. As an example, the green colour of the

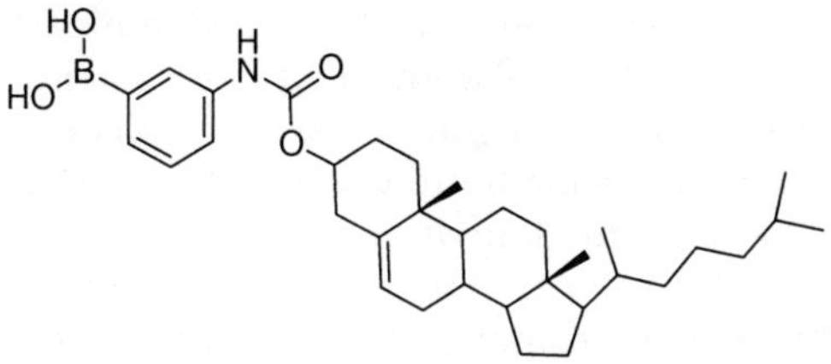

46

uncomplexed composite was changed to red upon introduction of D-glucose, and to blue by L-glucose. A direct correlation between the colour and the relative positioning of the bound boronic acid moieties to the saccharide ring has been proposed. For additional information the reader is directed to an excellent recent review on the use of cholesterol in molecular recognition [94].

As with the absorption spectroscopic techniques this system may find 'test paper' applications. Unfortunately, this system may have a serious drawback: its response time. For this system to recognise a saccharide in solution, the saccharide must diffuse from the aqueous layer into the cholesteric liquid crystal layer of the 'test paper'.

4.3
Heterogeneous Systems

Colour changes upon introduction of saccharides have been observed in molecular aggregates of **8** in a DMF:water mixed solvent system [36, 37]. The colour changes appear to be due to the deaggregation process induced by the saccharide binding.

5
Porphyrin-Based Receptors

A novel glucose-6-phosphate selective system **47** based on a boronic acid appended metalloporphyrin derivative has been devised [95, 96]. The two point binding of glucose-6-phosphate creates a rigid complex which gives a strong exciton coupling signal in the CD spectrum. It is believed that the strong binding by the primary binding site of glucose (1,2-diol) to the boronic acid followed

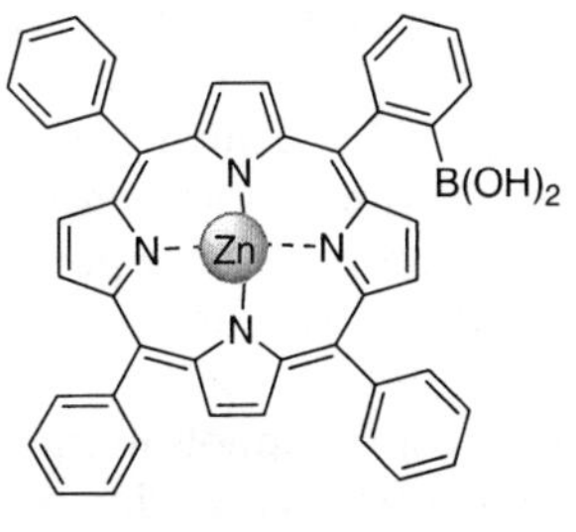

47

by the secondary interaction of the phosphate-metal centre results in the high affinity for glucose-6-phosphate. The replacement of the 1-hydroxy group, which is part of the primary binding site of glucose, by a phosphate unit in glucose-1-phosphate weakens the saccharide-boronic acid interaction and hence the overall strength of the complex. Binding of phosphate to the metal centre was revealed by ^{31}P NMR peak shifts. CD exciton coupling peaks were not found for either metal ion free **47** with glucose phosphate or for **47** with glucose itself, indicating the importance of the two-point binding.

A very interesting system consisting of a monoboronic acid zinc porphyrin **48** and monoboronic acid pyridine unit **49** has been discovered [97]. Self-assembly of the porphyrin and pyridine results in a 'diboronic acid'. This system is particularly interesting since the synthesis of monoboronic acids is much easier than that of diboronic acids. Modification of each constituent monoboronic acid is therefore much easier than that required in the direct synthesis of a diboronic acid. This system will allow for a large number of structurally similar self-assembled diboronic acids to be investigated with the minimum of synthetic effort.

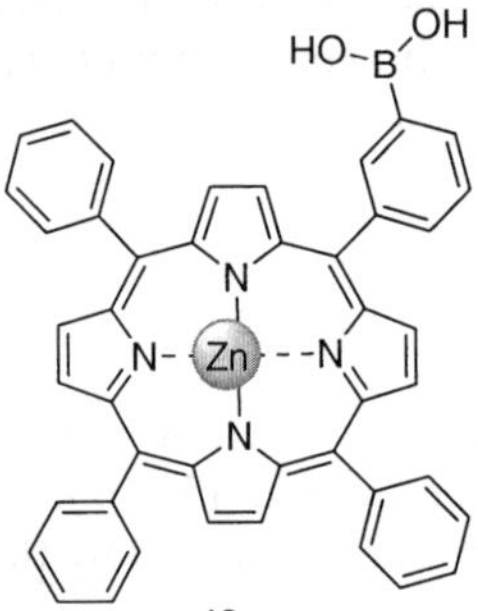

A novel porphyrin assembly has been constructed using **49** and dicatechol porphyrin [98]. Another self-assembled system is the μ-oxodimer of iron porphyrin (Scheme 10) [99, 100]. This tweezer-like molecule shows very high and selective binding with D-glucose and D-galactose.

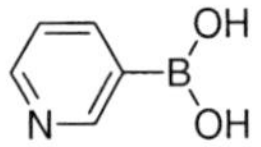

A novel dimeric system using anionic porphyrin **50** and cationic porphyrins **51** or **52** has been investigated [101]. The 1:1 dimer formed between **50** and **51** or **52** showed selective binding with glucose and xylose. Porphyrin **51** has also been used in saccharide controlled intercalation with DNA [102]. With no added saccharide **51** intercalates with DNA but when saccharides are added **51** dissociates from DNA. Compounds **51** and **52** have also been used with 1,5- or 2,6-anthraquinonedisulfonates (ADS) as a competitive system for the fluorescence detection of D-fructose [103]. 1,5- or 2,6-ADS binds with **51** or **52** and quenches the fluorescence; addition of D-fructose causes decomplexation and fluorescence recovery.

A D-lactulose selective system **53** based on a diboronic acid porphyrin has been developed [104]. The spacial disposition of the two boronic acids in **53** produces the perfect 'cleft' for the disaccharide D-lactulose.

Novel porphyrin dimers have been prepared using monoboronic acid porphyrin and saccharides [105, 106] These systems show saccharide controllable electron-transfer efficiency.

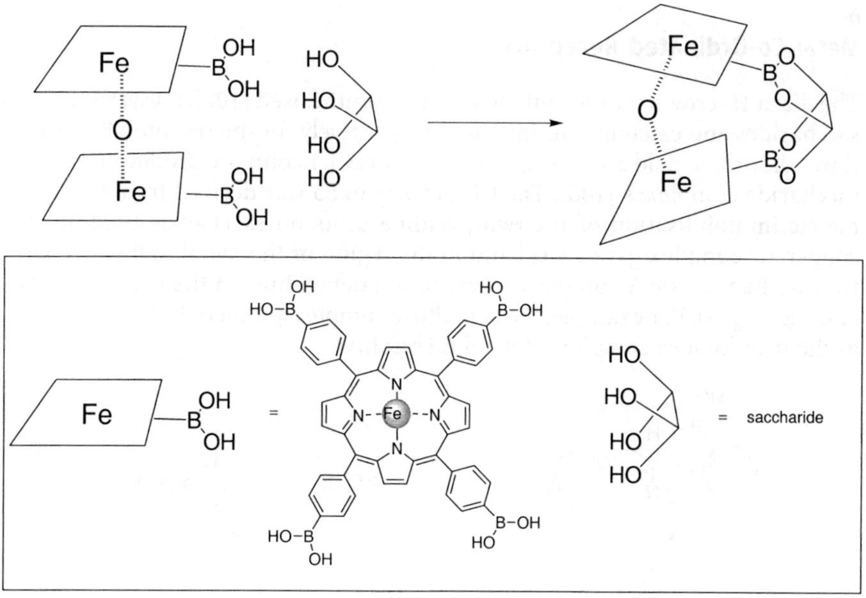

Scheme 10. Saccharide 'tweezer' formed from an iron porphyrin μ-oxodimer

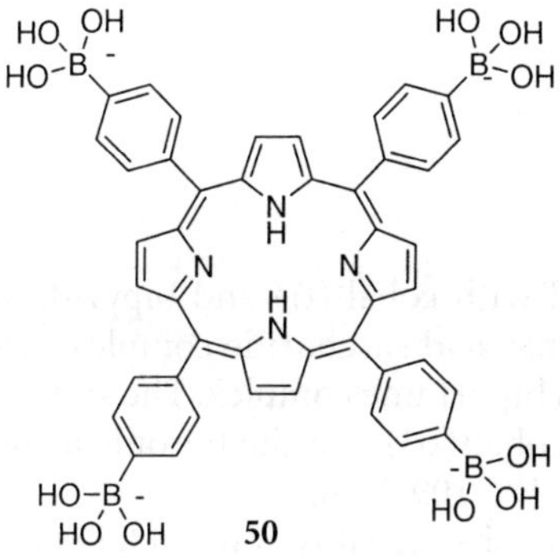

50

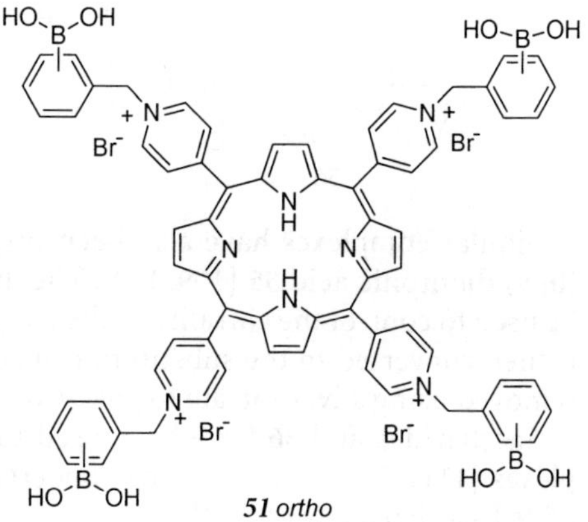

51 ortho
52 meta

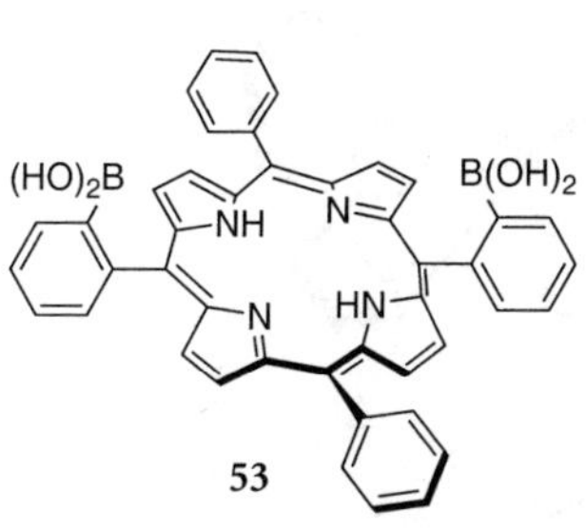

53

6
Metal Co-Ordinated Receptors

The diaza 18-crown-6 diboronic acid **54** was synthesised [107]. It was shown that saccharides and calcium ions interact competitively for the receptor. Bipyridine (bpy) diboronic acid **55** and its iron(II) (as $FeCl_2$) complex produce CD active saccharide complexes [108]. The CD activity of **55** was derived from the asymmetric immobilisation of the two-pyridine units on saccharide binding. The copper(II) complex gave a CD band in the region of the metal to ligand charge transfer band. Δ or Λ complexes were found depending on the complexed saccharide (Fig. 8). For example, the D-maltose complex adopted Λ chirality whereas the D-cellobiose complex adopted Δ chirality.

Similar complexes have also been prepared with cobalt(II) and bipyridine (bpy) diboronic acid **55** [109, 110]. The diboronic acid saccharide complex can be used to control the chirality of the cobalt(II) bipyridine complex. The system is then converted to the substitution-inactive cobalt(III) and the boronic acids removed using silver nitrate to give $[Co^{III}(bpy)_3]^{3+}$ [109, 110].

Saccharides and **56** have been used to transcribe chirality into metal complexes [111]. The helicity of the copper(I) complex of **28** is also controlled by added saccharide [112, 113].

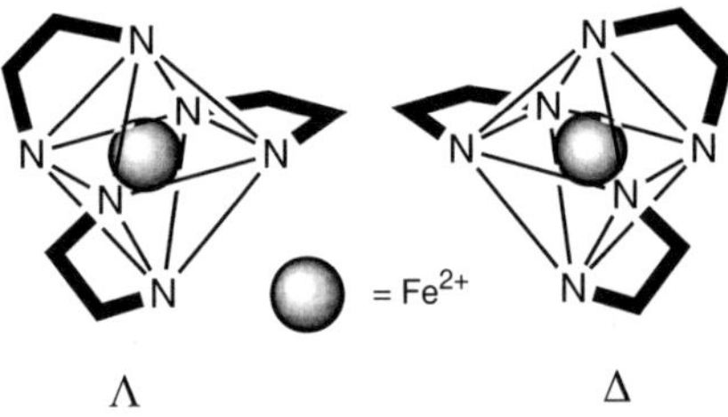

Fig. 8. Structure of bipyridine Δ or Λ chiral complexes

The chiral salen cobalt(II) complexes **57** and **58** have been used for the detection of saccharides [114]. Spectroscopic changes in the metal complexes were used to monitor saccharide complexation. Interestingly **59** showed twofold selectivity for L-allose over D-allose.

The rhenium(I) complex of **59** has been studied by Yam and Kai [115] and then reinvestigated in our group [116] for potential saccharide sensing properties. Deetz and Smith have investigated the ruthenium(II) complex **60** as a heteroditopic receptor for phosphate and saccharides [117]. It was shown that phosphorylated saccharides display enhanced binding and that binding of saccharides and phosphate ions shows positive co-operativity.

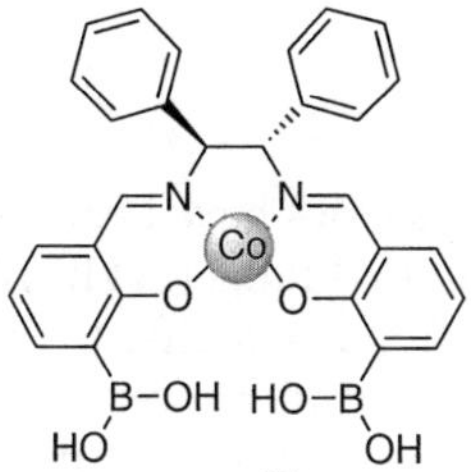

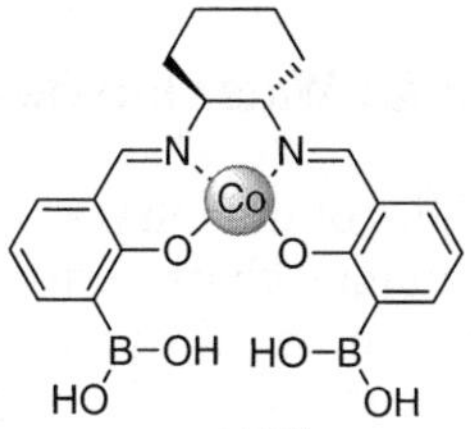

7
Electrochemical Receptors

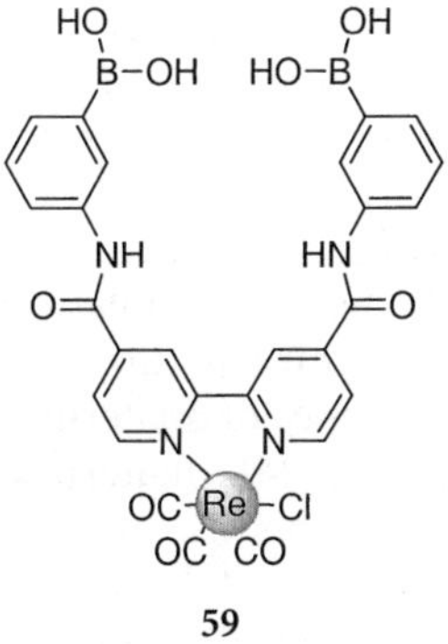

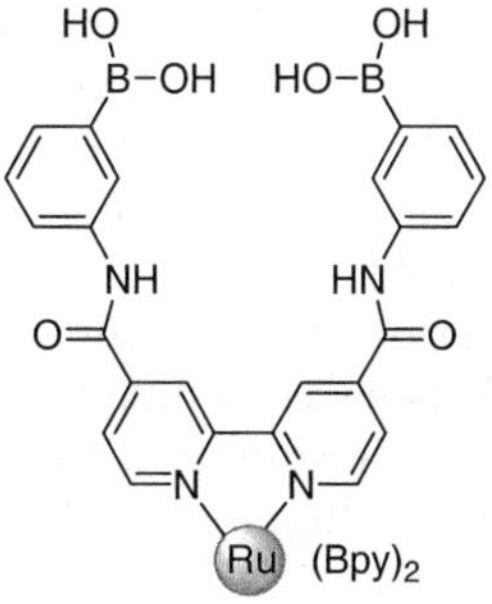

Electrochemical detection of saccharides by enzymatic decomposition of saccharides is well known [118]. The development of boronic acid-based electroactive saccharide receptors could provide selectivity for a range of saccharides. Chiral ferroceneboronic acid derivatives (−)-**61** and (+)-**61** have been synthesised and tested for chiral electrochemical detection of monosaccharides [119]. D-Fructose, D-mannitol, D-sorbitol, L-sorbitol and L-iditol gave stability constants with (+)-**61** and {(−)-**61**} of 15{14}, 28 {27}, 110 {70} and 76 {120} (mol^{-1} dm^3) respectively at pH 7.0 in 0.1 mol dm^{-3} phosphate buffer solution.

A recent paper by Moore and Wayner has explored the redox switching of carbohydrate binding with commercial ferrocene boronic acid [120]. From their

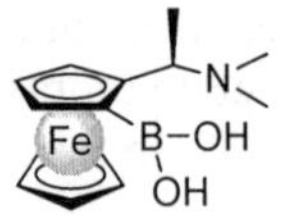

61

detailed investigations they have determined that binding constants of saccharides with the ferrocenium form are about two orders of magnitude greater than for the ferrocene form. The increased stability is ascribed to the lower pK_a of the ferrocenium (5.8) than ferrocene (10.8) boronic acid.

8
Receptors at the Air-Water Interface

Molecular assemblies of monolayers and their properties have been well established [121]. The unique characteristics of Langmuir-Blodgett (LB) films have drawn particular attention. The pressure of an LB film is sensitive to the activity of its individual constituents. The boronic acids **62** and **63** were prepared in order to demonstrate the effect of saccharide binding at the air/water interface [122]. The reactivity of the boronic acids was tested by solvent extraction methods (solid-solvent, neutral solvent-solvent and basic solvent-solvent). Extractabilities of both compounds were found to be in the order D-fructose > D-glucose > D-maltose > D-saccharose. Monolayers formed by **62** were found to be unstable based on both unreproducible pressure-area (π-A) isotherms and the crystalline nature of the monolayer. The meta-isomer, **63**, on the other hand, gave very reproducible results. The π-A isotherm of **63** was affected by the introduction of saccharide into the subphase at pH 10. The chiral cholesterylboronic acid derivative **46** behaved similarly [123]. Due to its chiral nature it could selectively recognise chiral isomers of fructose. The influence of quaternised amines on the saccharide binding has also been investigated [124]. Quaternised amines facilitate the saccharide detection by the monolayer at neutral pH. Assistance of closely located ammonium cations in the formation of boronate anion is believed to be the source of enhancement. The co-operative binding of saccharides by the diboronic acid derivatives **64(a–d)** on monolayers has also been investigated and found to be in agreement with its recognition pattern in homogeneous solutions [125, 126]. Molecular recognition in this system also seems to be facilitated by closely located ammonium cations [126].

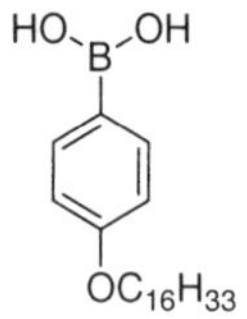

62 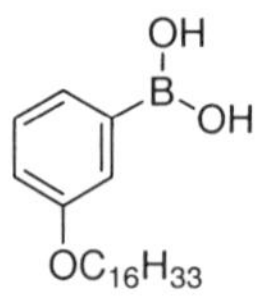   **63**

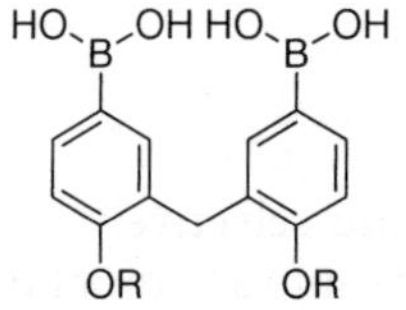

64 a: R = Methyl
b: R = 2-Octyldodecyl
c: R = 4-tert-Butylbenzyl
d: R = -CH$_2$CH=C(CH$_3$)CH$_2$CH$_2$CH=C(CH$_3$)$_2$
(Geranyl)

9
Polymeric and Imprinted Receptors

Okano and coworkers have used poly(vinyl alcohol) (PVA) polymers containing boronic acids and tertiary amines to prepare membrane coated platinum electrodes [127]. It was shown that glucose responsive swelling produced measurable current changes.

Smith has prepared a library of grafted polymers as potential sialic acid receptors [128]. The polymers were prepared using poly(allylamine) (PAA) to which 2% of boronic acid unit **65** was grafted. The final polymers also contained various amounts of 4-hydroxybenzoic acid, 4-imidazolacetic acid, octanoic acid and/or succinic anhydride.

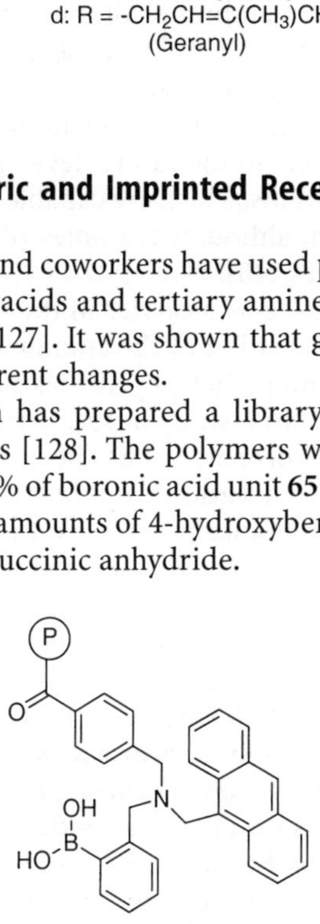

65

Polymers of poly(lysine) with boronic acids appended to the amine residue have been used as saccharide receptors [129–131]. On saccharide complexation these polymers are converted from neutral sp^2 boron to anionic sp^3 hybridised boron. The anionic polymer thus formed interact with added cyanine dye. Saccharide binding can then be 'read-out' by changes in the adsorption and ICD spectra of the cyanine dye molecule. Saccharide (backbone) polymers have been prepared by the polymerisation of diboronic acid monomers and saccharides [132, 133]. These systems have yet to be employed in chemosensor design.

A novel form of molecular imprinting in polyion complexes has been used to detect AMP using a QCM system [134, 135]. The molecular imprinting of [60] fullerene has also been used to create homogeneous chiral selective saccharide receptors [136–139].

10
Transport and Extraction

The aim of a saccharide extraction is to bind selectively with a saccharide in water and then be able to move the saccharide into a hydrophobic solvent (membrane). Saccharide transport is very similar to extraction; a saccharide must be bound in water then moved into a hydrophobic membrane. Once in the membrane the saccharide must be transported across the membrane and then released into the water on the other side of the membrane. Although the properties that are required for a good molecular extractor and transporter are similar, good transporters must balance extraction with release.

Efficient transport of saccharide related water-soluble artificial drugs into individual cells via the cell membrane are critical to the future development of drug design and delivery. Many biomimetic systems, which are capable of transporting neutral molecular species, are known, although examples of systems that can transport such species actively are rare [140].

Boronic acid and its derivatives have been used as a carrier in the transport of saccharide through membranes [141]. As formation of the anionic boronate is favoured in alkaline pH and disfavoured at lower pH, this provides a means of actively transporting glucose by a pH gradient. In a study of saccharide transport by phenylboronic acid derivatives, phenyl boronate ion has been accompanied by the lipophilic trioctylmethylammonium cation (TOMA) [141]. Not only saccharides but also saccharide-related biologically important molecular species such as uridine have been found to be effectively transported in this way [142, 143]. The negatively charged boronate complex accompanies lipophilic cation (trioctylmethylammonium chloride, TOMAC) during the transport. Paugam and Smith have used the assistance of F^- ions in saccharide transport [144]. Reaction of fluoride ions with boronic acid to form phenylfluoroboronate has been proposed as the reason for this observation. This provides a means of active transport of saccharide related molecular species at neutral pH with a fluoride ion gradient. Interestingly, other halogens do not assist in the transport. The incorporation of a cationic charge into the boronic acid could waive the requirement of an accompanying ammonium cation, such as pyridinium derivatives **66**. The strongly acidic pyridinium boronic acid **66c** transports saccharide

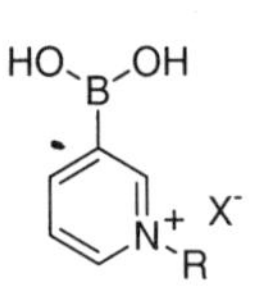

66 a. R = CH3   X = OMe
      b. R = n-C18H37   X = OTs

      c. R =

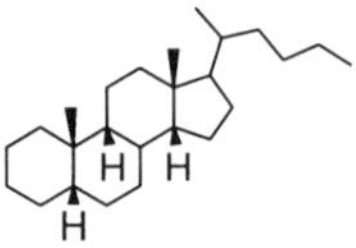

      X = OMs

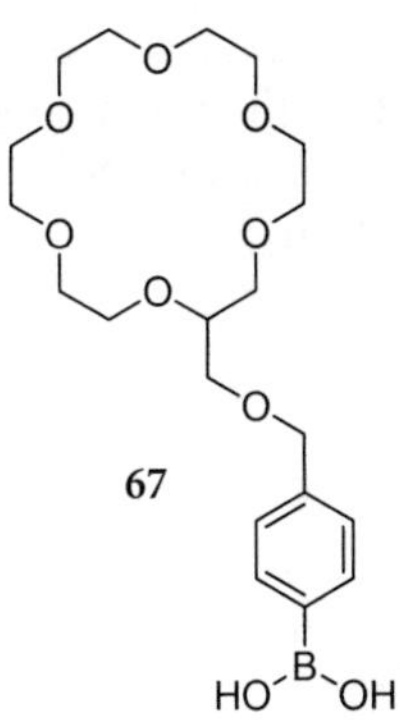

67

related species through membranes [145]. The transport of dopamine and related derivatives by the molecular receptor **67** has been reported [143]. The co-operative binding by two different receptor sites is possible in this case. Transport of amino acid derivatives by boronic acid also has been reported [146].

Smith and coworkers have investigated the ability of 21 monoboronic acids to transport saccharides through lipid bilayers [147]. It was found that lipophilic boronic acids are capable of facilitating the transport of monosaccharides through lipid bilayers, but that disaccharides are not transported. The mechanism of transport requires complexation of the saccharide as the tetrahedral boronate. However, the transported species is the neutral conjugate acid. Smith and coworkers have also investigated selective fructose transport through supported liquid membranes using both mono- and diboronic acids **68, 69a, 69b** and **69c** [148, 149]. They have also used a monoboronic acid **70** in combination with a diammonium cation **71** to facilitate the transport of bororibonucleoside-5′-phosphates [150].

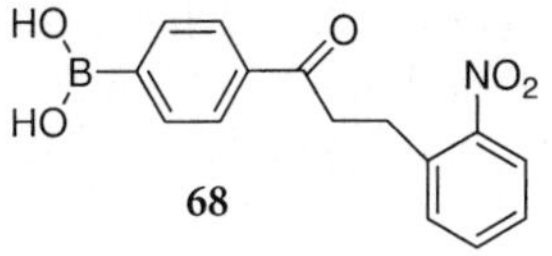

68

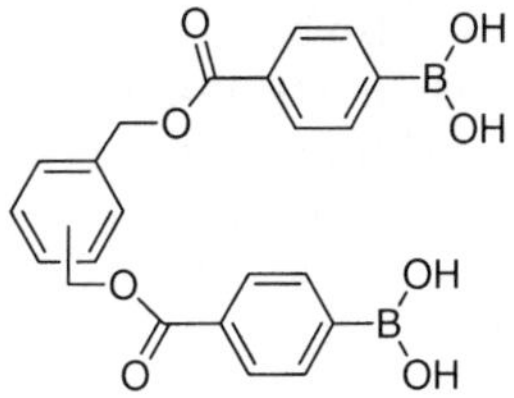

69 a: *ortho*
b: *meta*
c: *para*

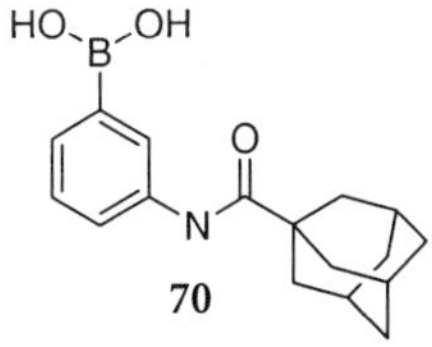

70

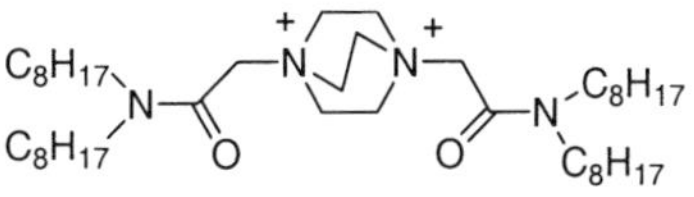

71

Boronic acids with linked ammonium ions have been used to aid saccharide extraction [151]. Using boronic acids in combination with crown ethers Smith and coworkers have developed a sodium-saccharide co-transporter **72** [152] and facilitated catecholamine transporters **73**, **74** and **75** [153].

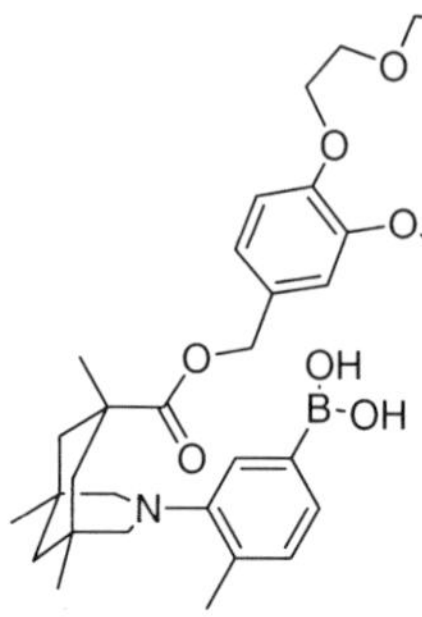

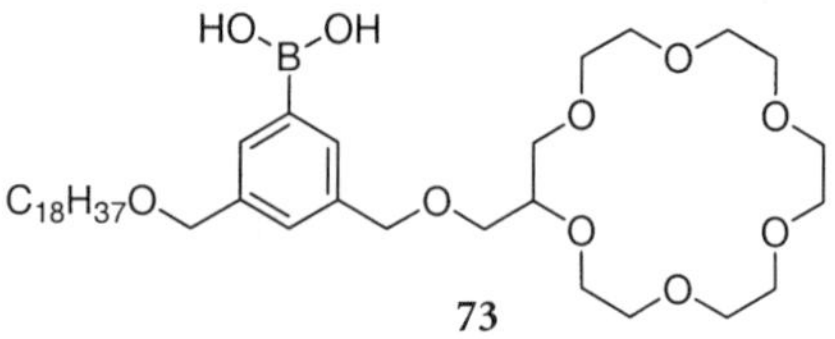

11
Fluoride Receptors

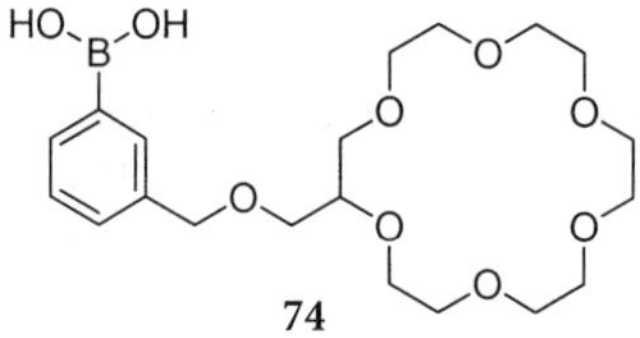

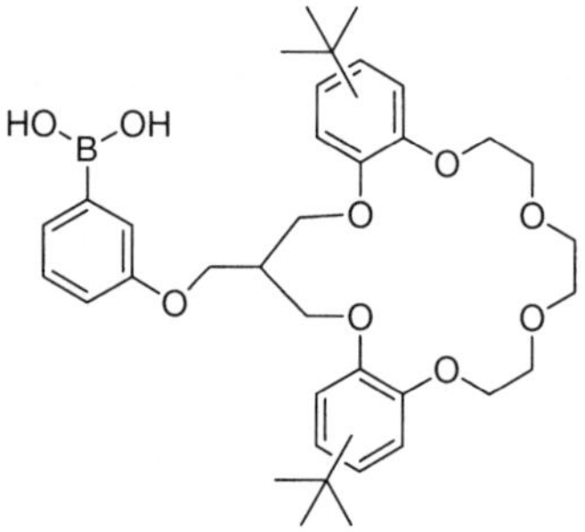

Boronic acids have also been used as anion receptors. The systems are based on the Lewis acid-base interaction between boron and anions. When boron binds with certain anions the hybridisation changes from sp^2 to sp^3 [154, 155]. A fluoride receptor based on commercial ferrocene boronic acid has been studied electrochemically [156] or by the colour change of a redox coupled dye molecule [157]. Ferrocene boronic acid **61** [119] has also been used to detect saccharides using the colour change of a redox coupled dye molecule. Commercial and synthetic **76** fluorescent fluoride sensors have been investigated [158]. Yuchi et al. have investigated the fluorescent and electrochemical properties of arylboronic acids [159]. Fabre and coworkers have used boronate-functionalised polypyrrole as a fluoride sensing material [160]. Shiratori et al. have used fluoride binding with boron to modulate the electron transfer pathways in zinc porphyrin systems [161, 162].

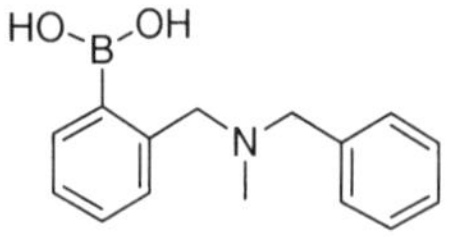

12
Conclusion

... in the end it will always be possible to understand nature, even in every new field of experience, but that we may make no a priori assumptions about the meaning of the word understand.

Werner K. Heisenberg

This chapter has hopefully introduced 'boronic acid' as a useful receptor in the design of 'chemosensors' for carbohydrates. It is hoped that the chemistry discussed will encourage other researchers to explore the uses of artificial receptors in chemosensor design. What we hope is that this chapter will encourage its readers to go out and solve many of the still unanswered problems in the design of chemosensors for carbohydrates. The solution to these problems will require the synthesis of new molecular receptors. These new receptors may contain a 'boronic acid', or may be a totally new receptor-type.

Acknowledgement. TDJ wishes to thank the Royal Society and the British Council for their financial support.

13
References

1. James TD, Kawabata H, Ludwig R, Murata K, Shinkai S (1995) Tetrahedron 51:555
2. James TD, Sandanayake KRAS, Shinkai S (1995) Supramol Chem 6:141
3. James TD, Sandanayake KRAS, Shinkai S (1996) Angew Chem, Int Ed Engl 35:1911
4. James TD, Linnane P, Shinkai S (1996) Chem Commun 281
5. Sandanayake KRAS, James TD, Shinkai S (1996) Pure Appl Chem 68:1207
6. Christophe M (1994) Tetrahedron 50:12521
7. Lappert MF (1994) Chem Rev 56:959
8. Michaelis A, Becker P (1880) Ber Dtsch Chem Ges Berlin 13:58
9. Michaelis A, Becker P (1880) Ber Dtsch Chem Ges Berlin 15:180
10. Kuivila HG, Keough AH, Soboczenski EJ (1954) J Org Chem 19:780
11. Roy GL, Laferriere AL, Edwards JO (1957) J Inorg Nucl Chem 114:106
12. Böesken J, Rossem V (1912) Recl Trav Chim Pays-Bas 30:392
13. Böesken J (1913) Ber Dtsch Chem Ges Berlin 46:2612
14. Lorand JP, Edwards JO (1959) J Org Chem 24:769
15. Norrild JC, Eggert H (1995) J Am Chem Soc 117:1479
16. Norrild JC, Eggert H (1996) J Chem Soc, Perkin Trans 2:2583
17. De Marchi S, Cecchin E, Basil A, Proto G, Donadon W, Jengo A, Schinella D, Jus A, Villalta D, De Paoli P, Santini G, Tesio F (1984) J Nephrol 4:280
18. Elsaa LJ, Rosenberg LE (1969) J Clinic Invest 48:1845
19. Baxter P, Goldhill J, Hardcastle PT, Taylor CJ (1990) Gut 31:817
20. Fedoak RN, Gershon MD, Field M (1989) Gasteroenterology 96:37
21. Yasuda H, Kurokawa T, Fuji Y, Yamashita A, Ishibashi S (1990) Biochim Biophys Acta 1021:114
22. Yamamoto T, Seino Y, Fukumoto H, Koh G, Yano H, Inagaki N, Yamada Y, Inoue K, Manabe T, Imura H (1990) Biochem Biophys Res Commun 170:223
23. Seaman W, Johnson JR (1931) J Am Chem Soc 53:711
24. Donald AD, Challenger F (1930) J Chem Soc 2171
25. Davis AP, Wareham RS (1999) Angew Chem, Int Ed Engl 38:2979

26. Shiomi Y, Saisho M, Tsukagoshi K, Shinkai S (1993) J Chem Soc, Perkin Trans 1:2111
27. Shiomi Y, Kondo K, Saisho M, Harada T, Tsukagoshi K, Shinkai S (1993) Supramol Chem 2:11
28. Tsukagoshi K, Shinkai S (1991) J Org Chem 4089
29. Kondo K, Shiomi Y, Saisho M, Harada T, Shinkai S (1992) Tetrahedron 8239
30. Sandanayake K, Nakashima K, Shinkai S (1994) J Chem Soc, Chem Commun 1621
31. James TD, Shiomi Y, Kondo K, Shinkai S (1993) XVIIIth International Symposium on Macrocyclic Chemistry, University of Twente, Enschede, The Netherlands
32. Deng G, James TD, Shinkai S (1994) J Am Chem Soc 4567
33. Weiss RG (1988) Tetrahedron 44:3413
34. James TD, Murata K, Harada T, Ueda K, Shinkai S (1994) Chem Lett 273
35. Imada T, Murakami H, Shinkai S (1994) J Chem Soc, Chem Commun 1557
36. Nagasaki T, Shinmori H, Shinkai S (1994) Tetrahedron Lett 2201
37. Shinmori H, Takeuchi M, Shinkai S (1995) Tetrahedron 51:1893
38. Arimori S, Takeuchi M, Shinkai S (1996) J Am Chem Soc 118:245
39. Arimori S, Takeuchi M, Shinkai S (1998) Supramol Sci 5:1
40. de Silva AP, Gunnlaugsson T, Rice TE (1996) Analyst (London) 121:1759
41. de Silva AP, Gunaratne HQN, Gunnlaugsson T, Lynch PLM (1996) New J Chem 20:871
42. de Silva AP, Gunaratne HQN, Gunnlaugsson T, McCoy CP, Maxwell PRS, Rademacher JT, Rice TE (1996) Pure Appl Chem 68:1443
43. de Silva AP, Gunaratne HQN, McCoy CP (1997) J Am Chem Soc 119:7891
44. de Silva AP, Gunaratne HQN, Gunnlaugsson T, Huxley AJM, McCoy CP, Rademacher JT, Rice TE (1997) Chem Rev 97:1515
45. de Silva AP, Gunnlaugsson T, McCoy CP (1997) J Chem Ed 74:53
46. Yoon J, Czarnik AW (1992) J Am Chem Soc 114:5874
47. Suenaga H, Mikami M, Sandanayake KRAS, Shinkai S (1995) Tetrahedron Lett 36:4825
48. Suenaga H, Yamamoto H, Shinkai S (1996) Pure Appl Chem 68:2179
49. Nagai Y, Kobayashi K, Toi H, Aoyama Y (1993) Bull Chem Soc Jpn 66:2965
50. James TD, Sandanayake K, Shinkai S (1994) J Chem Soc, Chem Commun 477
51. James TD, Sandanayake KRAS, Iguchi R, Shinkai S (1995) J Am Chem Soc 117:8982
52. Bielecki M, Eggert H, Norrild JC (1999) J Chem Soc, Perkin Trans 2:449
53. James TD, Sandanayake K, Shinkai S (1994) Angew Chem, Int Ed Engl 33:2207
54. Cooper CR, James TD (1998) Chem Lett 883
55. Irie M, Yorozu T, Hayashi K (1978) J Am Chem Soc 100:2236
56. Cram DJ (1986) Angew Chem, Int Ed Engl 25:1039
57. James TD, Sandanayake KRAS, Shinkai S (1995) Nature (London) 374:345
58. James TD, Shinmori H, Shinkai S (1997) Chem Commun 71
59. Linnane P, James TD, Imazu S, Shinaki S (1995) Tetrahedron Lett 36:8833
60. James TD, Shinmori H, Takeuchi M, Shinkai S (1996) Chem Commun 705
61. James TD, Shinkai S (1995) J Chem Soc, Chem Commun 1483
62. Sandanayake KRAS, James TD, Shinkai S (1995) Chem Lett 503
63. Linnane P, James TD, Shinkai S (1995) J Chem Soc, Chem Commun 1997
64. Ohseto F, Yamamoto H, Matsumoto H, Shinkai S (1995) Tetrahedron Lett 36:6911
65. Matsumoto H, Ori A, Inokuchi F, Shinkai S (1996) Chem Lett 301
66. Lu K, Wu YJ, Wang HX, Zhou ZX (1999) Polyhedron 18:1153
67. Cooper CR, James TD (1997) Chem Commun 1419
68. Cooper CR, James TD (2000) J Chem Soc, Perkin Trans 1:963
69. Desilva AP, Gunaratne HQN, McCoy CP (1993) Nature (London) 364:42
70. Iwata S, Tanaka K (1995) J Chem Soc, Chem Commun 1491
71. Takeuchi M, Yamamoto M, Shinkai S (1997) Chem Commun 1731
72. Yamamoto M, Takeuchi M, Shinkai S (1998) Tetrahedron 54:3125
73. Kijima H, Takeuchi M, Robertson A, Shinkai S, Cooper C, James TD (1999) Chem Commun 2011
74. Turro NJ (1978) Modern molecular photochemistry. Benjamin, Menlo Park, CA, USA
75. Sandanayake KRAS, Imazu S, James TD, Mikami M, Shinkai S (1995) Chem Lett 139

76. Takeuchi M, Mizuno T, Shinmori H, Nakashima M, Shinkai S (1996) Tetrahedron 52:1195
77. Takeuchi M, Yoda S, Imada T, Shinkai S (1997) Tetrahedron 53:8335
78. Eggert H, Frederiksen J, Morin C, Norrild JC (1999) J Org Chem 64:3846
79. Murakami H, Nagasaki T, Hamachi I, Shinkai S (1993) Tetrahedron Lett 6273
80. Murakami H, Nagasaki T, Hamachi I, Shinkai S (1994) J Chem Soc, Perkin Trans 2:975
81. Snyder HR, Weaver C (1948) J Am Chem Soc 70:232
82. Snyder HS, Meisel SL (1948) J Am Chem Soc 70:774
83. Russell AP (1991) Pat WO 91/04488
84. Shinmori H, Takeuchi M, Shinkai S (1996) J Chem Soc, Perkin Trans 2:1
85. Takeuchi M, Taguchi M, Shinmori H, Shinkai S (1996) Bull Chem Soc Jpn 69:2613
86. Davis CJ, Lewis PT, McCarroll ME, Read MW, Cueto R, Strongin RM (1999) Org Lett 1:331
87. Sandanayake KRAS, Shinkai S (1994) J Chem Soc, Chem Commun 1083
88. Ward CJ, Patel P, Ashton PR, James TD (2000) Chem Commun 229
89. Cabell LA, Monahan MK, Anslyn EV (1999) Tetrahedron Lett 40:7753
90. Lavigne JJ, Anslyn EV (1999) Angew Chem, Int Ed Engl 38:3666
91. Hartley JH, James TD (1999) Tetrahedron Lett 40:2597
92. James TD, Harada T, Shinkai S (1993) J Chem Soc, Chem Commun 1176
93. James TD, Harada T, Shinkai S (1993) J Chem Soc, Chem Commun 857
94. Shinkai S, Murata K (1998) J Mater Chem 8:485
95. Imada T, Kijima H, Takeuchi M, Shinkai S (1995) Tetrahedron Lett 36:2093
96. Imada T, Kijima H, Takeuchi M, Shinkai S (1996) Tetrahedron 52:2817
97. Takeuchi M, Kijima H, Hamachi I, Shinkai S (1997) Bull Chem Soc Jpn 70:699
98. Sarson LD, Ueda K, Takeuchi R, Shinkai S (1996) Chem Commun 619
99. Takeuchi M, Imada T, Shinkai S (1996) J Am Chem Soc 118:10,658
100. Takeuchi M, Imada T, Shinkai S (1998) Bull Chem Soc Jpn 71:1117
101. Arimori S, Takeuchi M, Shinkai S (1996) Chem Lett 77
102. Suenaga H, Arimori S, Shinkai S (1996) J Chem Soc, Perkin Trans 2:607
103. Arimori S, Murakami H, Takeuchi M, Shinkai S (1995) J Chem Soc, Chem Commun 961
104. Kijima H, Takeuchi M, Shinkai S (1998) Chem Lett 781
105. Takeuchi M, Chin Y, Imada T, Shinkai S (1996) Chem Commun 1867
106. Takeuchi M, Yoda S, Chin Y, Shinkai S (1999) Tetrahedron Lett 40:3745
107. Nakashima K, Shinkai S (1995) Chem Lett 443
108. Nakashima K, Shinkai S (1994) Chem Lett 1267
109. Mizuno T, Takeuchi M, Hamachi I, Nakashima K, Shinkai S (1997) Chem Commun 1793
110. Mizuno T, Takeuchi M, Hamachi I, Nakashima K, Shinkai S (1998) J Chem Soc, Perkin Trans 2:2281
111. Nuding G, Nakashima K, Iguchi R, Ishii T, Shinkai S (1998) Tetrahedron Lett 39:9473
112. Yamamoto M, Takeuchi M, Shinkai S (1998) Tetrahedron Lett 39:1189
113. Yamamoto M, Takeuchi M, Shinkai S, Tani F, Naruta Y (2000) J Chem Soc, Perkin Trans 2:9
114. Mizuno T, Takeuchi M, Shinkai S (1999) Tetrahedron 55:9455
115. Yam VWW, Kai ASF (1998) Chem Commun 109
116. Mizuno T, Fukumatsu T, Takeuchi M, Shinkai S (2000) J Chem Soc, Perkin Trans 1:407
117. Deetz MJ, Smith BD (1998) Tetrahedron Lett 39:6841
118. Schuhmann W, Schmidt H-L (1992) Adv Biosensors 2:79
119. Ori A, Shinkai S (1995) J Chem Soc, Chem Commun 1771
120. Moore ANJ, Wayner DDM (1999) Can J Chem 77:681
121. Laschewsky A (1989) Angew Chem, Int Ed Engl 28:1574
122. Shinkai S, Tsukagoshi K, Ishikawa Y, Kunitake T (1991) J Chem Soc, Chem Commun 1039
123. Ludwig R, Harada T, Ueda K, James TD, Shinkai S (1994) J Chem Soc, Perkin Trans 2:697
124. Ludwig R, Ariga K, Shinkai S (1993) Chem Lett 1413
125. Ludwig R, Shiomi Y, Shinkai S (1994) Langmuir 10:3195
126. Dusemund C, Mikami M, Shinkai S (1995) Chem Lett 157
127. Kikuchi A, Suzuki K, Okabayashi O, Hoshino H, Kataoka K, Sakurai Y, Okano T (1996) Anal Chem 68:823

128. Patterson S, Smith BD, Taylor RE (1998) Tetrahedron Lett 39:3111
129. Nagasaki T, Kimura T, Arimori S, Shinkai S (1994) Chem Lett 1495
130. Kimura T, Arimori S, Takeuchi M, Nagasaki T, Shinkai S (1995) J Chem Soc, Perkin Trans 2:1889
131. Kimura T, Takeuchi M, Nagasaki T, Shinkai S (1995) Tetrahedron Lett 36:559
132. Mikami M, Shinkai S (1995) J Chem Soc, Chem Commun 153
133. Mikami M, Shinkai S (1995) Chem Lett 603
134. Kanekiyo Y, Ono Y, Inoue K, Sano M, Shinkai S (1999) J Chem Soc, Perkin Trans 2:557
135. Kanekiyo Y, Inoue K, Ono Y, Sano M, Shinkai S, Reinhoudt DN (1999) J Chem Soc, Perkin Trans 2:2719
136. Ishii T, Nakashima K, Shinkai S, Araki K (1998) Tetrahedron 54:8679
137. Ishii T, Nakashima K, Shinkai S (1998) Chem Commun 1047
138. Ishii T, Iguchi R, Shinkai S (1999) Tetrahedron 55:3883
139. Ishii T, Nakashima K, Shinkai S, Ikeda A (1999) J Org Chem 64:984
140. Araki T, Tsukube H (1990) Liquid membranes: chemical applications. CRC Press, Boca Raton, FL, USA
141. Shinbo T, Nishimura K, Yamaguchi T, Sugiura M (1986) J Chem Soc, Chem Commun 349
142. Grotjohn BF, Czarnik AW (1989) Tetrahedron Lett 30:2325
143. Paugam MF, Valencia LS, Boggess B, Smith BD (1994) J Am Chem Soc 116:11,203
144. Paugam MF, Smith BD (1993) Tetrahedron Lett 34:3723
145. Mohler LK, Czarnik AW (1993) J Am Chem Soc 115:2998
146. Mohler LK, Czarnik AW (1993) J Am Chem Soc 115:7037
147. Westmark PR, Gardiner SJ, Smith BD (1996) J Am Chem Soc 118:11,093
148. Paugam MF, Riggs JA, Smith BD (1996) Chem Commun 2539
149. Gardiner SJ, Smith BD, Duggan PJ, Karpa MJ, Griffin GJ (1999) Tetrahedron 55:2857
150. Riggs JA, Hossler KA, Smith BD, Karpa MJ, Griffin G, Duggan PJ (1996) Tetrahedron Lett 37:6303
151. Takeuchi M, Koumoto K, Goto M, Shinkai S (1996) Tetrahedron 52:12,931
152. Bien JT, Shang MY, Smith BD (1995) J Org Chem 60:2147
153. Paugam MF, Bien JT, Smith BD, Chrisstoffels LAJ, de Jong F, Reinhoudt DN (1996) J Am Chem Soc 118:9820
154. Jacobson S, Pizer R (1993) J Am Chem Soc 115:11,216
155. Worm K, Schmidtchen FP, Schier A, Schafer A, Hesse M (1994) Angew Chem, Int Ed Engl 33:327
156. Dusemund C, Sandanayake KRAS, Shinkai S (1995) J Chem Soc, Chem Commun 333
157. Yamamoto H, Ori A, Ueda K, Dusemund C, Shinkai S (1996) Chem Commun 407
158. Cooper CR, Spencer N, James TD (1998) Chem Commun 1365
159. Yuichi A, Sakurai J, Tatebe A, Hattori H, Wada H (1999) Anal Chim Acta 387:189
160. Nicolas M, Fabre B, Simonet J (1999) Chem Commun 1881
161. Shiratori H, Ohno T, Nozaki K, Yamazaki I, Nishimura Y, Osuka A (1998) Chem Commun 1539
162. Shiratori H, Ohno T, Nozaki K, Osuka A (1999) Chem Commun 2181

Artificial Multivalent Sugar Ligands to Understand and Manipulate Carbohydrate-Protein Interactions

Thisbe K. Lindhorst

Institut für Organische Chemie, Christian-Albrechts-Universität zu Kiel, Otto-Hahn-Platz 4, 24098 Kiel, Germany
E-mail: tklind@oc.uni-kiel.de

Multivalency plays an important functional role in carbohydrate-protein interactions. Understanding of the molecular principles underlying multivalency effects is an important goal of biological chemistry. Consequently, a large number of different synthetic mulitvalent glycoligands have been designed to interfere effectively with carbohydrate-protein interactions and to facilitate the investigation of the multiple interactions occurring during these molecular recognition events. Control of inflammation processes and of microbial adhesion are important examples where multivalent carbohydrate ligands might be developed into useful therapeutics such as in the context of an anti-adhesion therapy. Furthermore, cell–cell interactions can also be manipulated by a chemical bioengineering approach which utilizes synthetic substrates in the biosynthetic pathways leading to the assembly of the complex carbohydrate environment on cell surfaces.

Keywords. Neoglycoconjugates, Glycodendrimers, Glycobiology, Multivalency, Bacterial adhesion, Oligosaccharide mimetics, Cell surface modification, Chemical bioengineering

1
Introduction

Cells are covered by a large number of different carbohydrate structures which are exposed to the exterior of the cell by membrane proteins and lipids (glycoproteins, glycolipids, GPI anchors). The carbohydrate environment on the surface of an eukaryotic cell is a complex milieu, in which the various carbohydrate molecules are arranged in a diverse and heterogeneous manner. The oligosaccharide portions can be evenly distributed across the cell surface or they are concentrated within microdomains. Their expression and mode of exposure is the subject of dynamic processes in which the cell surface characteristics undergo changes, for example, during cell differentiation and development, cell degeneration and other pathologic modifications [1–4].

The cell surface serves as a site for docking of other cells, molecules and pathogens. In this processes, parts of the oligosaccharide layer of the cell surface (glycocalyx) serve as multivalent ligands for cell-surface receptors, which bind to these carbohydrate ligands in a more or less specific recognition process. These receptors are called 'lectins' and 'selectins', respectively [5]. The stereochemistry and three-dimensional arrangement of functional groups provided by complex carbohydrates is of crucial importance for the recognition process, receptor binding and the biological signals which are eventually triggered by these molecular interactions. Thus the study of carbohydrate-protein interactions forms a key focus of glycosciences [6–9].

While the molecular information encoded in nucleic acids is fully understood nowadays, an analogous code has not yet been discovered in glycobiology. It can be expected that the underlying principles are much more diverse and flexible. While Nature affords a considerable microheterogeneity of glycoconjugates resulting from the biosynthesis of cell surface carbohydrates, only very small structural changes in the exposed saccharide structures can cause major biological effects, on the other hand. This is exemplified by the structural differences which form the basis of the different blood groups A and B (Fig. 1), where a hydroxyl group is substituted for an acetylamido group in the terminal oligosaccharide portions of erythrocytes.

As carbohydrate-protein interactions are of fundamental importance for a large number of physiological events as well as for many disease-states, there is a great interest in unraveling the secrets of molecular recognition in these processes, also with regard to control and regulation which necessarily has to occur. Based on a better understanding of carbohydrate-protein interactions, many advantageous applications for treatment of disease, their prevention and

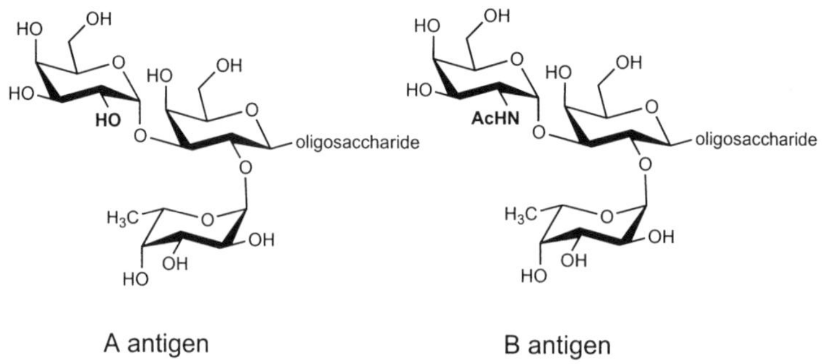

A antigen   B antigen

Fig. 1. A small structural variation in the terminal saccharide units of the glycoconjugates found on erythrocytes is the molecular basis for the differences of blood groups A and B

diagnosis should be possible. In both regards, for the investigation and for the manipulation of carbohydrate-protein interactions, synthetic glycoconjugates are of great value. Two approaches for obtaining artificial glycoconjugates may be distinguished, engineering by chemical and biochemical methods; the both alternatives are discussed in this contribution.

Interestingly, monovalent carbohydrate-protein interactions are typically associated with weak binding constants. Strength of binding and also its specificity is improved by multivalent interactions, which are found quite regularly in biosystems. It has even been postulated, that whenever systems, in which multiple ligands and receptors are involved, are concerned, multivalency of interactions is a general principle [10]. Consequently, a large amount of work has been dedicated to the design and synthesis of multivalent carbohydrate ligands and oligosaccharide analogs, following the philosophy of glycomimetic research [11]. Multivalent glycoassemblies of various kinds have been designed to interfere effectively with and control multivalent carbohydrate-protein interactions as well as to support the understanding of the principles of multivalency in molecular recognition.

In this contribution, the principal chemical architectures which have been introduced so far are briefly surveyed followed by a number of interesting and recent examples of biological applications. Then, biochemical engineering of cell surface carbohydrates is discussed and some controversial and difficult aspects of the issue of multivalency are touched on in the closing section.

2
Multivalency in Carbohydrate-Protein Interactions

It is frequently found that interactions between monovalent carbohydrate ligands and a single binding site (carbohydrate recognition domain, CRD) of a complementary receptor protein are weak with dissociation constants in the millimolar range [12]. Nature effects tight binding by multivalency which has been recognized as an important functional principle of carbohydrate-protein interactions during recent years [10, 13]. Multivalency in carbohydrate recognition is easily achieved, as the involved ligands are most frequently presented as multiple copies of recognition elements and this usually finds an analogy in a cluster of receptor binding sites. For the enthalpy gain during receptor-ligand complexation, a number of criteria have to be taken into account, such as dipole–dipole interactions, dispersive forces (London forces), and specific forces such as hydrogen bonding and n-σ bonding. In aqueous solution, however, the loss of favorable interactions of both ligand and the receptor with the solvent have to be overcompensated in the binding event because binding two solutes necessarily involves the loss of favorable interactions with solvent as discussed in detail by Toone et al. [14]. Consequently, binding is largely driven by hydrophobic effects. This and the quite shallow binding sites of lectins account for the weak binding affinities found in carbohydrate recognition.

Enhancement of affinity upon multivalent binding (Fig. 2) requires that the change in Gibbs free energy ΔG upon binding of an N-valent ligand is greater than N times the change in Gibbs free energy upon binding of each of

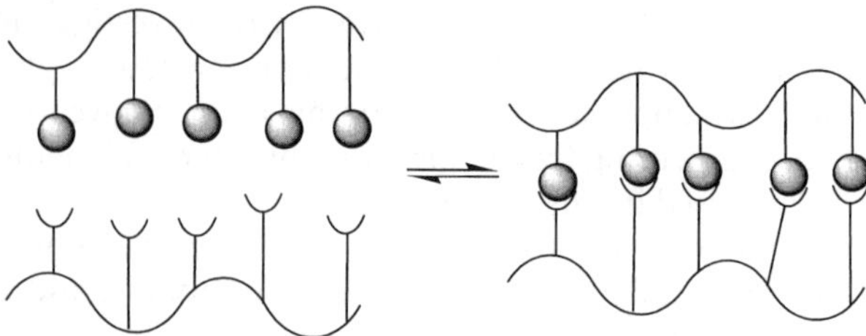

Fig. 2. Polyvalent binding between a multivalent array of receptors and multivalent ligands

the constituent monomeric ligands to an individual lectin binding site. In other words, an avidity effect is only observed when a higher binding energy is found in multivalent binding than it can be expected of the sum of individual affinities. To describe the thermodynamics of polyvalent interactions more precisely, Whitesides et al. have used the quantities α and β, where α is the cooperativity factor and β is introduced as an enhancement factor [10]. However, determination of α values requires knowledge about the number of ligands which are actually bound. Because in most instances this information is not known, β can be used for the simplest description of the phenomenology, with $K_{poly} = \beta K_{mono}$.

Multivalent interactions have several advantages and are often used by Nature to control a wide variety of cellular processes including cell surface recognition events, immune response, tumor metastasis, fertilization, and microbial adhesion. An understanding of the mechanistic principles that underlie multivalent binding events would facilitate the generation of new classes of carbohydrate-based therapeutic agents because the low affinity of saccharide ligands for their lectin receptors is, of course, one major impediment for the development of carbohydrate derivatives as therapeutics. In addition to modifying the monomeric ligands to enhance the free energy which is available from desolvation during protein-carbohydrate complexation, multivalent carbohydrate ligands have become important targets to investigate cell surface recognition. Many different types of multivalent glycoligands have been synthesized and have often shown remarkable enhancements in activity compared to an equivalent concentration of the corresponding monomeric ligands.

The first synthetic oligovalent glycoligands which showed enhanced affinities for their receptors were introduced by Y. C. Lee [15–18]. With these simple cluster glycosides a logarithmic increase of affinity for multimeric hepatic lectins was observed in a hemagglutination assay upon a linear increase of scaffolded carbohydrate ligands which were varied from one to three [19] (Fig. 3). For this observation the term 'cluster effect' was coined.

Nowadays, Y. C. Lee's finding together with a number of additional exciting observations, made in the context of multivalency, are a major incentive for the design and preparation of synthetic multivalent ligands to serve for the exploration and modification of cell-surface interactions.

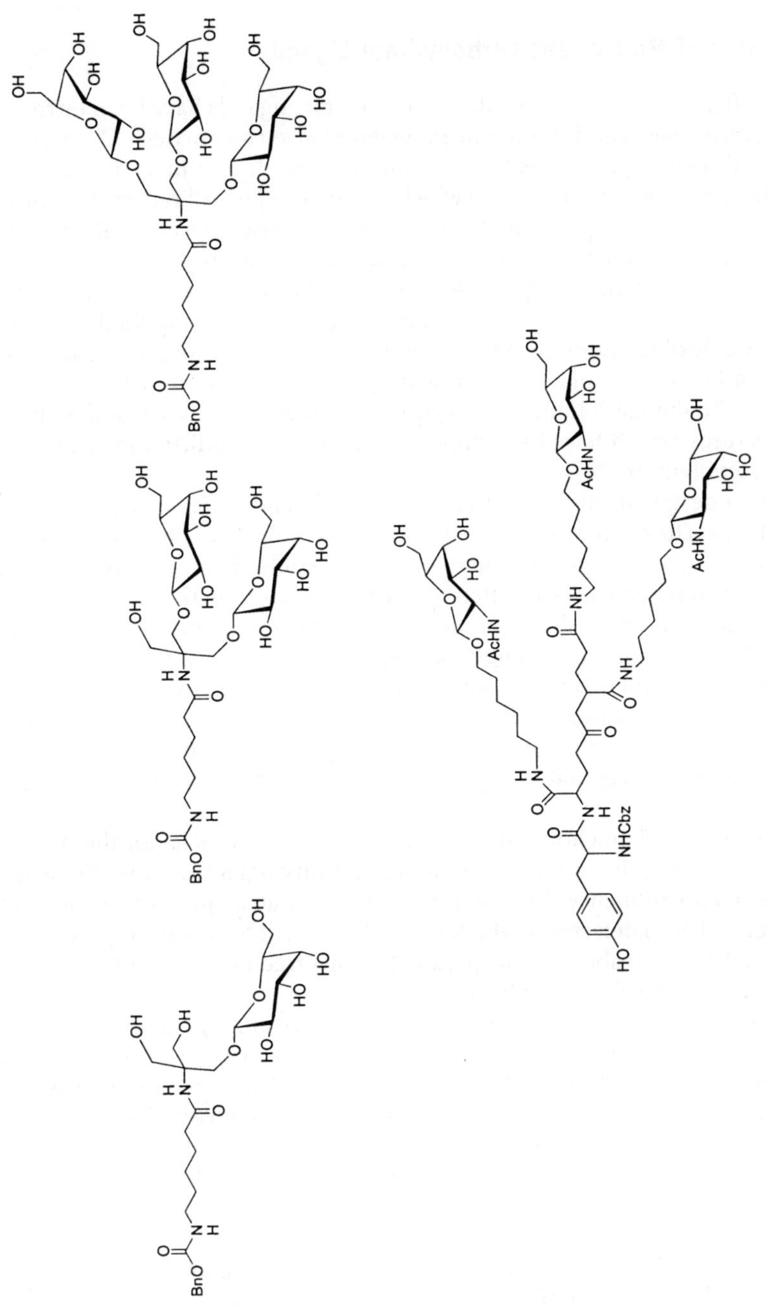

Fig. 3. Structures of cluster glycosides with which the 'cluster effect' was first observed

3
Chemistry of Multivalent Carbohydrate Ligands

Many different types of multivalent carbohydrate ligands have been introduced during recent years and this has been reviewed in various articles [1, 13, 20 – 23]. All the different approaches have in common that they aim at mimicking the complex carbohydrate structures which are involved in cell recognition processes. These endeavors span from the synthesis of oligosaccharide analogs over the mimicry of oligoantennary glycoconjugate saccharides to glycomimetics which assemble parts of the complex, polyvalent carbohydrate environment on cell surfaces. In all cases, the natural example structures are simplified and substituted by a flexible design involving a multivalent scaffold (or backbone), a minimum carbohydrate epitope to serve as ligand and spacer moieties to link the ligands with the scaffolding core. Figure 4 provides an overview of the different architectures, which have been applied so far or await utilization such as dendronized polymers [24].

The so-designed multivalent carbohydrate ligands offer a large number of advantages. They can often be synthesized much more easily than the natural example structures. They are more easily accessible than the natural material which is heterogeneous and often too scarce. The synthetic routes chosen for the preparation of multivalent neoglycoconjugates allow easy variation of scaffold symmetries, linker lengths and spacer properties, as well as the systematic variation of their size and sugar valencies. Moreover, natural and non-natural saccharide building blocks can be assembled with natural or non-natural linkages. The addition of functional groups, pharmacophores and biolabels is facilitated and conceptual extensions such as drug delivery approaches can be implied.

However, the right choice of scaffolds and linkers is crucial for the success of a multivalent neoglycoconjugate as a high-affinity ligand, because these structural elements influence the size, three-dimensional shape, and valency of the product and even determine whether a molecule will be mono- or polydisperse. A rational decision about all the parameters involved in the design of a multivalent ligand was hardly possible until today.

Relatively small oligofunctional molecules were among the first ones which were used as cores for scaffolding of sugar epitopes (Fig. 3). Tris(hydroxymethyl)aminomethane (TRIS) [15] and tripeptides [16] formed the basis of the ancestors of today's glycoclusters and cluster glycosides. Very similar molecules have recently been conjugated to lipid moieties. The resulting glycolipids, which were decorated with cluster galactosides, showed high affinity for the hepatic asialoglycoprotein receptor and were able to be loaded into liposomes which eventually showed uptake by the liver [25].

Meanwhile, many other small, tri- and tetrafunctional non-carbohydrate core molecules have been used for glycocluster synthesis such as tris(2-aminoethyl)amine [26], a number of pentaerythritol derivatives [27, 28], nitromethane-derived molecules [29] or cyclams [30], for example (Fig. 5). Based on a lysine oligopeptide, cluster mannosides were able to be obtained which showed binding to the human mannose receptor with nanomolar affinities [31] (Fig. 6).

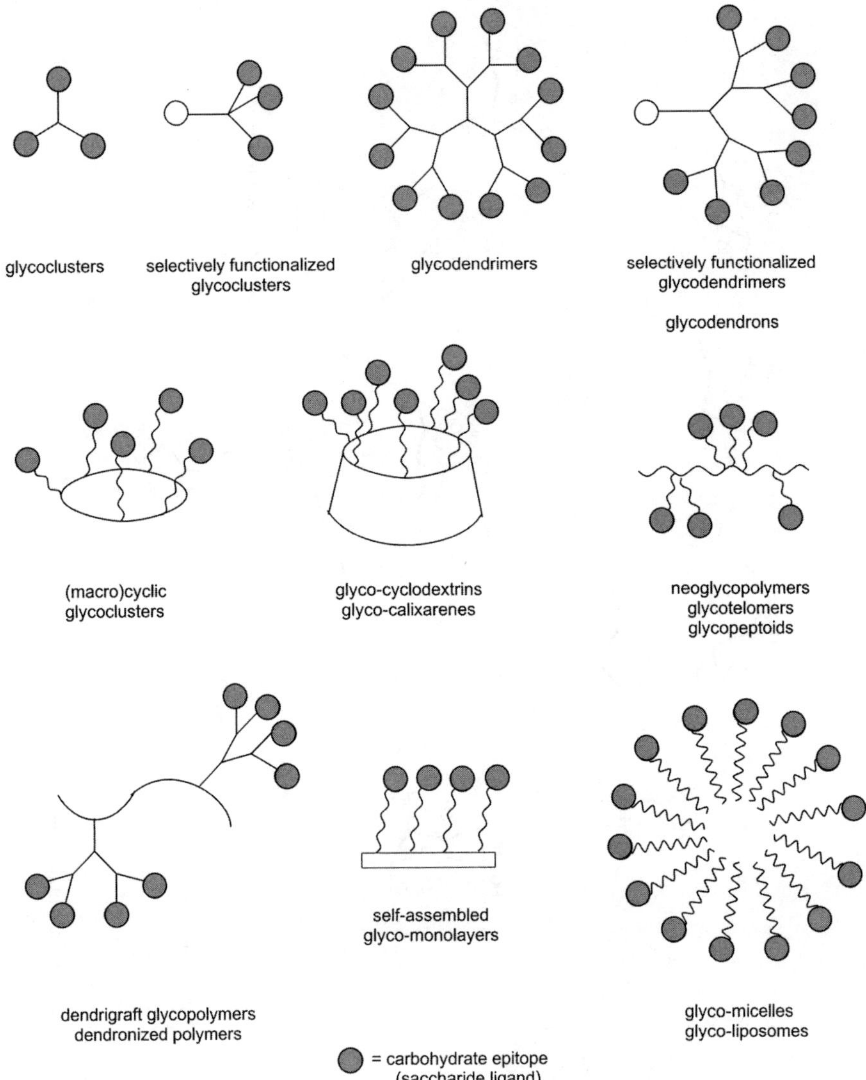

Fig. 4. Principal possibilities to architecture multivalent glycoligands

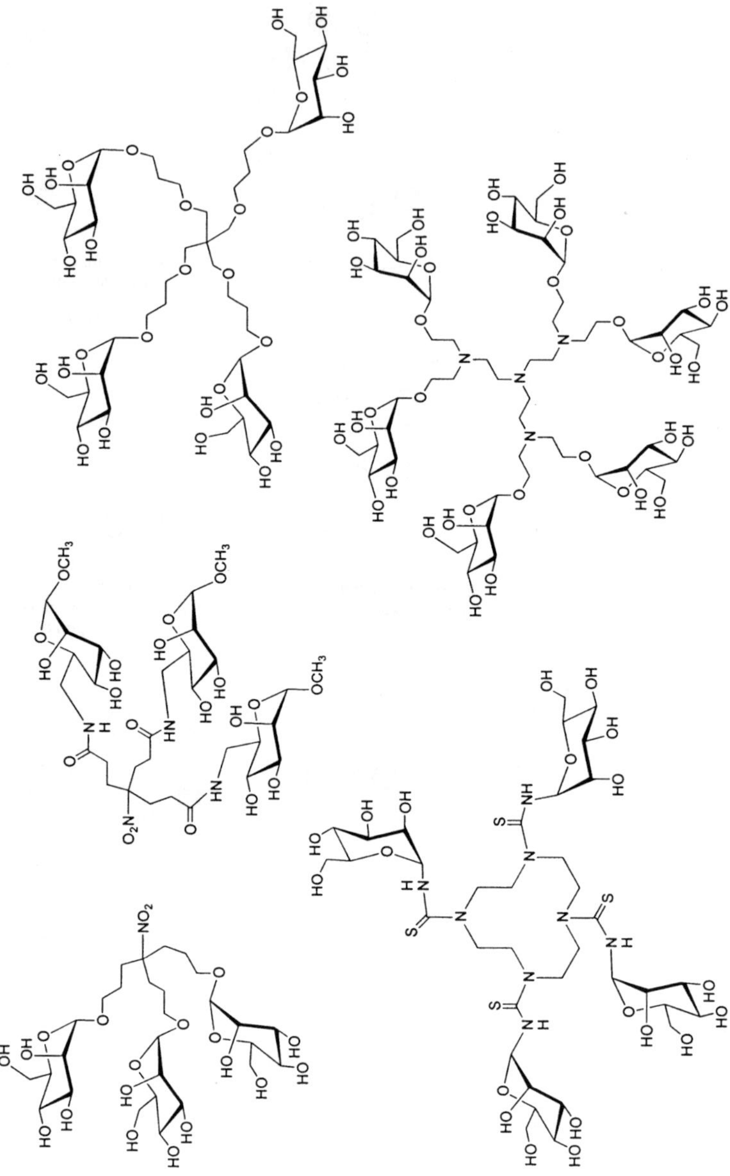

Fig. 5. Examples of various small molecular weight glycoclusters

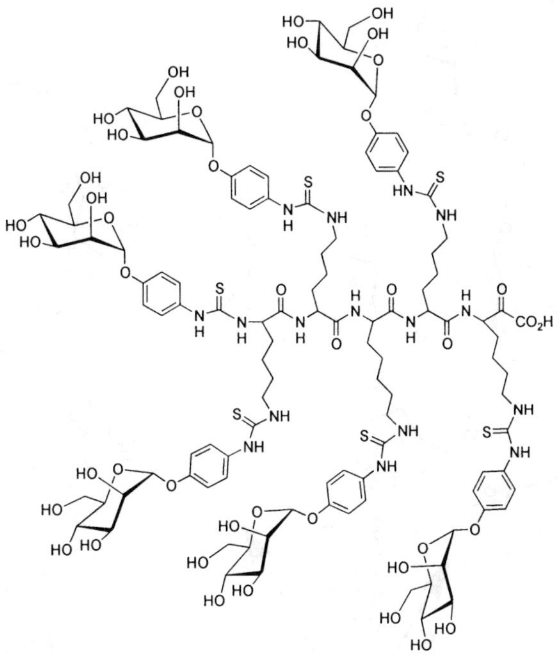

Fig. 6. Lysine-based cluster mannoside which binds to human mannose receptor at nanomolar concentrations

Recently, cyclic peptides [32] and carbohydrate derivatives were introduced as scaffolds for glycocluster synthesis. Uniformly modified glucose derivatives, so-called 'octopus glycosides' [33] were used to prepare carbohydrate-centered glycoclusters which can be regarded as oligosaccharide analogs (Fig. 7A). The same synthetic route starting from selectively functionalized octopus glycosides led to tethered carbohydrate-centered cluster glycosides with a handle for further modification, functionalization or enlargement [34] (Fig. 7B).

Extending the concept of using carbohydrates as scaffolds for the synthesis of oligosaccharide mimetics, a mannoside derivative has been synthesized which served as an orthogonally protected trifunctional scaffold for the successive assembly of oligosaccharide mimetics [35]. Peptide coupling and thiourea-bridging reactions were used in the linkage steps instead of glycosylation reactions (Fig. 8). This orthogonally functionalized scaffold also offers the advantage of synthesizing oligosaccharide mimetics of a mixed type, by introduction of sugar building blocks of different configuration.

To allow the synthesis of larger multivalent carbohydrate ligands, many different backbones have been utilized, including membranes and liposomes [36–39], polymers, and telomers, respectively. Polymers and the smaller telomers have first been prepared by polymerization and copolymerization reactions, respectively (Fig. 9) [10, 13, 40–42].

Neoglycopolymers have been synthesized by radical copolymerization involving alkenyl glycosides [43] or *N*-acryloylated glycosides [44–52]. Glycopo-

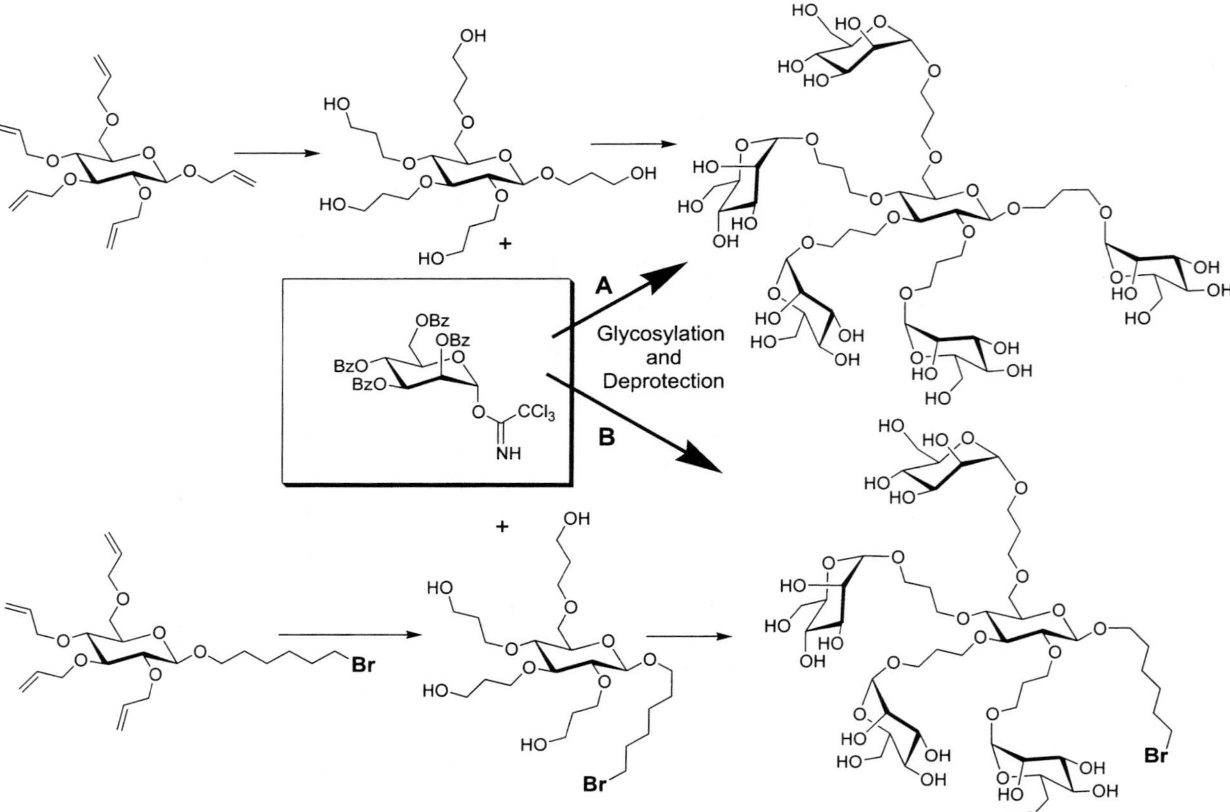

Fig. 7. A: Synthesis of carbohydrate centered cluster glycosides; B: Synthesis of selectively functionalized cluster glycosides

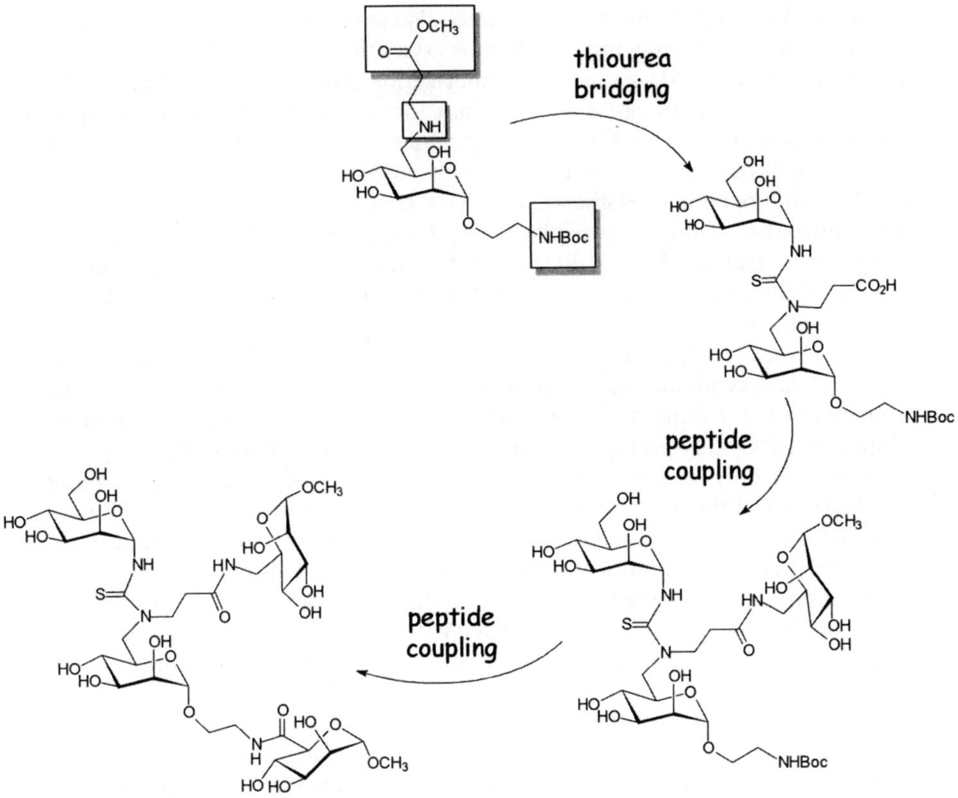

Fig. 8. Variable synthesis of oligosaccharide mimetics based on an orthogonally protected tri-functional carbohydrate scaffold molecule

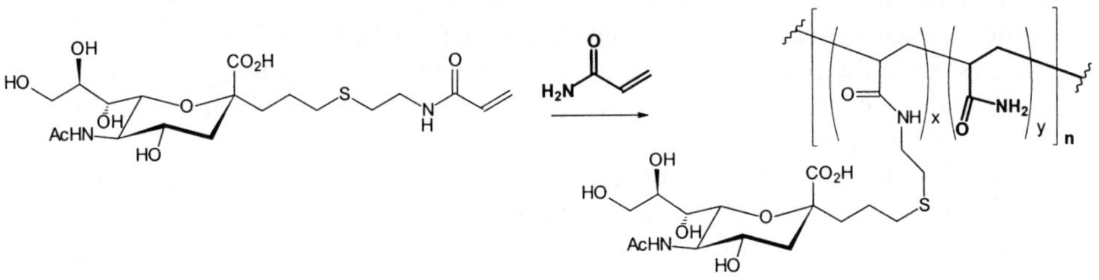

Fig. 9. Preparation of glycopolymers

lymers and sequential glycopeptides were also shown to be amenable to enzymatic modification, allowing the chemoenzymatic synthesis of these kinds of multivalent ligands [53]. As in the polymerization process very broad molecular weight distributions are obtained, especially for acrylate polymerizations, sugar-functionalization of a preformed and purified homopolymer was used as an alternative, as demonstrated with oligosaccharide-polylysine conjugates [54, 55].

Then, ring-opening metathesis polymerization (ROMP) has been utilized for the synthesis of multivalent displays of carbohydrate ligands using Ruthenium carbenes as catalysts [56–59]. ROMP can be a living polymerization and thus, if the rate of initiation is faster than that of propagation, material of defined length can be generated by adjusting the monomer/initiator ratio. Therefore, material with relative narrow polydispersities can be obtained. ROMP has been developed into a valuable synthetic method allowing easy variation of valencies and epitope density of the resulting molecules, thus leading to structure-function relationships in various biological systems [60]. This chemistry has been extensively used and elaborated by Kiessling's group [1, 20, 61, 62], allowing large-scale oligomer production and the generation of libraries of oligomers either by ROMP with glycosyl-modified cycloalkenes or by the introduction of sugar moieties in a post-polymerization modification reaction. The method also implies the possibility of end-labeling by a bifunctional capping agent which is added in excess to terminate the ruthenium-initiated ROMP reactions [63] (Fig. 10). So functionalized polymers were able to be used for conjugation with fluorescein for example.

In order to allow the synthesis of strictly monodisperse multivalent glycoconjugates, dendrimers have been selected as molecular scaffolds. The glycodendrimers so obtained are also very helpful for investigating carbohydrate-protein interactions. Their valency and size can be easily adjusted and even the chemical nature of the spacers used can be varied. Both, a divergent and a convergent strategy can be favorably followed for their synthesis. Many different initiator cores and dendritic architectures have been used for the synthesis of glycodendrimers which has been extensively reviewed [23, 64–68]. Typical glycodendrimers are carbohydrate-coated non-carbohydrate dendrimers, however, interestingly, the principles of dendrimer growth could also be applied to the synthesis of hyperbranched glycopeptide dendrons from carbohydrate building blocks [69] (Fig. 11).

In addition to the multivalent carbohydrate ligands mentioned so far, there is an increasing number of further examples, especially those involving supramolecular chemistry. Molecules which offer a cavity such as calixarenes and cyclcodextrins have been appended with carbohydrate wedges to allow site-directed drug delivery for example [70–73]. Supramolecular and self-assembled arrangements, respectively, have also been explored for the investigation of cell recognition phenomena, such as micelles and liposomes [38, 39, 74] and self-assembled monolayers [75, 76]. In the following section, the biological effects which have been obtained using multivalent glycoligands will be highlighted.

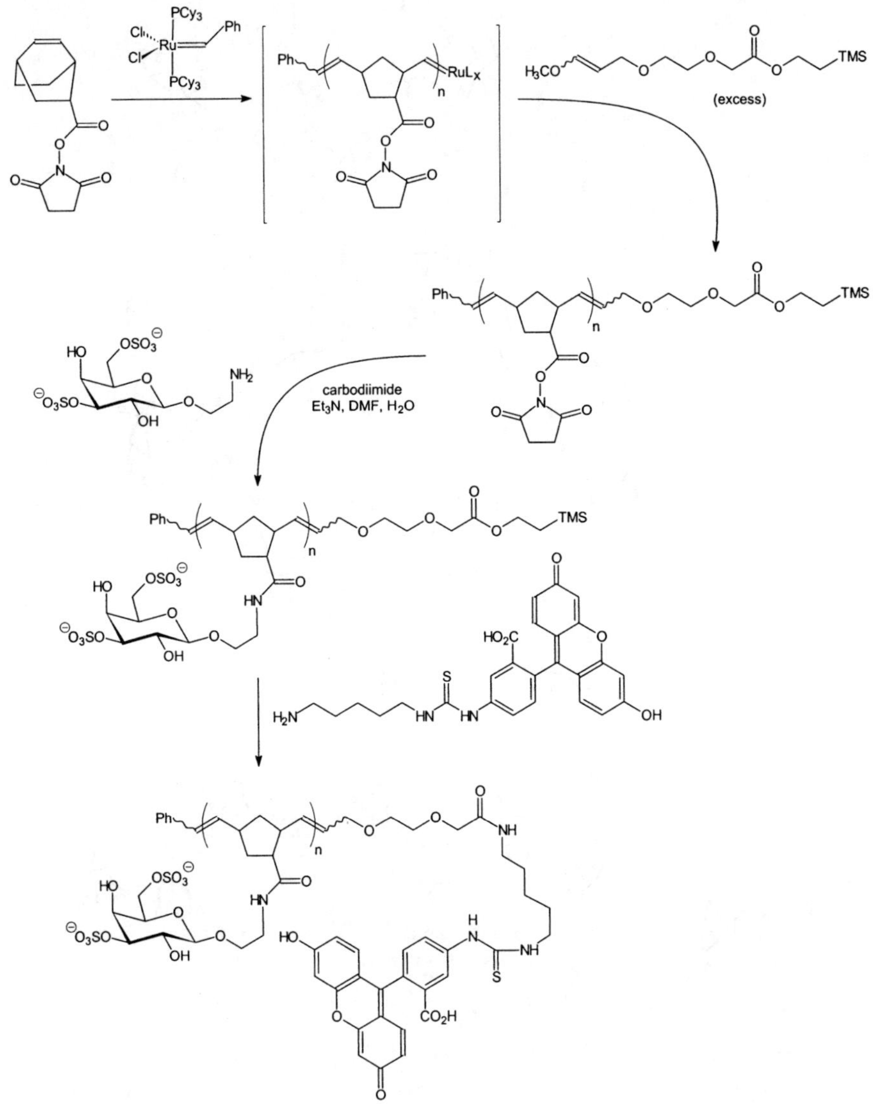

Fig. 10. Ring-opening metathesis polymerization (ROMP) and end-labeling

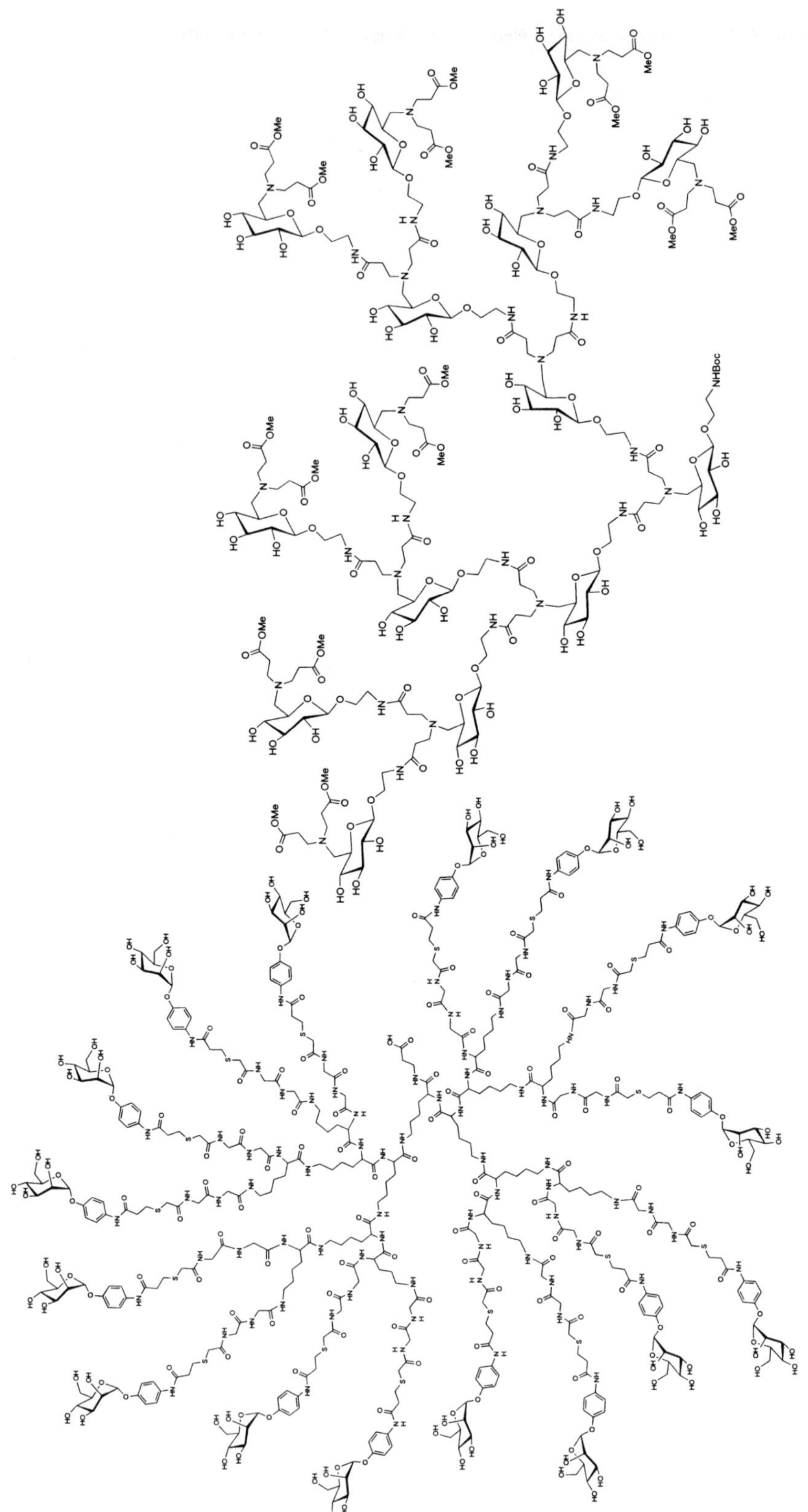

Fig. 11. Examples for glycodendrimers

4
Biology of Multivalent Carbohydrate Ligands

Synthetic multivalent ligands can be used to interfere in the biology of cell-surface interactions in two different ways. In one case, multivalent ligands can be used to prevent receptor–ligand binding, thus acting as an inhibitor, in another case multivalent ligands may be used to induce cellular response, thus assuming the function of an effector [1]. In other words, multivalent ligands may be designed and utilized as antagonists or agonists, respectively. In addition, the structural variation of multivalent glycomimetics is used to explore the structural preferences of a receptor under investigation in the context of structure-activity relationships. As the synthesis of artificial multivalent glycomimetics is relatively easy compared to the preparation of the naturally occurring compounds, a large arsenal of molecules was used to probe structure and function of the receptors under investigation. However, the syntheses are often not dominated by rational design, even more as this is difficult due to a large number of unknown parameters in each system. A number of interesting examples from the rich literature about biological effects with multivalent neoglycoconjugates has been selected here to be discussed in this section.

One important example of functional multivalency in biology are microbial infections, which require an adhesion event as a prerequisite. Often microbial adhesion is mediated by multiple carbohydrate-protein interactions. Blocking of these initiating interactions may lead to an anti-adhesion therapy against microbial infections based on carbohydrates, a concept which was first suggested by Sharon et al. some 20 years ago [77].

The best investigated microbial adhesion system is the adhesion of the influenza virus to its host cells. All types of influenza virus contact host cells through the interaction of the trimeric influenza virus hemagglutinin with *N*-acetyl neuraminic acid residues on the host-cell surface. Multiple interactions of many hemagglutinin molecules and the sialic acid molecules exposed on the host cell surface establish adhesion [78]. This interaction can be inhibited by monovalent sialic acid derivatives at millimolar concentrations. To prevent the binding of the influenza virus to its host cells at lower concentrations, neoglycopolymers were used which were capable of inhibiting the adhesion of the influenza virus hemagglutinin at 10^6 times lower concentrations [10]. Small glycoclusters and dimers were also tested with success [79, 80] and more recently, further improvements in inhibitor design in this system were reported [81].

Because the polyacrylamide backbones used in neoglycopolymers are cytotoxic, dendritic polymers were investigated as nontoxic alternative backbones. A series of differently shaped dendrigraft polymers [82] was prepared and tested for their ability to inhibit virus hemagglutinin [83]. Comb-branched and dendrigraft inhibitors showed up to the 5×10^4-fold inhibitory potency compared to the monovalent ligand with Sendai virus, however, with the influenza virus strain H2N2 an only twofold increase was observed. This clearly demonstrates the importance of ligand architecture and the three-dimensional shape of the multivalent carbohydrate ligands for biological function.

Multivalent carbohydrate ligands have also been employed in the control of inflammation processes which are mediated by leukocytes. Excessive infiltration of leukocytes from blood vessels into surrounding tissues can cause acute or chronic reactions as observed in reperfusion injuries, stroke, psoriasis, rheumatoid arthritis, or respiratory diseases [84]. An early step in the inflammatory cascade that finally leads to leukocyte extravasation is mediated by selectin-carbohydrate interactions [85], which are a prerequisite for the inflammatory cascade to take place [86]. Therefore, blocking of the initial carbohydrate-protein interactions by suitable carbohydrate derivatives may allow us to control the diseases related to leukocyte over-recruitment. A large number of monomeric ligands have been designed as selectin antagonists based on the tetrasaccharide sialyl Lewis[X] as the lead structure [87, 88]. Moreover, multivalent glycomimetics have been envisaged as ligands of the selectins to prevent leukocyte binding to endothelial cells [60, 89] and to eventually develop a class of carbohydrate-based anti-inflammatory therapeutics.

Again, multivalency plays an obvious role in this system as the chemoattractant chemokines interact with glycosaminoglycans to form multivalent arrays that recruit leukocytes through their multiple interactions with receptors on the cell surface. Multivalent 3′,6-disulfo Lewis[X] derivatives, which were prepared by ring-opening metathesis polymerization (ROMP) (Fig. 12) and were shown to inhibit L-selectin can also promote L-selectin downregulation from the surface of human neutrophils [60, 90].

Thus, these compounds combine an effector with a inhibitor function. Therefore, in addition to inhibiting L-selectin, an alternative approach to control inflammation with multivalent ligands may be envisioned, which exploits alterations in L-selectin concentration on the cell surfaces, resulting from a proteolytic cleavage event.

Receptor clustering has also been implicated in controlling the chemotaxis of bacteria toward attractants [91, 92]. Recently it was shown by Kiessling et al. that galactose- and glucose-bearing multivalent carbohydrate ligands were effective chemoattractants for the bacteria *E. coli* and *Bacillus subtilis*. The chemotactic response elicited by a ligand depended on its valency, ligands with higher carbohydrate valencies being the most potent ones [93]. This result suggests, that multivalent glycoligands lead to clustering of chemoreceptors, thus initiating and modulating receptor response.

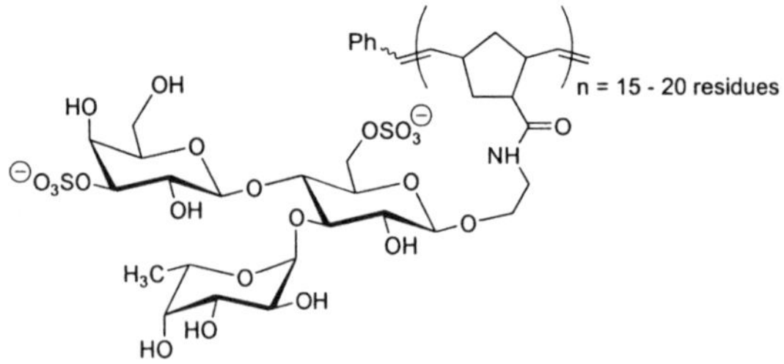

Fig. 12. Oligomeric 3′,6-disulfo Lewis[X] displays, prepared by ring-opening metathesis polymerization (ROMP)

Glycodendrimers can often, but not always, substitute for neoglycopolymers. The first example of a glycodendrimer at all was a hyperbranched oligolysine molecule, terminated with α-thiosialic acid residues, introduced by Roy et al. (Fig. 13) [94]. A glycodendrimer of the type depicted (Fig. 13) bearing 16 sialic acid residues was shown to be as potent as the analogous neoglycopolymers in the inhibition of hemagglutination of the influenza virus. Glycodendrimers containing 3′-sulfo-Lewis^{X} residues served as potent antagonists of E- and L-selectin with IC_{50} values in the low micromolar range [95]. Mannose-coated glycodendrimers which were obtained by convergent synthesis by Stoddart's group [96], however, performed rather poorly when they were tested as ligands in the inhibition of binding of the concanavalin A lectin to yeast mannan. Consequently,

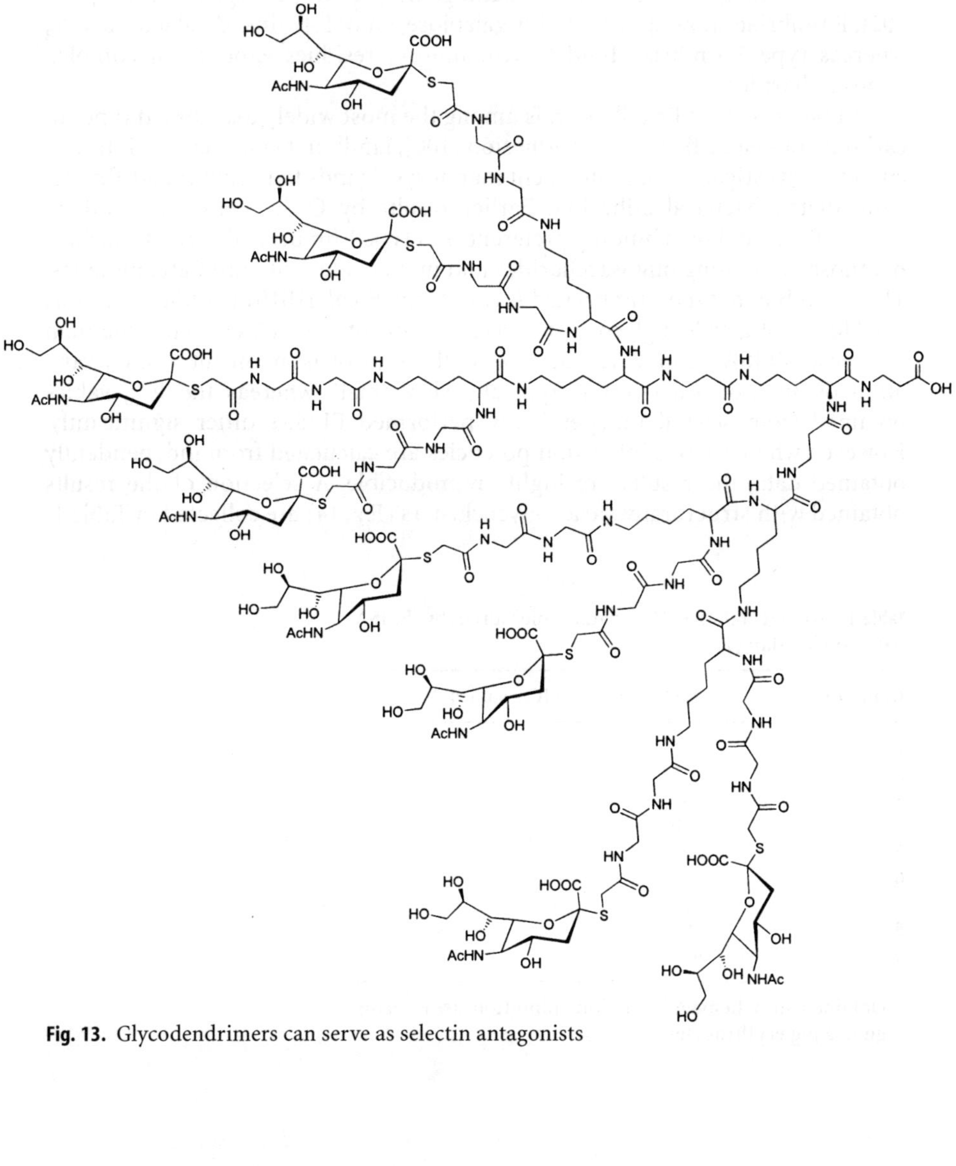

Fig. 13. Glycodendrimers can serve as selectin antagonists

many attempts have meanwhile been made to systematically vary the shape and valency of glycodendrimers and glycoclusters and this has led to numerous interesting results in many adhesion systems including concanavalin A [97–99].

Bacterial adhesion is also carbohydrate-dependent and is mediated by protein appendages on the bacterial surface, called 'pili' or 'fimbriae' [100]. They contain multiple lectin domains, which are often also referred to as 'adhesins'. Polyvalent interactions between bacterial adhesins and cell surface carbohydrates allow attachment to the host cells, a process which is a prerequisite for infection.

The fimbriae found on bacteria are classified according to their sugar specificities. *Escherichia coli* mainly utilizes two different types of fimbriae, the so-called P fimbriae, containing the PapG adhesive protein, and type 1 fimbriae, containing FimH as the adhesin. Both types of fimbriae have been shown to contribute essentially to virulence and pathogeniety in uropathogenic *E. coli* [101, 102]. P fimbriae show specificity for galabiose, an $\alpha(1,4)$-linked galactobioside, whereas type 1 fimbriae bind to α-mannosyl residues exposed in complex oligosaccharides.

As mannose-specific adhesion is among the most widely distributed types of carbohydrate-specific bacterial adhesion [100], Lindhorst et al. became interested in the investigation of multivalent mannosyl-ligands to inhibit type 1 fimbriae-mediated bacterial adhesion. Earlier results by Ofek, Sharon and others [103–105] have shown binding preferences of type 1 fimbriated *E. coli* to various mannose-containing oligosaccharides and mannosides with aromatic moieties. The recombinant type 1 fimbriated *E. coli* strain, *E. coli* HB101 (pPKl4) [106] was used to test oligovalent glycoclusters as inhibitors of their adhesion to a mannan layer in an ELISA setup [27, 29, 30, 107]. In this type of assay, the intra-assay variation of an individual test is typically very small, whereas the IC_{50} values obtained from several independently performed ELISAs differ significantly. However, when relative inhibition potencies are calculated from independently obtained data, the results are highly reproducible. A selection of the results obtained with structurally varied glycoclusters (Fig. 14) are collected in Table 1.

Table 1. Average relative IC_{50} values measured by ELISA based on MeMan (**1**)

Inhibitor	Average relative IC_{50} values
1	1
2	35[a]
3	32
4	90
5	200
6	257
7	318
8	99
9	950

[a] Obtained in a hemagglutination inhibition assay using guinea pig erythrocytes.

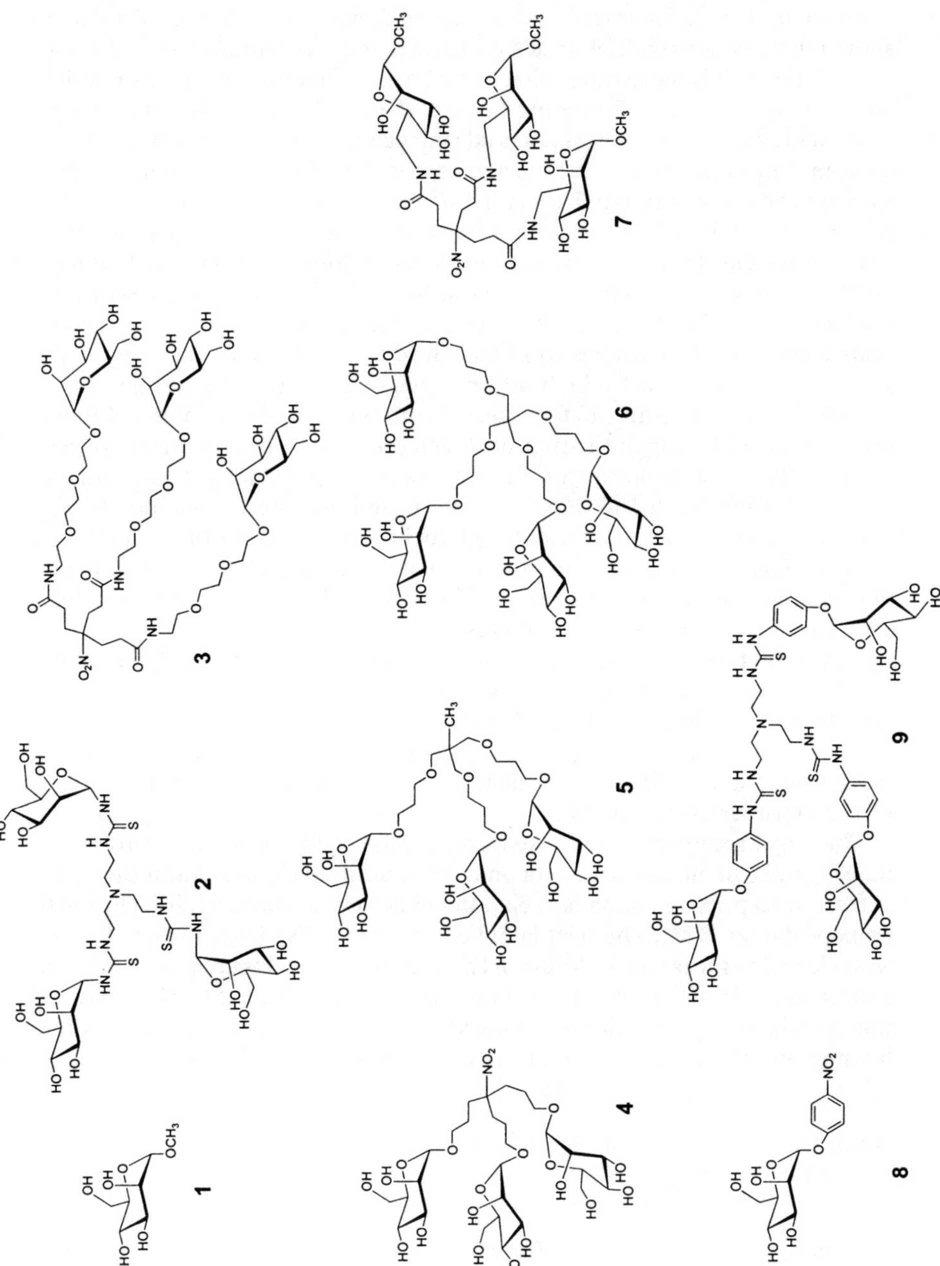

Fig. 14. Glycoclusters tested as inhibitors of type 1 fimbriae-mediated adhesion of *E. coli*

So far, no conclusive interpretation can be drawn form the obtained data to allow a more rational inhibitor design. Clearly, the incorporation of hydrophobic moieties as it is seen with p-nitrophenyl α-D-mannoside (**8**) and its trivalent analog **9**, adds favorably to binding. Some of the trivalent glycoclusters obviously resemble a conformation which is advantageous for the interactions with the adhesin. However, parts of the clusters might also bind to secondary binding sites of the receptor protein. The crystal structure of the FimCFimH protein has recently been solved [108] showing a binding pocket of the size of a monosaccharide and this does not provide an explanation for the collected testing data. Further structural work [109, 110] suggests that additional binding sites distributed over the lectin surface are of importance for the interactions with carbohydrate ligands. Further studies will finally lead to a better understanding of the structural requirements for high-affinity antagonists of type 1 fimbriae.

Quite similar structures to those tested with type 1 fimbriae inhibited P fimbriae-mediated hemagglutination of *Streptococcus suis* at nanomolar concentrations [111]. Clustering of galabioside moieties, the known ligand for the adhesive P fimbriae, led to dimeric, trimeric and tetrameric compounds (Fig. 15). The IC_{50} values obtained with the glycoclusters **12**, **13**, and **14** are compared to the inhibitory potencies of monovalent ligands **10** and **11** with and without a phenyl moiety in the aglycon part (Table 2). A molecular rational for the high affinities observed is not known at present.

If the structure of an oligovalent receptor protein is known from X-ray analysis, this may form the basis for successful tailoring of multivalent ligands. It was recently shown, that tailored pentavalent carbohydrate ligands can lead to a much higher increase in binding energy than it was possible by a more random-like introduction of multivalency such as in the synthesis of glycopolymers and glycodendrimers, respectively.

The target receptors were chosen from a class of AB_5 bacterial toxins, called the Shiga-like toxins, consisting of an enzymatically active A subunit that gains entry to susceptible mammalian cells after oligosaccharide recognition by the B_5 homopentamer [112]. The heat-labile enterotoxin (LT9) from *Escherichia coli* was selected as the target molecule in Fan's group. LT is a close relative to cholera toxin, a member of the AB_5 family of bacterial toxins [113]. The five B subsites of this protein are symmetrically arranged and represent identical carbohydrate binding sites that recognize ganglioside GM1 headgroups (Fig. 16).

Table 2. Inhibition of agglutination of human erythrocytes by *S. suis* bacteria inhibitory concentration for complete inhibition of hemagglutination of *S. suis*

Inhibitor	Type P_N	Type P_O
10	1800	1300
11	300	300
12	6	3
13	25	16
14	3	2

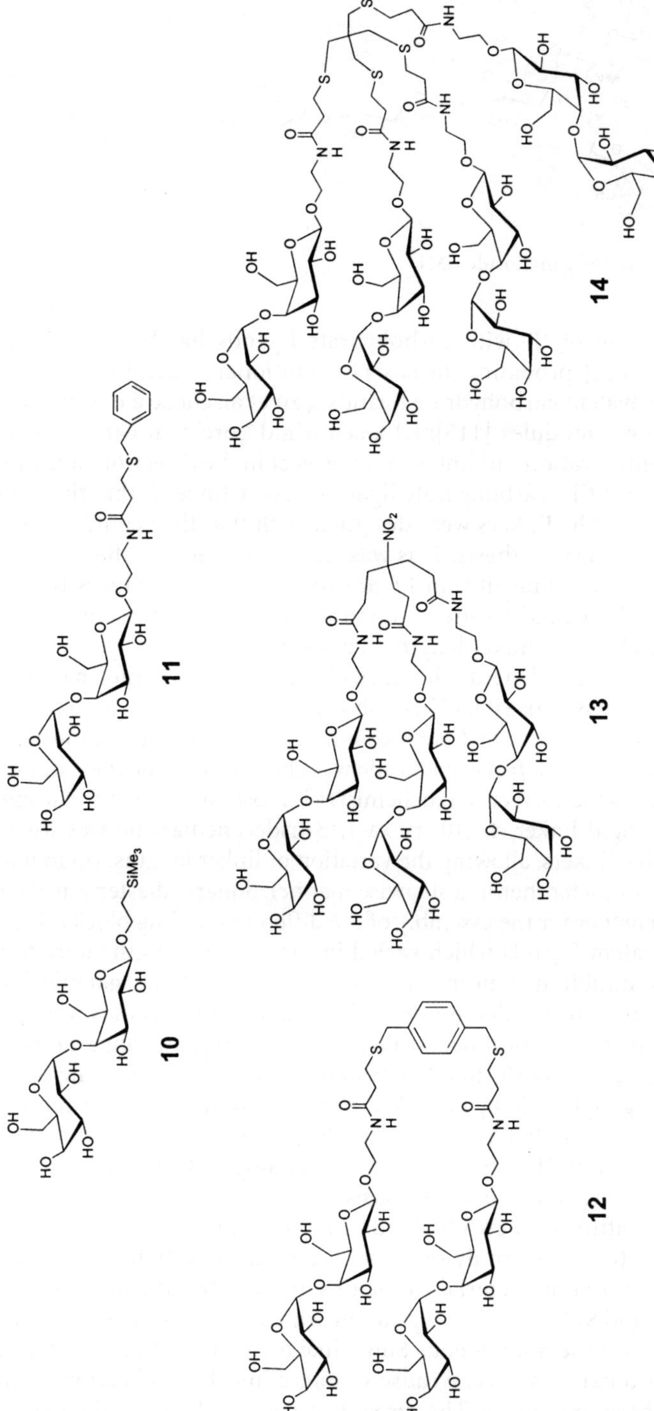

Fig. 15. Glycoclusters which inhibited hemagglutination of *Streptococcus suis* at nanomolar concentrations

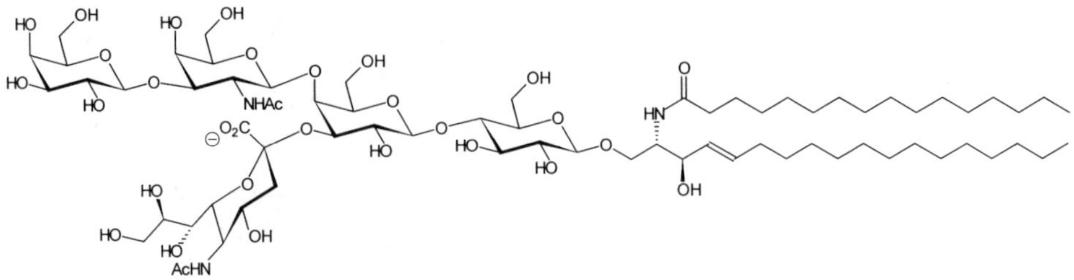

Fig. 16. Structure of the ganglioside GM1

The interaction of LT with carbohydrate ligands has been determined in atomic details [114], providing the basis for a rational structure-based design of synthetic pentavalent carbohydrate ligands. Fan et al. chose a design which was made up of three modules [115]: (i) a semirigid 'core' that can adopt a 5-fold symmetric configuration, (ii) linkers that project in the direction of the receptor binding sites, and (iii) carbohydrate ligands, called 'fingers', that fit snugly into the binding sites. The linkers were designed such that they could be oligomerized in an interactive synthesis. This was important because the dimensions of the carbohydrate-binding sites in LT are so large, with distances between the toxin's carbohydrate binding sites of up to 45 Å. The synthetic setup allowed one to explore the effects of linker length on affinity.

For the core pentacyclen was chosen, which can adopt a conformation close to a 5-fold symmetry. Because it had been shown earlier that galactosides bind in the same manner to LT's carbohydrate recognition sites as the much more complex ganglioside GM1 [114], a β-N-galactoside was chosen to mimic the natural carbohydrate ligand. As the correct stereochemistry is most difficult to meet using a conformationally rigid linker, 4,7,10-trioxa-1,13-tridecanediamine was used as the basis for flexible linkers, allowing the variation of linker lengths. Up to four units were attached to each other in a stepwise manner. Squaric diester was the principal coupling partner for the assembly of the different building blocks (Fig. 17).

The pentavalent ligands which varied in their linker lengths were tested for their ability to inhibit the binding of LT B pentamer (LT-B5) to ganglioside using a ELISA protocol. The results showed a significant affinity gain compared to the monovalent ligand. The best pentavalent ligand having the longest linkers (**21**), showed an IC_{50} that was 10^5-fold better than galactose, approaching the affinity of the natural ganglioside ligand GM1, which is about five times more potent. The gain of affinity could be attributed to the pentavalent design by comparison of the IC_{50} values of **21** with a monovalent analogue **17** revealing a 2000-fold affinity gain on a valancy-corrected basis.

Even higher affinities were obtained by Bundle's group with Shiga-like toxins from *E. coli* [116]. This family of structurally and functionally related toxins belong to the subset of bacterial AB_5 toxins and can be classified into two subgroups, SLT-I and SLT-II. [117]. Shiga toxin, which is produced by *Shigella dysenteriae* type 1, and the homologous Shiga-like toxins (SLTs) of *E. coli*, which are also called Verotoxins (VTs) can cause serious clinical complications in humans infected by these organisms. The diseases caused by Shiga and cholera toxins

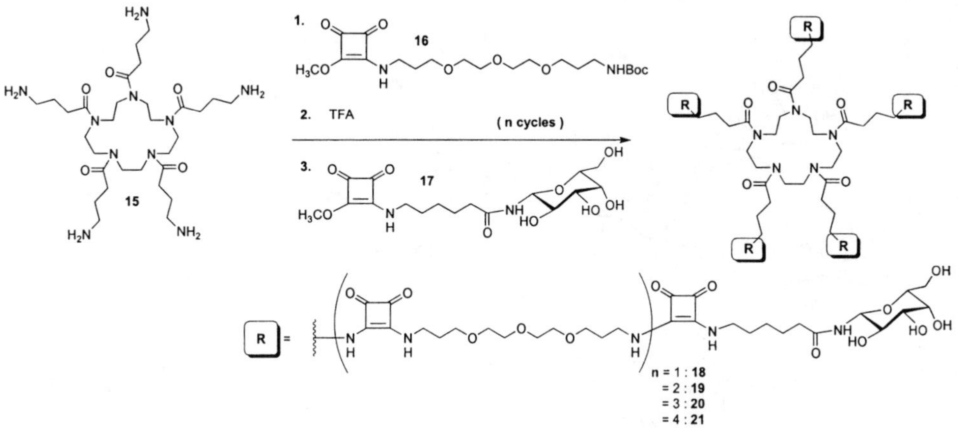

Fig. 17. Synthesis of a designed pentavalent glycoligand for heat-labile enterotoxin

account for the loss of millions of lives each year. Blocking of the initial and obligatory oligosaccharide-toxin recognition would pave the ground for a novel therapeutic approach. A suitable carbohydrate derivative providing tight binding to the B_5 pentamer structure, again, should be of polyvalent character to allow simultaneous binding to the 5 subsites thus overcoming the low affinity of every single interaction by multivalency. The crystal structure of the B_5 subunit of *E. coli* O157:H7 Shiga-like toxin I (SLT-I) in complex with an analogue of its carbohydrate receptor Gb_3 (Fig. 18) [118] was taken as the basis for the design of a pentavalent carbohydrate ligand assembling the trisaccharide epitope.

Trisaccharide **22** was selected as Gb_3 analog based on preliminary binding studies with the toxin. First a tethered dimer of this trisaccharide ligand (**23**) was synthesized leading to a 40-fold higher affinity than the univalent trisaccharide. This bridged dimer was then used as building block in the synthesis of a pentameric cluster using a pentafunctionalized carbohydrate core molecule [33] as the scaffold (Fig. 19). The linkers emerging from the glucose core were designed to span ~30 Å from the central pore of the toxin. Squaric acid diester was used to couple the trisaccharide dimer to the pentavalent core molecule. The resulting starfish-like molecule **24**, hence named STARFISH, exhibited more than a million-fold increase in inhibition of SLT-I over the univalent trisaccha-

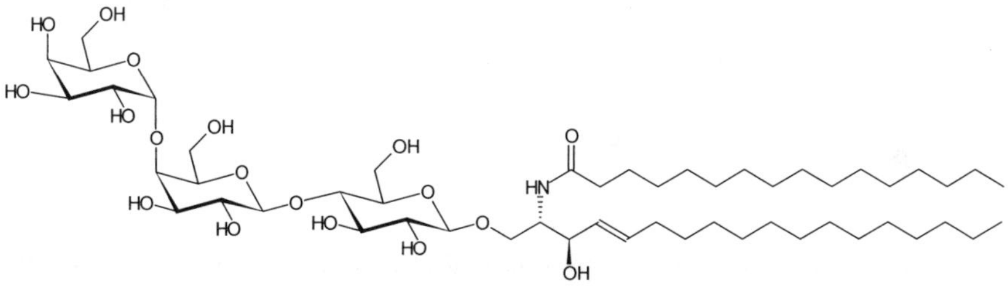

Fig. 18. Structure of the ganglioside Gb_3

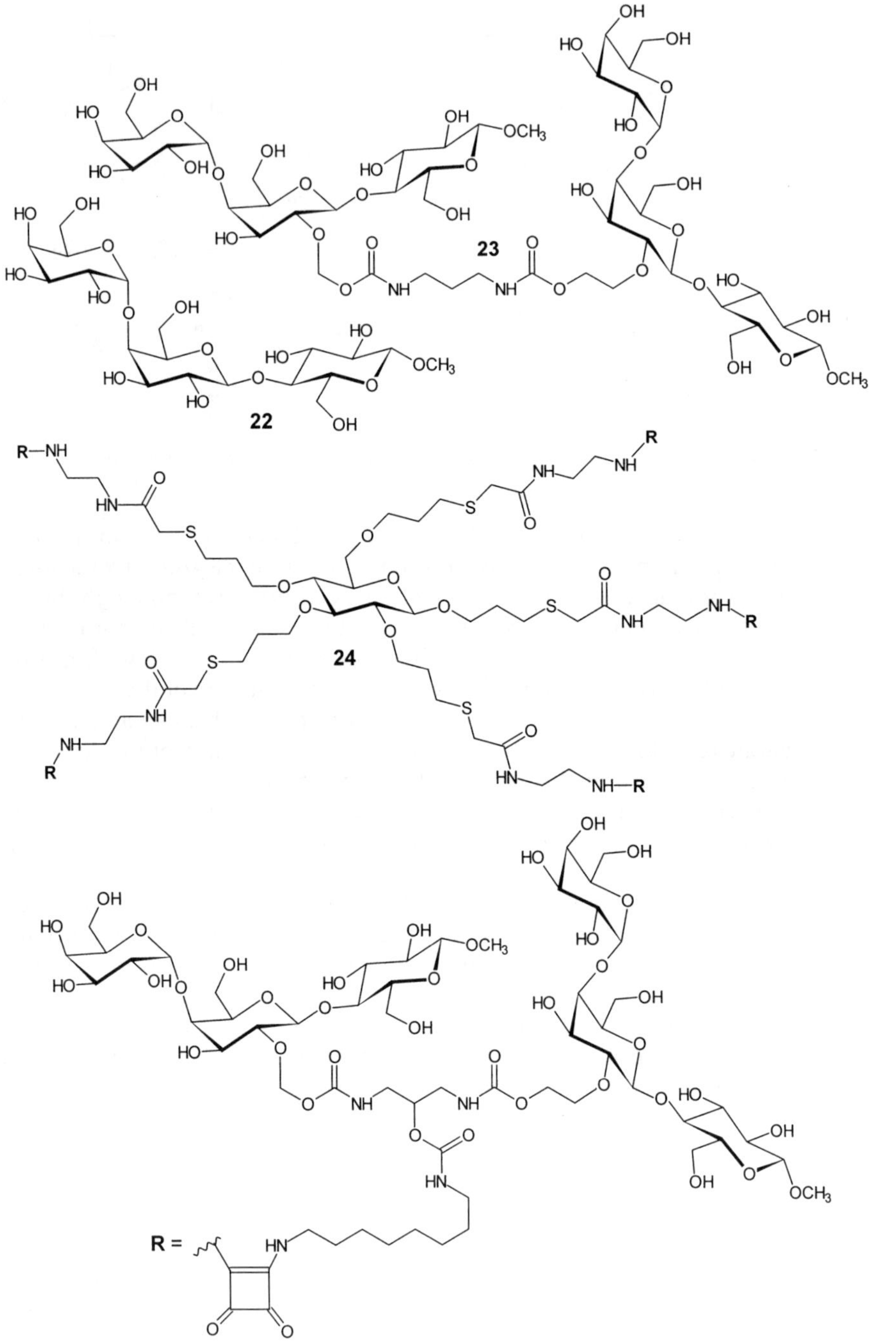

Fig. 19. STARFISH: Synthesis of a designed pentavalent glycoligand for Shiga-like toxin

ride 22. The in-vitro inhibitory activity is 1 to 10-million-fold higher than that of univalent ligands. This subnanomolar concentration lies within the range desired for an anti-adhesive therapeutic. It could be shown that the STARFISH inhibitor protects susceptible cells against prolonged exposure to SLT-I as well as SLT-II.

Interestingly, a crystallographic study of the complex formed between the B-subunit pentamer of SLT-I and STARFISH revealed a mode of binding that differs from that which was envisaged by the 'rational design' approach which was originally chosen. The crystal structure shows that one STARFISH molecule binds not one, but bridges two B-subunit pentamers, which are then positioned to bind across the face of a second toxin molecule.

5
Modification of Cell Surfaces

Scientists dealing with the chemistry and biochemistry of the various classes of biopolymers are in quite different starting positions. While oligopeptides and oligonucleotides consist of bifunctional monomers, oligosaccharides are assembled from a large variety of oligofunctional monosaccharides, each linkage step requiring regio- and stereoselectivity. Consequently, oligopeptide and oligonucleotide synthesis was able to be developed into a fully automated process many years ago, whereas the analogous approach with sugars remains a difficult adventure even today [119, 120]. On the other hand, the characteristics of the biosynthesis of these molecules offer some advantages for glycobiologists. Protein and nucleic acid biosynthesis is a template-driven process which has to be precisely controlled. Oligosaccharide synthesis, however, is governed by a large number of enzymes, glycosyl transferases and glycosidases, leading to less accurate biosynthetic results as it is reflected by the microheterogeniety [121] found in cell surface carbohydrates. When glycosyl transferases, which have been designed by Nature to form glycosidic bonds regio- and stereoselectively, are utilized in synthetic chemistry, they show some useful tolerance in regioselectivity and can even be persuaded to accept a more or less large variety of sugar derivatives as substrates in the catalyzed reaction. This also points to the intriguing possibility of using Nature's biosynthetic machinery for bio-engineering of the cell surface oligosaccharides and glycoconjugates in a living system, feeding unnatural synthetically modified substrates to the biosynthetic pathway. This is an option in glycosciences which will hardly find an analogy in protein and nucleic acid research.

Such biochemical engineering could indeed be demonstrated by the Reutter group. They showed that unnatural mannosamine derivatives, in which the *N*-acetyl group of ManNAc was substituted with *N*-propanoyl, *N*-butanoyl, or *N*-pentanoyl (Fig. 20), were converted to the corresponding sialosides and incorporated into glycoconjugates in cell culture and in rats [122–124].

Sialic acids are biosynthesized from *N*-acetylmannosamine (ManNAc) as the six-carbon precursor and phosphoenolpyruvate in an aldolase-catalyzed reaction (Fig. 21) [125]. When the unnatural ManNAc derivatives depicted in Fig. 20, ManProp, ManBut, and ManCrot, were used as substitutes for the native mono-

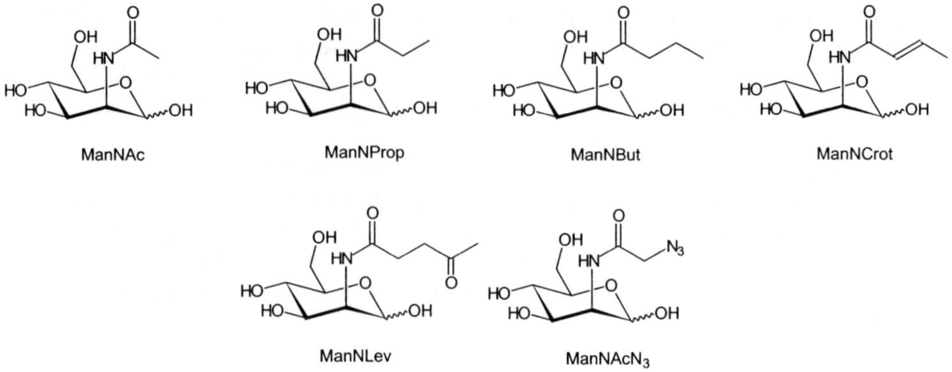

Fig. 20. The natural (ManNAc) and unnatural substrates for sialic acid biosynthesis

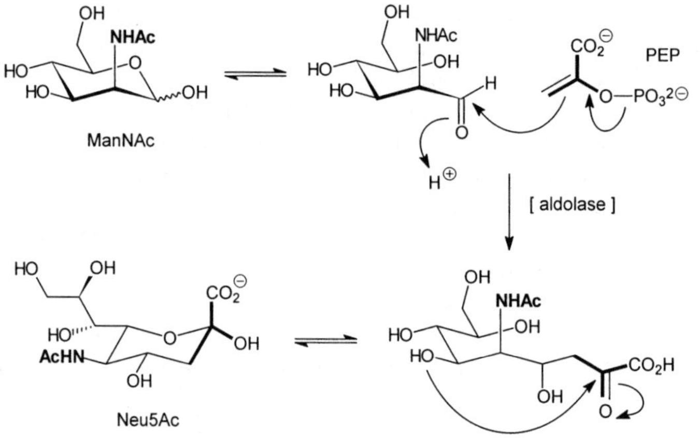

Fig. 21. Aldolase-catalyzed biosynthesis of sialic acid

saccharides they could intercept with the biosynthetic pathway (Fig. 22), finally leading to changes in cell surface glycosylation.

The work of Reutter et al. paved the way for a chemical strategy which has been used in biology by the group of Bertozzi. Bertozzi et al. reasoned that the incorporation of functionalized *N*-acyl residues in sialic acid would lead to changes in cell surface glycosylation in vivo which allows eventual chemical remodeling of the cell surface carbohydrates. This biochemical engineering technology [126] would enable direct intervention in the biological processes governed by cell surface carbohydrates.

It could be shown that levulinoyl- as well as azidoacetyl-modified ManNAc (ManLev and ManNAcN$_3$ in Fig. 20) were also tolerated by all enzymes involved in sialoside biosynthesis (Fig. 23). First, *N*-levulinoyl mannosamine (ManLev) carrying an extra ketone functionality, which is not present in ManNAc, the natural substrate of the *N*-acetylneuraminic acid aldolase, could be introduced onto the cell surface of cells using the biosynthesis pathway of sialic acid biosynthe-

N-Propanoyl-D-mannosamine

ATP — N-Acetylmannosamine kinase

N-Propanoyl-D-mannosamine-6-P

PEP — N-Acetylneuraminic acid aldolase

N-Propanoylneuraminic acid -9-P

N-Acetylneuraminic acid-9-phosphatase
P

N-Propanoylneuraminic acid

CTP — CMP-N-Acetylneuraminic acid synthetase

CMP-N-Propanoylneuraminic acid

acceptor saccharide — Sialyltransferase

N-Propanoylneuraminic acid-saccharide + CMP

Fig. 22. Proposed metabolic route of ManNProp *in vitro* and *in vivo*; P = phosphate, PEP = phosphoenolpyruvate, ATP = adenine triphosphate, CMP = cytidine monophosphate

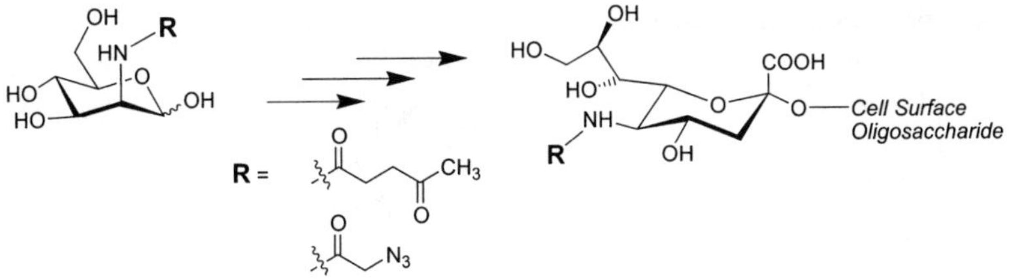

Fig. 23. "Electrophilic" sialic acid moieties can be incorporated into cell surface glycoconjugates by biochemical engineering

sis and assembly of sialylated oligosaccharides [127]. This was proved by the chemical reaction of the so engineered cells with biotinamidocaproyl hydrazide which was eventually detected and quantified on the cell surface by flow cytometry using FITC avidin/fluorescein isothiocyanate. Ketone groups are absent on the natural, chemically nucleophilic, cell surface environment and can, therefore, be chemoselectively ligated with hydrazide, hydroxylamino, and thiosemicarbazide groups under physiological conditions. However, the ketone ligation reactions have only limited intracellular use, as there are endogenous keto-metabolites such as pyruvic acid and oxaloacetate competing in the respective reactions. An alternative to a ketone-modified cell surface could be introduced by feeding N-azidoacetylmannosamine (ManNAcN$_3$) to the sialic acid biosynthetic machinery instead of ManLev. This was also tolerated and led to the delivery of azide functionalities to cell surfaces [128].

To allow chemical modification of the "azido-functionalized" cell surface, a modified Staudinger reaction was developed which permitted its use in an aqueous cellular environment. The original Staudinger reaction [129, 130] occurs between a phosphine and an azide to produce an aza-ylide. In the presence of water however, this intermediate hydrolyzes spontaneously to yield a primary amine and the corresponding phophine oxide. Therefore, a phosphine was designed which contains an appropriately situated electrophilic trap, such as a methyl ester within the phospine structure. This enables intramolecular cyclization once the nucleophilic aza-ylide is formed, producing a stable amide and preventing hydrolysis of the aza-ylide (Fig. 24).

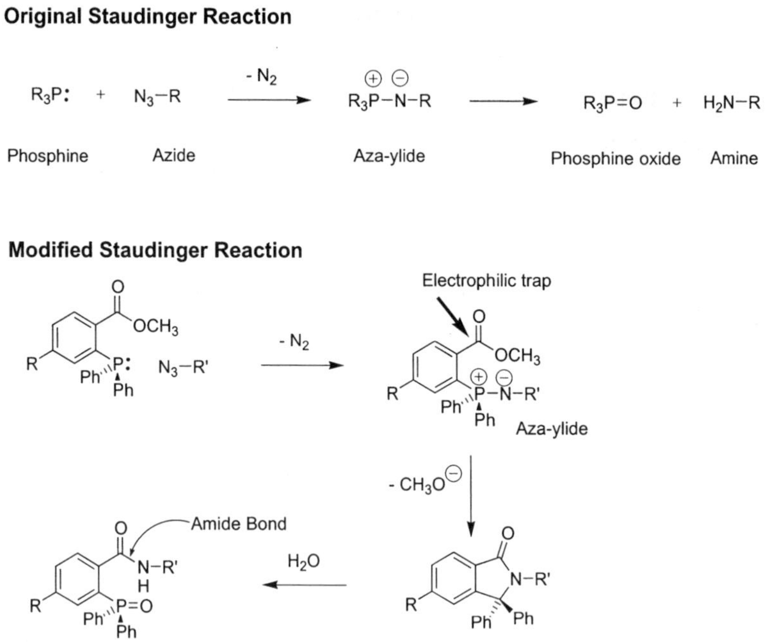

Original Staudinger Reaction

Modified Staudinger Reaction

Fig. 24. Classic and modified Staudinger reaction for biochemical engineering

The cell surface Staudinger process is superior to the cell surface ketone reaction, regarding the azide's reactivity in a cellular context and during the incorporation process where it is better tolerated than ManLev.

The possibility of manipulating the chemical reactivity of cell surfaces with biosynthetic processes opens the door to a large number of possible applications. These include the chemical construction of new glycosylation patterns on cells [131], new approaches to tumor cell targeting [132] and novel receptors for facilitating viral-mediated gene transfer [133]. Moreover, sialic acid residues which are crucial in cell communication are overexpressed on a number of human cancers, offering the expression of unnatural, reactive sialic acids as possible mechanism to differentiate cancer cells from normal cells in a new targeting strategy. Thus the ketone as well as the azido group provide molecular handles for the chemical engineering of the cell surface, with the option of utilizing the two approaches in tandem. Physiological consequences await exploration and might also lead to options for useful applications. Reutter et al. showed, that N-propanoylneuraminic acid, which was incorporated into the cell surface glycoconjugates by biochemical engineering, modulated the calcium responses of individual oligodendrocytes differentially when GABA was applied [134]. This led to the proposal that neurotransmitter-induced calcium transients in oligodendrocytes are modulated after unnatural sialylation of ionotropic GABA receptors which are sialylated glycoproteins. Eventually alteration in the sialyation of glycoproteins may provide a means for intracellular or intercellular information coding.

6
Problems and Perspectives

The investigation of receptor-ligand interactions has numerous facets. It certainly has to deal with the dispersive interactions during ligand binding to a single receptor binding site. In addition to questions about ligand constitution and configuration which are necessary for the specific formation of non-covalent complexes, the importance of conformation has become the major issue of molecular recognition during the last two decades. The elucidation of energy minimum conformations of ligands in the non-bound state and the observation of conformational changes upon complexation by various methods has brought the discussion beyond the view of a simple lock-and-key model.

Utilizing the obvious advantages of conformational preorganization and restrained conformational flexibility has allowed the design of high-affinity synthetic ligands in some instances [135]. Aiming at the synthesis of rigid ligands, however, is a risky adventure not necessarily crowned by success. If the conformation of the receptor binding site which is complementary to the three-dimensional structure of a high-affinity ligand, is unknown, the conformational space which has to be explored in ligand design is quite large, depending on the amount of information which is available about the receptor and the binding site, respectively. As exemplified in Figure 25, a ligand structure with a high amount of conformational degrees of freedom (such as A and B in Fig. 25) might be able to adopt the conformation which is required for binding the receptor R (A in Fig. 25). The binding constant, however, will be weak as the entropic penalty

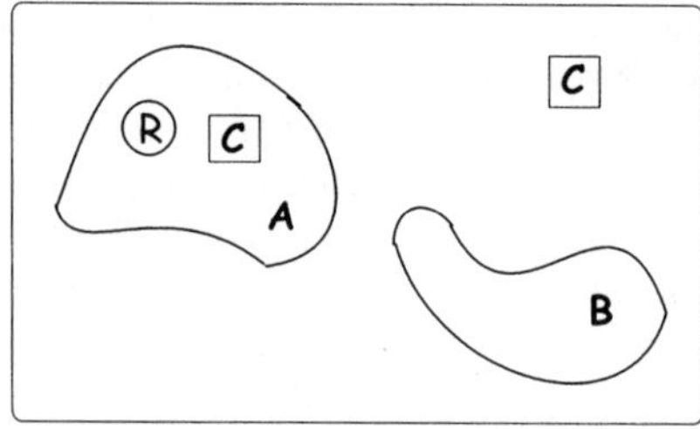

Fig. 25. Exploring conformation space in ligand receptor binding

which has to be paid upon binding is high. A rigid ligand covers only a small area of the conformational space (type C in Fig. 25), however, its chances to meet the structure complementary to the receptor conformation R are low and binding will be zero, even if the correct conformation was only missed very narrowly [136].

In addition to the questions which have to be asked with regard to the fit of a ligand into the receptor binding site, the situation can be more complex due to additional binding sites of the receptor protein. These may be utilized to improve binding and even to preorganize some parts of a more complex ligand for receptor binding. The same problems, as they were just discussed, apply for an optimal ligand design in this scenario. The situation becomes even more complicated when multiple binding events are a central part of molecular recognition. Considering multivalency in carbohydrate-lectin and carbohydrate-selectin interactions confronts one with problems even beyond the already difficult questions about the general forces and interactions that provide both affinity and specificity during a monovalent association process in aqueous solution.

It is obvious that Nature has designed multivalency to improve affinity and specificity as well as to induce supramolecular rearrangements followed by biological signal induction and transduction [10]. Multivalent carbohydrate ligands are fascinating tools for exploring multivalency in carbohydrate-protein interactions as well as effectively interfering in these processes. This was demonstrated first some 20 years ago using cluster glycosides as ligands for the hepatic asialoglycoprotein receptor [18]. Even today, the origin of the cluster glycoside effect found in this system is not completely understood at a molecular level. It can be reasoned by an entropy gain in analogy to the chelate effect [133]. This, however, is only reasonable when a conformationally quite restricted multivalent ligand is applied, providing a perfect fit for a multiple binding site arrangement. As this is a rare scenario and difficult to design, cluster effects as found by Y. C. Lee are hardly regularly observed with synthetic ligands. Multivalent high affinity interactions must be largely influenced by entropy because the enthalpy gained in every single monovalent interaction cannot account for the avidity effects which have been observed.

Nevertheless, multivalent carbohydrate ligands are among the most useful tools for probing and investigating multivalency in carbohydrate-protein interactions and investigating cellular interactions. Thus, the principles of carbohydrate-protein interactions may be explored starting from the least complex model compound and adding complexity from there. On the other hand, chemists are doomed to a random design, when the structural information about the receptors is only limited. Linkers have to be designed so, that they are long enough to facilitate optimal placement of ligands within binding sites, too short or too rigid linkers will add unfavorably to interaction enthalpies. On the other hand, flexibility of linkers leads to a conformational entropic penalty upon binding, which should rather be minimized. Furthermore, upon binding, the linker will almost certainly contact the surface of the protein at the periphery of the binding site and thus, the chemical nature of the linker is important for the multimeric ligand design. A prediction of favorable or unfavorable contributions to ΔH, however, is almost impossible. While binding is largely driven by hydrophobic effects, dipolar and ionic interactions and hydrogen bonding play important roles in determining the specificity of solute-solute interactions and the precise structure of the bound complex as they are directional interactions. Even though they are not likely to contribute strongly to the overall net negative free energy of complexation they have therefore, to be taken into account [14].

It has also to be considered that multivalent ligands can bind multiple binding sites by a number of different mechanisms, and this makes the exploration of the underlying principles much more difficult than when only one binding site is investigated. In addition, the spatial arrangement and flexibility of the assembled carbohydrate epitopes form a number of additional variables of which it is uncertain how they have to be orchestrated for optimum cooperative binding. The discussion about possible interpretations of how multivalent ligands function is not closed. Multivalent ligands may simultaneously occupy several carbohydrate recognition domains of the respective receptors, which may occur as oligomers or monomers that contain several CRDs. Simultaneous binding may also occur through a process which leads to clustering of receptors on a cell surface and eventual activation of signaling pathways. On the other hand, it is also possible, that enhanced binding affinity is obtained not by occupation of more than one binding sites, but through higher local ligand concentrations as provided by multivalent ligand displays. Moreover, the mechanistic principles that underlie multivalent binding events in vivo can be different, depending on what Nature has designed for the various biological circumstances.

For the various in vitro testing systems which have been used to evaluate the antiadhesive properties of multivalent glycoligands, interpretations have to be made with caution [138] because the nature of the assay is crucial for the results obtained. Gel-like layers or other higher-order aggregates such as lattices may form and thus lead to the observation of a quasi "artificial" enhancement of affinity [14].

The physical chemistry of the effects observed are far from being fully understood. Additional biophysical methods may help to solve the questions about the thermodynamics and kinetics of multivalent ligand-receptor interactions. Nevertheless, multivalency is one of the most promising strategies for achieving high-affinity binding and for the development of therapeutical useful carbohydrate derivatives.

7
References

1. Kiessling LL, Gestwicki JE, Strong LE (2000) Curr Opin Chem Biol 4:696
2. Varki A, Cummings R, Esko J, Freeze H, Hart G, Marth J (eds) (1999) Essentials of Glycobiology. Cold Spring Harbor Laboratory Press, La Jolla
3. Bertozzi CR, Kiessling LL (2001) Science 291:2357
4. Karlsson K-A (1995) Curr Opin Struct Biol 5:6221
5. Lis H, Sharon N (1998) Chem Rev 98:637
6. Varki A (1993) Glycobiology 3:97
7. Kobata A (1993) Acc Chem Res 26:319
8. Dwek RA (1996) Chem Rev 96:663
9. Rudd PM, Elliott T, Cresswell P, Wilson IA, Dwek RA (2001) Science 291:2370
10. Mammen M, Choi SK, Whitesides GM (1998) Angew Chem 110:2908; Angew Chem Int Ed 37:2754
11. Sears P, Wong C-H (1999) Angew Chem 111:2446; Angew Chem Int Ed 38:2300
12. Toone EJ (1994) Curr Opin Struct Biol 4:719
13. Kiessling LL, Pohl NL (1996) Chem Biol 3:71
14. Burkhalter NF, Dimick SM, Toone EJ (2000) In: Ernst B, Hart GW, Sinaÿ P (eds) Carbohydrates in Chemistry and Biology 2:863
15. Lee YC (1978) Carbohydr Res 67:509
16. Lee RT, Lee YC (1987) Glycoconjugate J 4:317
17. Lee RT, Lin P, Lee YC (1984) Biochemistry 23:4255
18. Lee YC, Lee RT (1995) Acc Chem Res 28:321
19. Connolly T, Townsend RR, Kawaguchi K, Bell WR, Lee YC (1982) J Biol Chem 257:939
20. Kiessling LL, String LE, Gestwicki JE (2000) Annu Rep Med Chem 35:321
21. Lindhorst TK (1996) Nachr Chem Tech Lab 44:1073
22. Sears P, Wong C-H (1998) Chem Commun:1161
23. Jayaraman N, Nepogodiev SA, Stoddart JF (1997) Chem Eur J 3:1193
24. Zistler A, Koch S, Schlüter AD (1999) J Chem Soc, Perkin Trans 1:501
25. Sliedregt LAJM, Rensen PCN, Rump ET, van Santbrink PJ, Bijsterbosch MK, Valentijn ARPM, van der Marel GA, van Boom JH, vam Berkel TJC, Biessen EAL (1999) J Med Chem 42:609
26. Lindhorst TK, Kieburg C (1996) Angew Chem 108:2083; Angew Chem Int Ed Engl 35:1953
27. Lindhorst TK, Dubber M, Krallmann-Wenzel U, Ehlers S (2000) Eur J Org Chem:2027
28. Langer P, Ince SJ, Ley SV 1998, J Chem Soc, Perkin Trans 1:3913
29. Lindhorst TK, Kötter S, Krallmann-Wenzel U, Ehlers S (2001) J Chem Soc, Perkin Trans 1:823
30. König B, Fricke T, Waßmann A, Krallmann-Wenzel U, Lindhorst TK (1998) Tetrahedron Lett 39:2307
31. Biessen EAL, Noorman F, van Teijlingen ME, Kuiper J, Barrett-Bergshoeff M, Bijsterbosch MK, Rijken DC, van Berkel TJC (1996) J Biol Chem 271:28024
32. Wittmann V, Seeberger S (2000) Angew Chem 112:4508; Angew Chem Int Ed 39:4348
33. Dubber M, Lindhorst TK (1998) Carbohydr Res 310:350
34. Dubber M, Lindhorst TK (2000) J Org Chem 65:5275
35. Patel A, Lindhorst TK (2000) J Org Chem 66:2674
36. Spevak W, Foxall C, Charych DH, Dasgupta F, Nagy JO (1996) J Med Chem 39:1018
37. Spevak W, Nagy JO, Charych DH, Schaefer ME, Gilbert JH (1993) J Am Chem Soc 115:1146
38. Kingery-Wood JE, Williams KW, Sigal GB Whitesides GM (1992) J Am Chem Soc 114:7303
39. Spevak W, Nagy JO, Charych DH, Schaefer ME, Gilbert JH, Bednarski MD (1993) J Am Chem Soc 115:1146
40. Bovin NV, Gabius JH (1995) Chem Soc Rev:413

41. Roy R (1996) TIGG 8:79
42. Roy R (1996) In: Khan SH, O'Neill RA (eds) Modern Methods in Carbohydrate Synthesis. Harwood Academic Publishers, Amsterdam, p 378
43. Kobayashi K, Akaike T, Usui T (1994) Methods Enzymol 242:226
44. Lee RT, Lee YC (1974) Carbohydr Res 34:151
45. Lee RT Lee YC (1982) Methods Enzymol 83:299
46. Roy R, Laferriére CA (1988) Carbohydr Res 177:C1
47. Roy R, Andersson FO, Harms G, Kelm S, Schauer R (1992) Angew Chem 104:1551; Angew Chem Int Ed Engl 31:1478
48. Roy R, Park WK, Srivastava OM, Foxall LV (1996) Bioorg Med Chem Lett 6:1399
49. Byramova NE, Mochalova LV, Belyanchikov IM, Matrosovich MN, Bovin NV (1991) J Carbohydr Res 10:691
50. Chernyak AY, Kononov LO, Kochetkov NK (1994) J Carbohydr Res 13:383
51. Lees WJ, Spaltenstein A, Kingery-Wood JE, Whitesides GM (1994) J Med Chem 37:3419
52. Mammen M, Dahmann G, Whitesides GM (1995) J Med Chem 38:4179
53. Sallas F, Nishimura S-I (2000) J Chem Soc, Perkin Trans 1:2091
54. Thoma G, Magnani J, Öhrlein R, Ernst B, Schwarzenbach F, Duthaler RO (1997) J Am Chem Soc 119:7414
55. Thoma G, Patton JT, Magnani JL, Ernst B, Öhrlein R, Duthaler RO (1999) J Am Chem Soc 121:5919
56. Fraser C, Grubbs RH (1995) Macromolecules 28:7248
57. Lynn DM, Kanaoka S, Grubbs RH (1996) J Am Chem Soc 118:784
58. Dias EL, Nguyen ST, Grubbs RH (1997) J Am Chem Soc 119:3887
59. Nomura K, Schrock RR (1996) Macromolecules 29:540
60. Gordon EJ, Sanders WJ, Kiessling LL (1998) Nature 392:30
61. Mann DA, Kanai M, Maly DJ, Kiessling LL (1998) J Am Chem Soc 120:10575
62. Pohl NL, Kiessling LL (1999) Synthesis SI:1515
63. Strong LE, Kiessling LL (1999) J Am Chem Soc 121:6193
64. Roy R (1997) In: Driguez H, Thiem J (eds) Top Curr Chem 187:Glycoscience, Synthesis of Substrate Analogs and Mimetics, p 241
65. Roy R (1996) Curr Opin Struct Biol 6:692
66. Roy R (1999) Carbohydr Eur 27:34
67. Zanini D, Roy R (1998) Architectonic Neoglycoconjugates:Effects of Shapes and Valencies in Multiple Carbohydrate-Protein Interactions. In: Chapleur Y (ed) Carbohydrate Mimics, Concepts and Methods. Wiley-VCH, Weinheim, p 385
68. Lindhorst TK (2001) In: Vögtle F (ed) Top Curr Chem:Dendrimers IV, in press
69. Sadalapure K, Lindhorst TK (2000) Angew Chem 112:2066; Angew Chem Int Ed 39:20
70. Roy R, Kim JM (1999) Angew Chem 111:380; Angew Chem Int Ed 38:369
71. Roy R, Das SK, Santoyo-González F, Hernández-Mateo F, Dam TK, Brewer CF (2000) Chem Eur J 6:1757
72. Dondoni A, Marra A, Scherrmann M-C, Casnati A, Sansone F, Ungaro R (1997) Chem Eur J 3:1774
73. Ortiz-Mellet C, Benito JM, García Fernández JM, Law H, Chmurski K, Defaye J, O'Sullivan ML, Caro HN (1998) Chem Eur J 4:2523
74. Stoll MS, Hounsell EF, Lawson AM, Chai W, Feizi T (1990) Eur J Biochem 189:499
75. Mrksich M (2000) Chem Soc Rev 29:267
76. Mrksich M (2001) this issue of Top Curr Chem
77. Aronson M et al. (1979) J Infect Dis 139:329
78. Mammen M, Dahmann G, Whitesides GM (1995) J Med Chem 38:4179
79. Sabesan S, Duus JØ, Domaille P, Kelm S, Paulson JC (1991) J Am Chem Soc 113:5865
80. Glick GD, Knowles JR (1991) J Am Chem Soc 113:4701
81. Kamitakahara H, Suzuki T, Nishigori N, Suzuki Y, Kanie O, Wong C-H (1998) Angew Chem 110:1611; Angew Chem Int Ed 37:1524
82. Tomalia DA, Esfand R (1997) Chem Ind 416

83. Reuter JD, Myc A, Hayes MM, Gan Z, Roy R, Qin D, Yin R, Piehler LT, Esfand R, Tomalia DA, Baker JR, Jr (1999) Bioconj Chem 10:271
84. Cines DB, Pollak ES, Buck CA, Loscalzo J, Zimmermann GA, McEver RP, Pober JS, Wick TM, Konkle BA, Schwartz BS, Barnathan ES, McCrae KR, Hug BA, Schmidt A-M, Stern DM (1998) Blood, 91:3527
85. Varki A (1994) Proc Nat Acad Sci USA 91:7390
86. Thoma G, Magnani JL, Patton JT, Ernst B, Jahnke W (2001) Angew Chem 113:1995; Angew Chem Int Ed 40:1941
87. Simanek EE, McGarvey GJ, Jablonski JA, Wong C-H (1998) Chem Rev 98:833
88. Bertozzi CR (1995) Chem Biol 2:703
89. Kretzschmar G, Sprengard U, Kunz H, Bartnik E, Schmidt W, Toepfer A, Hörsch B, Krause M, Seiffge D (1995) Tetrahedron 51:13015, 86
90. Gordon EJ, Strong LE, Kiessling LL (1998) Bioorg Med Chem Lett 6:1293
91. Bray D, Levin MD, Morton-Firth CJ (1998) Nature 393:85
92. Liu Y, Levit M, Lurz R, Surette MG, Stock JB (1997) EMBO J 16:7231
93. Gestwicki JE, Strong LE, Kiessling LL (2000) Chem Biol 7:583
94. Roy R, Zanini D, Meunier SJ, Romanowska A (1993) J Chem Soc, Chem Commun:1896
95. Roy R, Park WKC, Zanini D, Foxall C, Srivastava OP (1997) Carbohydr Lett 2:259
96. Ashton PR, Hounsell EF, Jayaraman N, Nilsen TM, Spencer N, Stoddart JF (1998) J Org Chem 63:3429
97. Roy R, Pagé D, Perez S F, Bencomo VV (1998) Glycoconjugate J 15:251
98. Roy R, Baek MG, Xia Z, Rittenhouse-Diakun K (1999) Glycoconjugate J 16:S53
99. Zanini D, Roy R (1998) J Org Chem 63:3486
100. Sharon N (1987) FEBS Lett 217:145
101. Connell H, Agace W, Klemm P, Schembri M, Mårlid S, Svanborg C (1996) Proc Natl Acad Sci USA 93:9827
102. Struve C, Krogfelt KA (1999) Microbiology 145:2683
103. Firon N, Ofek I, Sharon N (1983) Carbohydr Res 120:235
104. Neeser J-R, Koellreutter B, Wuersch P(1986) Infect Immun 52:428
105. Firon N, Ashkenazi S, Mirleman D, Ofek I, Sharon N (1987) Infect Immun 55:472
106. Klemm P, Jørgensen, van Die I, de Ree H, Bergmans H (1985) Mol Gen Genet 199:410
107. Lindhorst TK, Kieburg C, Krallmann-Wenzel U (1998) Glycoconjugate J 15:605
108. Choudhury D, ThompsonA, Stojanoff V, Langermann S, Pinkner J, Hultgreen SJ, Knight SD (1999) Science 285:1061
109. Sauer FG, Barnhart M, Choudhury D, Knight SD, Waksman G, Hultgren SJ (2000) Curr Opin Struct Biol 10:548
110. Knight SD, Berglund J, Choudhury D (2000) Curr Opin Chem Biol 4:653
111. Hansen HC, Haataja S, Finne J, Magnusson G (1997) J Am Chem Soc 119:6974
112. Lindberg AA, Brown JE, Stromberg N, Westling-Ryd M, Schultz JE, Karlsson KA (1987) J Biol Chem 262:1779
113. Merrit EA, Hol WGJ (1995) Curr Opin Struct Biol 5:165
114. Merrit EA, Sarfaty S, Feil IK, Hol WGJ (1997) Structure 5:1485
115. Fan E, Zhang Z, Minke WE, Hou Z, Verlinde CLMJ, Hol WGJ (2000) J Am Chem Soc 122:2663
116. Kitov PI, Sadowska JM, Mulvey G, Armstrong GD, Ling H, Pannu NS, Read RJ, Bundle DR (2000) Nature 403:669
117. Fan E, Merritt EA, Verlinde CLMJ, Hol WGJ (2000) Curr Opin Struct Biol 10:680
118. Ling H, Boodhoo A, Hazes B, Cummings MD, Armstrong GD, Brunton JL, Read RJ (1998) Biochemistry 37:1777
119. Seeberger PH, Haase W-C (2000) Chem Rev 100:4349
120. Sears P, Wong C-H (2001) Science 291:2344
121. Vliegenthart JFG, Montreuil J (1995) In: Montreuil J, Vliegenthart JFG, Schachter H (eds) Glycoproteins, New Compr Biochem, 29a:p 13
122. Kayser H, Zeitler R, Kannicht C, Grunow D, Nuck R, Reutter W (1992) J Biol Chem 267:16934

123. Keppler OT, Stehling P, Herrmann M, Kayser H, Grunow D, Reutter W, Pawlia M (1995) J Biol Chem 270:1308
124. Keppler OT, Horstkorte R, Pawlita M, Schmidt C, Reutter W (2001) Glycobiology 11:11R
125. Varki A (1991) FASEB J 5:226
126. Reutter W, Keppler O, Pawlita M, Schüler C, Horstkorte R, Schmidt C, Hoppe B, Lucka L, Stehling P, Mickeleit M, Biochemical Engineering by New N-Acylmannosamines of Sialic Acid Creates New Biological Characteristics and Technical Tools of its N-Acyl Side Chain. In Sialobiology and other forms of glycosylation (1999) Inoue Y, Lee YC, Fat I (eds). Osaka:Gakushin Publishing Company, pp 281
127. Mahal LK, Yarema KJ, Bertozzi CR (1997) Science 276:1125
128. Saxon E, Bertozzi CR (2000) Science 287:2007
129. Staudinger H, Meyer J (1919) Helv Chim Acta 2:635
130. Gololobov YG, Kasukhin LF (1992) Tetrahedron 48:1353
131. Yarema KJ, Mahal LK, Bruehl RE, Rodriguez EC, Bertozzi CR (1998) J Biol Chem 273:31168
132. Lemieux GA, Yarema KJ, Jacobs CL, Bertozzi CR (1999) J Am Chem Soc 121:4278
133. Lee JH, Baker TJ, Mahal LK, Zabner J, Bertozzi CR, Wierner DF, Welsh MJ (1999) J Biol Chem 274:21878
134. Schmidt C, Ohlemeyer C, Kettenmann H, Reutter W, Horstkorte R (2000) FEBS Lett 478:276
135. Lohof E, Planker E, Mang C, Burkhart F, Dechantsreiter MA, Haubner R, Wester H-J, Schwaiger M, Hölzemann G, Goodman SL, Kessler H (2000) Angew Chem 112:2868; Angew Chem Int Ed 39:2761
136. Kessler H (1982) Angew Chem 94:509; Angew Chem Int Ed Engl 21:512
137. Breslow R, Belvedere S, Gershell L, Leung D (2000) Pure Appl Chem 72:333
138. Dimick SM, Powell SC, McMahon SA, Moothoo DN, Naismith JH, Toone EJ (1999) J Am Chem Soc 121:10286

Author Index Volume 201–218

Printing (Computer to Film): Saladruck Berlin
Binding: Stürtz AG, Würzburg